10th International Conference of DAAAM Baltic INDUSTRIAL ENGINEERING

Tallinn, Estonia
12-13 May 2015

Editors:

Tauno Otto

ISBN: 978-1-5108-2263-4

Printed from e-media with permission by:

Curran Associates, Inc.
57 Morehouse Lane
Red Hook, NY 12571

Some format issues inherent in the e-media version may also appear in this print version.

Printed by Curran Associates, Inc. (2016)

For permission requests, please contact DAAAM International Vienna
at the address below.

DAAAM International Vienna
Karlsplatz 13/311, A-1040
Wien, Austria, European Union

Phone: 0043-1-58801-31121
Fax: 0043-1-58801-31199

secretary@daaam.info

Additional copies of this publication are available from:

Curran Associates, Inc.
57 Morehouse Lane
Red Hook, NY 12571 USA
Phone: 845-758-0400
Fax: 845-758-2633
Email: curran@proceedings.com
Web: www.proceedings.com

DANUBE ADRIA ASSOCIATION FOR AUTOMATION & MANUFACTURING
DAAAM International Vienna
DAAAM Baltic

PROCEEDINGS

OF THE 10TH INTERNATIONAL CONFERENCE OF DAAAM BALTIC
INDUSTRIAL ENGINEERING
12-13th MAY 2015, TALLINN, ESTONIA

ORGANIZED UNDER THE AUSPICES
OF DAAAM INTERNATIONAL VIENNA

in co-operation with:
BALTECH Consortium
Estonian Academy of Sciences
Federation of Estonian Engineering Industry
Association of Estonian Mechanical Engineers
Innovative Manufacturing Engineering Systems Competence Centre IMECC
Tallinn City Enterprise Board
DAAAD
MECTORY
ITL

EDITED BY
T. OTTO

TALLINN UNIVERSITY OF TECHNOLOGY

The Chairman of DAAAM Baltic Scientific Committee expresses special personal thanks to Professor Branko Katalinic for encouraging to organise and for advising during the preparation of the conference. Also special thanks for the support during the organising the conference to the following persons: A. Hermaste, R. Kulbas, K. Karjust.

EDITORS' NOTE

This publication was reproduced by the photo process, using the soft copies supplied by their authors. The editor and the DAAAM Baltic are not responsible either for the statements or for the opinion expressed in this publication.

Proceedings of 10th International Conference of DAAAM Baltic INDUSTRIAL ENGINEERING
Editor T. Otto
12–13 May 2015, Tallinn, Estonia

Phone: +372 620 3257
E-mail: daaam@ttu.ee

Foreword

Over the past years, DAAAM-Baltic has been an international forum for researchers and engineers to present their research results in the areas of industrial engineering, manufacturing and automation. It provides an avenue for discussion and exchange of new ideas addressing new techniques and methods for product development and manufacturing engineering.

This DAAAM-Baltic-2015 conference in the DAAAM-Baltic Conference series will continue the mission of the DAAAM, which is to give for young and active scientists of the Baltic Sea region an opportunity to introduce their works and find partners.

The main idea of starting DAAAM-Baltic meetings was to organize regional conferences of DAAAM International events (DAAAM – Danube-Adrian Association for Automation & Manufacturing). DAAAM International is association for international scientific and academic cooperation in the fields of intelligent automation and modern production. Estonia has been represented as a member of International DAAAM Committee since 1994. In year 2015 we have 10th Conference of DAAAM-Baltic and 19 years of history of publishing the Proceedings. The Proceedings of DAAAM-Baltic have been included in ISI Web of Science of Thomson Reuters since 2004 and in Scopus since 2012.

The Conference addresses the issues of managing globalization in the internet age. We emphasize importance of European initiatives towards Factories of the Future and technology platform Manufuture, and hope to give our contribution by bringing scientists, entrepreneurs and industry specialists together for exchange of future visions based on scientific research and case studies.

In 2015 we listened over fifty oral presentations from 8 countries. The conference itself was organized in the frameworks of ICT Week and Der deutsche Frühling, introducing Industry 4.0 developments all over Europe. All scientific papers have been carefully selected and full texts of presentations peer-reviewed to address key topics of the conference. The proceedings include a variety of topics: New Manufacturing Technologies; Factories of the Future; Industrial Internet; Industry 4.0; Mechatronics Design; Control Systems; Robotics; Supply Chain Management and Logistics; Quality Management; Productivity Management; and Human Resource Management.

We wish to thank all authors, referees, members of the Organizing Committee and Program Committee, as well as supportive organizations for their efforts which made this conference possible. DAAAM-Baltic would not be possible without contributions from members of the scientific community of the Baltic-Sea region.

We look forward to a very exciting and stimulating conference, and hope that you will join us in next DAAAM-Baltic in 2016.

Tallinn; May 2015

Tauno Otto
Chairman of the Scientific Committee DAAAM-Baltic 2015

Steering Committee

Tauno Otto - *chair*

Kristo Karjust

Rein Küttner

Andres Siirde

Leo Mõtus

Mart Tamre

Renno Veinthal

Gunnar Klaus Prause

International Program Committee

M.Airila (FIN)

J.Balic (SLV)

F.Boór (HUN)

S.Bockus (LIT)

J.Brnić (CRO)

M.Eerme (EST)

H.Jähn (GER)

T.Karaulova (EST)

B.Katalinic (AUT)

T.Kjellberg (SWE)

K. Korbe (EST)

P.Kulu (EST)

P.Kuosmanen (FIN)

J. Lavrentjev (EST)

J.Martikainen (FIN)

V.Mironov (LAT)

J. Preden (EST)

J.-G.Persson (SWE)

J.Riives (EST)

A.Schulz (LAT)

F.Sergejev (EST)

T. Torims (LAT)

M.Veveris (UK)

A.Voet (BEL)

A.Zawada-Tomkiewicz (POL)

CONTENTS

I PRODUCTION ENGINEERING & MANAGEMENT

10th International DAAAM Baltic Conference
"INDUSTRIAL ENGINEERING"
12-13 May 2015, Tallinn, Estonia

ANALYSIS OF THE GUIDENANCE OF MECHANICAL PRESSES WITH REGARD TO NEW FORMING TECHNOLOGIES

Chval, Z.; Cechura, M.; Raz, K.

Abstract: *Number of technological operations performed on forming machines is increasing worldwide. It is caused by more accurate operations and therefore products with smaller tolerances and more precise shape are produced. Also material waste during technological operations and production is lower. These facts are resulting in various tasks for machine designers. Old forging machines must be upgraded and new solutions and machine designs should be found. A new machine must meet requirements of accuracy and also ram guidance quality. This paper is comparing different types of ram guidance on mechanical forging press in terms of stress and deformation.*

Key words: mechanical press, forging press, guidenance, forming technologies.

1. INTRODUCTION

Presumption for the quality machine design is good knowledge of used technologies. If the technology is determined, it is necessary to analyse the technological process for determining the load of machine parameters of individual operations (boundary conditions for virtual prototyping).

This may be done:

• Calculation based on given technological parameters and assumptions according to many years of practical experience.

• Monitoring of technological process and taking into account the obtained results.

Usually, it happens that both methods are combined.

Research of crank presses deals with the analysis of ram guidance method and analysis of various ram design solutions with respect to its rigidity and size of deformations.

2. INFLUENCE OF RAM GUIDANCE TO QUALITY OF FORGING

Accuracy and quality of ram guidance directly affect the accuracy and quality of forging. Therefore, many companies solved this problem and developed various solutions. The most commonly are used tetrahedral and an octagonal guidance. Octagonal guidance is considered as the best guidance. [1,2].

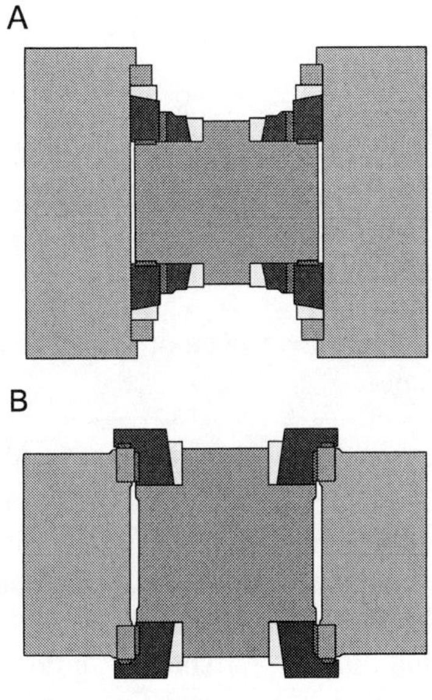

Fig. 1. A and B variants of the octagonal ram guidance.

Variant of ram guidance - A is designed for presses with a wide frame compared to the press frame. In this variant clearance of guidance is defined on the ram and on the press frame too. Variant B is designed for presses with the same wide of press frame and the ram. Clearance of guidance is defined only on the ram.

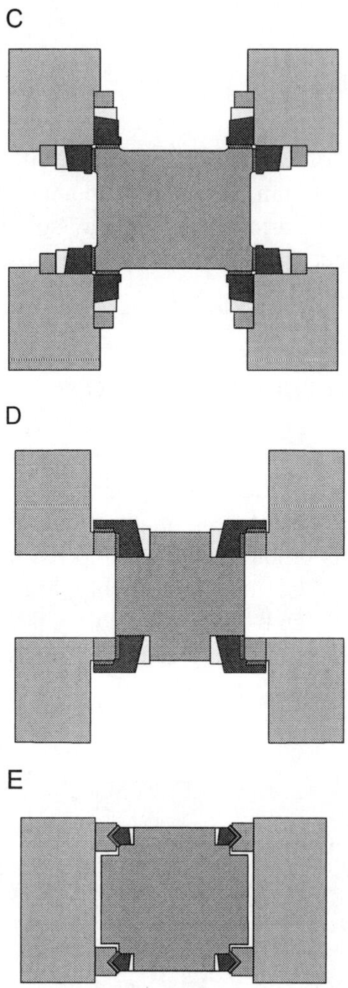

Fig. 2. C, D and E variants of the octagonal ram guidance.

Variant C and D is intended for machines with four columns, variant C is designed to eccentrically applied force in the x and y axes. Variant D is designed to eccentrically applied force only in the y-axis.

E variant is a prismatic guidance. Clearance of guidance is defined only on the ram.

3. VIRTUAL ANALYSIS OF THE EFFECT OF DIFFERENT TYPES OF RAM GUIDANCE TO THE STRESS AND DISPLACEMENT

3.1 Boundary conditions of calculation

Ram was modeled as a half symmetric model. Fixation in the vertical direction was carried out in part (taking account of eccentricity) of the bottom surface of the tool inserted through the contact to ram. In other directions the ram is caught by guidance (using contact).

Clearance in the guidance in all cases was set to 0.5 mm.

Guidance is considered as a rigid without the influence of frame deformation - the red part of models.

For all types of guidance the contact area is identical in both directions.

Nominal force is applied through pin of ram.

Fig. 3. Example of ram model with boundary conditions.

3.2 Calculation results

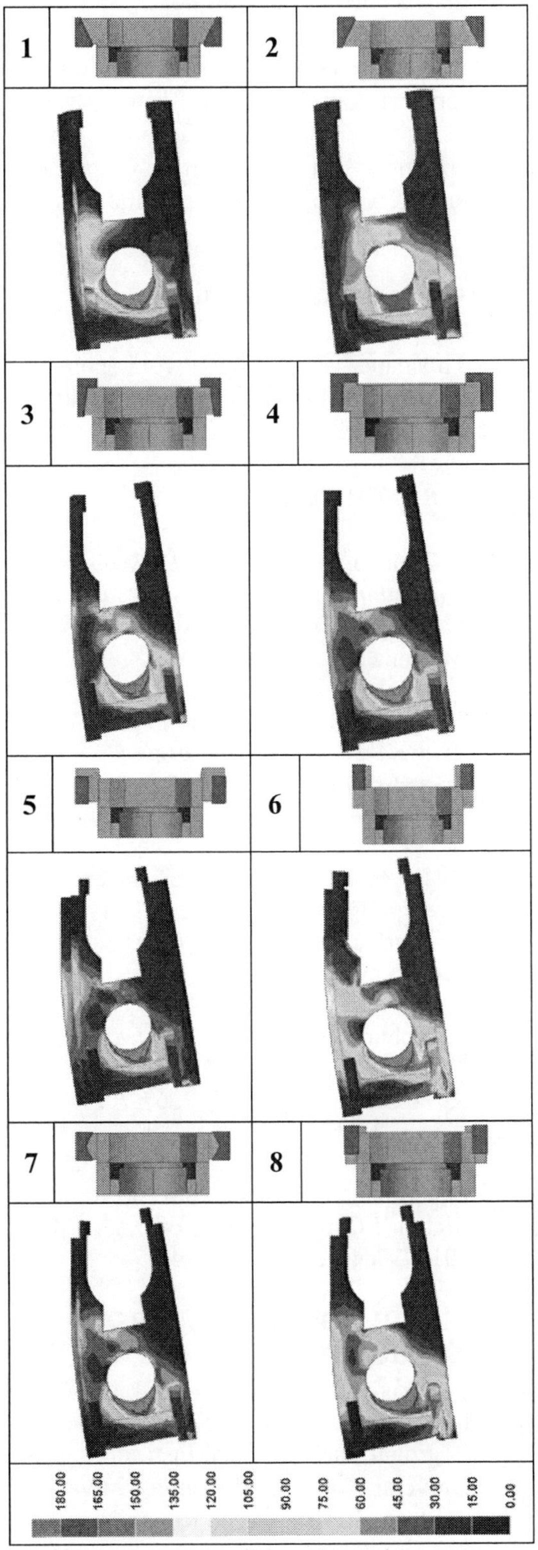

Table 1. Stress conditions in various types of rams (MPa).

In all cases the greatest stress in ram is under the pin. But the distribution stress is different depending to displacement of the ram in its guidance (tilting).
Case no. 2 leads to jamming of the pin and therefore the stress under the pin is decreased.
The smallest size of the stress fields in the ram, and the size of stress values show results no. 5. For all other variants are stress peaks in the corners between the ram and the guide bars.

Types list of ram guidance and displacement conditions of the ram.

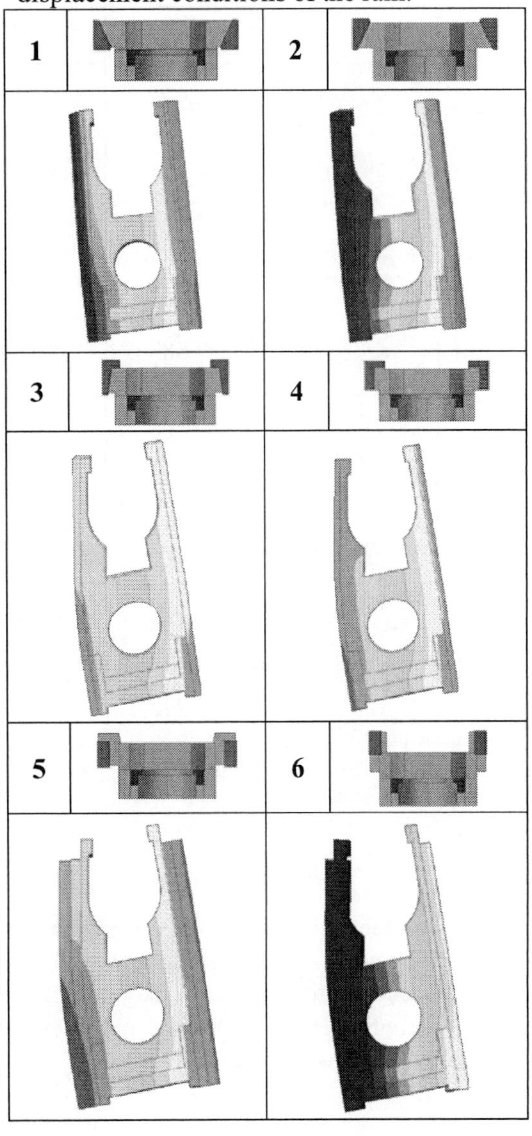

7		**8**	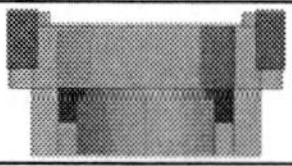
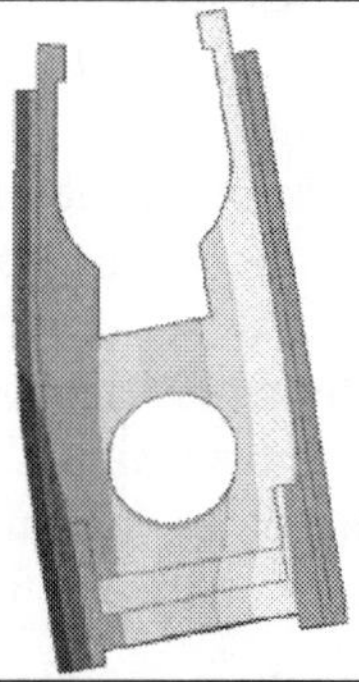		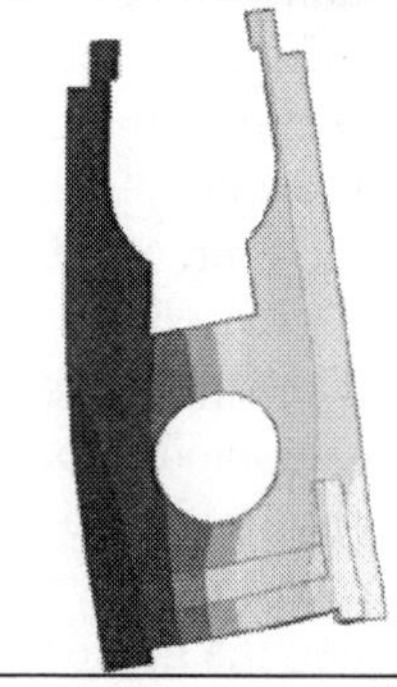	

Table 2. Displacement in various types of rams (mm, -3 to +1).

	Guidance type	average displacement	maximum displacement
1	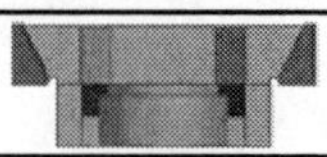	0,995	2,094
2	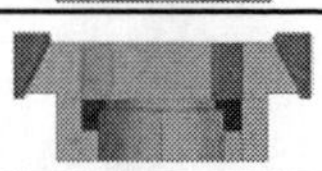	1,121	2,322
3	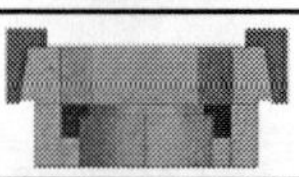	0,803	1,584
4	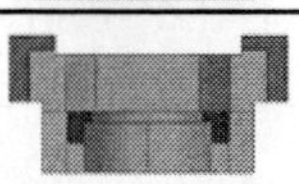	0,888	1,796
5	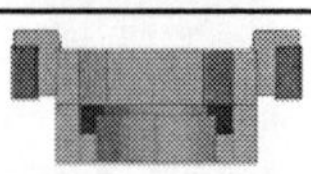	0,868	1,734
6	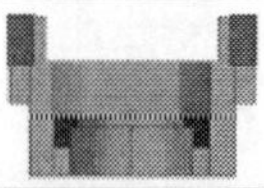	1,785	2,997
7	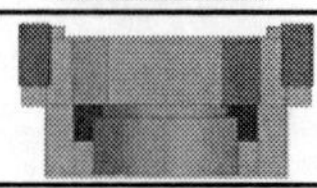	1,608	2,656
8		0,942	1,926

Table 3. Values of displacement in various types of rams (mm).

5. CONCLUSION

From the presented results is evident that the choice of ram guidance has a great influence to stress magnitude, its distribution, and especially to the size of displacement of the ram.

The analysis shows that the smallest displacement is in ram no. 3, 4, 5, and the smallest stress is in ram no. 5.

Proper selection of design solution of the ram can significantly influence stress and deformation of the ram. This obviously has an impact on the accuracy of production and economic cost of the proposed solution.

5. ACKNOWLEDGMENT

This paper is based upon work sponsored by project "Regionální technologický institut" reg. no. CZ.1.05/2.1.00/03.0093

6. REFERENCES

1. Cechura, M., Smolik, J. *Developement and Innovations of Existing Design Solutions of Forming Machines.* Univestity of West Bohemia, Plzen, 2012.

2. He, W., Zeng, P., Lin, F. *Finite Element Analysis for 40 MN Die Forging Press.* Advanced Materials Research, 2012, 690-693.

2. Hlavac, J., Cechura, M., *Boundary Conditions Setting Questions for Virtual Simulation of Mechanical Presses*, Technologia, September 2009/9-10, 61.

3. Cechura, M., Chval, Z., *Convectional Versus Multiple Operating Press*, Kovarenstvi, May 2013/17, 67-70.

4. Piscan, I., Janssens, T., Predincea, N. Experimental Validation of FEM for frictional Contacts. Annals of DAAAM for 2011, Vienna, 2011, 741-742.

7. ADDITIONAL DATA ABOUT AUTHORS

Ing. Zdenek Chval, Ph.D., researcher,
Regional Technological Institute,
University of West Bohemia, Univerzitni 8, Plzen, 306 14, Czech Republic.
Email: zdchval@rti.zcu.cz.
Tel: +420 377 638 741

10th International DAAAM Baltic Conference
"INDUSTRIAL ENGINEERING"
12-13 May 2015, Tallinn, Estonia

INFLUENCE OF COOLING METHODS ON TOOL LIFE DURING MACHINING WITH iMACHINING STRATEGY

Fulemova, J.; Hnatik, J.; Kozmin, P. & Sklenicka, J.

Abstract: *The choice of right method of cooling can influence not only the machining process, but also its results. Generally it is known that effect of different methods of cooling on machining process can be different at the same cutting conditions. It can be expressed by tool life, surface roughness, cutting temperature, cutting forces, etc. Currently, there are used more ecological technologies during machining that include for example dry machining with air blowing into cutting area. This article deals with influence of external cooling and dry machining of S235JR steel on tool life during milling with a four-teeth milling cutter during iMachining strategy. Further there are compared tools, delivered by three different tool producers, in terms of economic viewpoint.*
Key words: milling, cooling method, tool life, iMachining.

1. INTRODUCTION

Actual trends in an engineering industry are insisting on quality and efficiency of manufacturing, on its technical, economical and ecological level. To ensure these demands there are technologies like HSC (High Speed Cutting), HPC (High Performance Cutting), HFM (High Feed Milling), dry machining, etc. The benefit of these technologies is reducing of continuous production time and machine time of manufacturing at keeping quality and accuracy of a product [1, 2]. Machining technology, which is called iMachining is ranked among HPC and was used during progressive milling of a machine part. The workpiece material was carbon steel S235JR.

1.1 iMachining

As mentioned above, iMachining is ranked among HPC strategies. This technology is based on high dept of cut (a_p) and lower width of cut (a_e) and that is why it is possible to achieve great material removal. The algorithm for counting the tool path iMachining tries to keep constant load of the tool. This is achieved by keeping the tool engagement angle at the set boundaries, it means a_p is constant and a_e is changing. If the supposed tool loading were nevertheless higher, the feed rate would be reduced. Constant cutting forces are not guaranteed by keeping constant a_p and φ at outside and inside corners. For that reason the algorithm takes into account effective feed per tooth. In case of entrance and exit of the tool into/from the workpiece, there is not possible to keep constant φ, the algorithm automatically changes feed rate [2, 3].

1.2 Pros and Cons of this strategy

Uniform cutting forces allow fully utilize the possibilities of the tool without risk of its overloading. It brings expressive increasing of machining productivity and also there is an increase tool life and more uniform tool wear with regard to high depth of cut. Machining by iMachining strategy increases productivity on account of cycles reducing; it means time saving 70% and more. Certain problem is machining of semi-products which were made by technology of forging, plasma cutting, casting, etc. because they have very hard surface layer. iMachining strategy does not allow up-milling, which is especially suitable for mentioned kinds of semi-products. Huge amount of chips is produced during iMachining. Chips have needle shape, see in the Fig. 1.

Fig. 1. Needle shape of chips after iMachining strategy

In some cases it can cause flooding of a tool and its very fast breakage. During standard milling strategy the tool wear and the risk of its breakage is very often indicated in advance by increased noise of machining and higher values of vibrations. In contrast to it, in case of iMachining, after tool flooding, the destruction of tool is very fast and without preceding symptoms. This phenomenon is so fast that the machine operator is not able to sufficiently response. That is why it is necessary to ensure very good chip flow, namely by external cooling, internal pressurized cooling, blowing of the tool by external pressurized air or using a vortex tube [2, 3].

1.3 Research background

A choice the method of cooling dramatically influences results of cutting process. The effect of different cooling methods on cutting process can be profoundly different at the same conditions. Mentioned effect is possible to observe on tool life, cutting forces, cutting temperature, quality of machined surface, etc. Currently there are used following cooling methods: external cooling, internal pressurized cooling, MQL, external cooling by air or frozen air, cryogenic cooling, etc. [4,5].

Usage external cooling is dependent upon material of workpiece, machining technology, cutting conditions and type of machining. Generally, for milling is suitable dry machining or pressurized air as blowing. The higher temperature at the place of cut the less suitable is using cooling liquid, especially during roughing. Although cutting temperatures are not constant during milling so these changes do not exceed individual possibilities of sintered carbide grades [6, 7, 8]. Da Silva et al. [9] investigated the influence of different cooling method on tool life. They observed milling of carbon steel by sintered carbide with TiAlN thin layer at cutting speed 200 and 226m/min and changing feed rate 0.14 and 0.22mm/rev. Conclusion of their work is following. Except v_c = 200m/min and f = 0.22mm/rev. there is not strong different, in light of tool life, among external cooling and air blowing. At mentioned cutting conditions, the tool cooled by air has around 25% higher tool life than the tool cooled by external cooling. This conclusion is partly at variance with general recommendations. Many papers is mainly focused on the influence of cooling method during finishing milling of tool steels, Al alloys, Ti alloys, stainless steels, etc. on tool life. Choice of suitable cooling method for theirs machining differ from low alloyed carbon steels. That is why this article is focused on rough (productive) milling of carbon steel S235JR. This material is very often used in engineering industry. The main task of this paper is explaining the influence of external cooling and air blowing on tool life. Results of this article can be used for practical application and for everyone who is going to machine this kind of workpiece material. An integral part of this work is certain recommendations which will lead to more productive machining of S235JR. Last step will be comparison of used tool in terms of economic viewpoint.

2. EXPERIMENT

Experiment was done on the machining centre DMU 65monoBLOCK. The workpiece material was carbon steel S235JR. The size of workpiece was 110x50x70 mm and the amount was 1,000pieces. The cutting tool was four-tooth VHM solid milling cutter, this tool has teeth in helix and the diameter 16mm; it is universal tool which is suitable both roughing and finishing operations. Chemical composition and mechanical properties are written in Tab. 1. Tested tools, cutting conditions and other parameters are written in Tab. 2.

Chemical composition [%]					
C	Mn	Si	P	S	N
0.20	1.4	-	0.035	0.035	0.012
0.23	1.5	-	0.045	0.045	0.014
Mechanical properties					
$R_{p0.2}$[MPa]		R_m [MPa]		A[%]	KV[J]
195		350		22	27
		500			

Table 1. Chemical composition and mechanical properties of steel S235JR

The milling strategy was iMachining. Shape of the semi-product and the tool paths, which are generated by iMachining strategy, are in the Fig. 2.

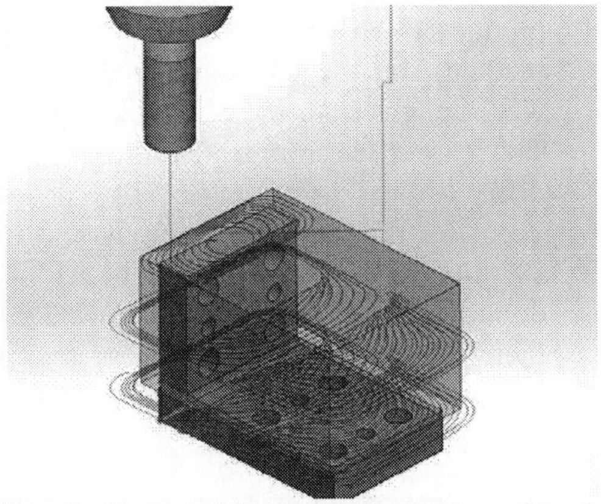

Fig. 2. Shape of the semi-products and tool paths

The tools were clamped into a weldon tool holder. This tool holder is designed for roughing operations for tools where is a risk of its pulling out or slipping. The workpiece was clamped into a jaw-vice. Cooling method was following:

- external cooling, emulsion Blaser Blasocut 35Combi, concentration 6% (in the Tab. 2 marked as wet)
- external, air (in the Tab. 2 marked as air)

	Tool A						Tool B	Tool C		Tool D	
	1.	2.	3.	4.	5.	6.	1.	1.	2.	1.	2.
D [mm]	16										
z	4										
v_c [m/min]	249	249	206								
n[min^{-1}]	4951	4951	4100								
f_{rev}[mm]	0.14	0.16	0.14								
v_f[mm/min]	2758	3099	2233								
a_e[mm]	2.34	1.78	1.57								
a_p[mm]	23										
cooling	wet				air					wet	air
tool life [min]	26	80	106	93	198	237	293	316	323	73	237
thin layer	TiAlN							TiAlN		TiN/TiCN	

Table 2. Cutting tools and their cutting conditions

2.1 Set up the experiment

Before experimental machining it was necessary to set up proper cutting conditions (cutting conditions were selected from catalogue of each tool producer) and to choose proper cooling method. Tools which are marked A and B were bought from the one tool producer. Tools are made from sintered carbide and differ from each other by design, deposited thin layer and a purchase price. Debugging of cutting conditions was done at the tools A no. 1, 2 and 3 and their results are not taken in relative comparison. The used tool can be seen in the Fig. 3.

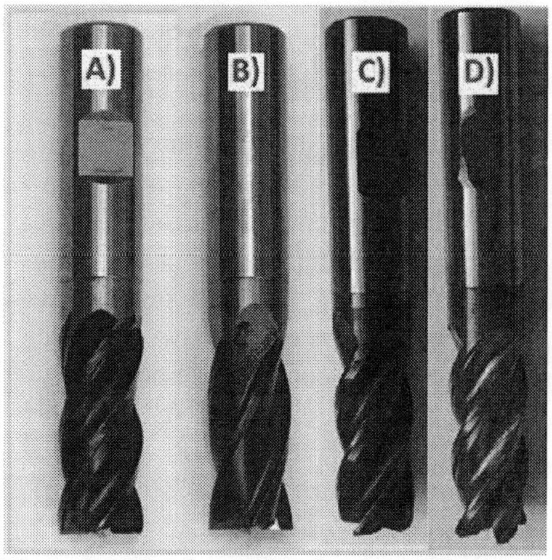

Fig. 3. Used tools for experimental machining

3. EXPERIMENTAL RESULTS

Because the tools are tools for high productive machining/cutting so the main criteria for explanation of their benefit are volume of removed material and purchase price of the tools.

3.1 Volume of removed material

In the graph Fig. 4 there is shown the size of tool wear, which was measured on the flank face and which is marked as VB_{max}. Tool wear was measured on all four teeth and along the helix, till the value of maximal axial depth of cut (a_p).
Average value of tool wear is marked by ▒ and the value of spread of measured points is marked by ▌. The spread of measured points is influenced by the way of tool wear. The main way of tool wear is edge chipping, see in the Fig. 5.

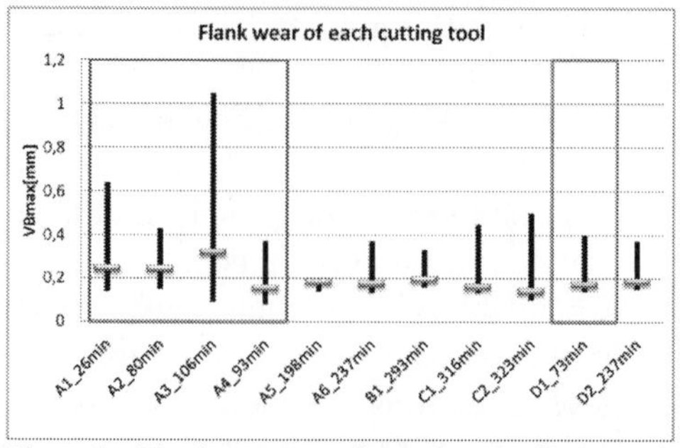

Fig. 4. Tool wear of each cutting tool (tools in the red square = external cooling)

The edge chipping occured along the whole cutting edge. At some tools and at some edge places, the edge chipping was more intensive. Average value of tool wear was 0.2mm. During experimental machining the tools were visually checked. If the machine operator indicated intensive edge chipping he exchanged the milling cutter. C milling cutters reached the higher tool life, but also the highest spread of measured points of tool wear. Comparable results were reached with tools A6 and D2 (the same tool life and the same values of tool wear). B1 milling cutter reached the tool life „only“ 293min. This tool would have reached better results if there had not been finished production on account of reaching demanded amount of produced parts (it means 1,000pieces). It is possible to say that tool B would reach the same results as C tools.

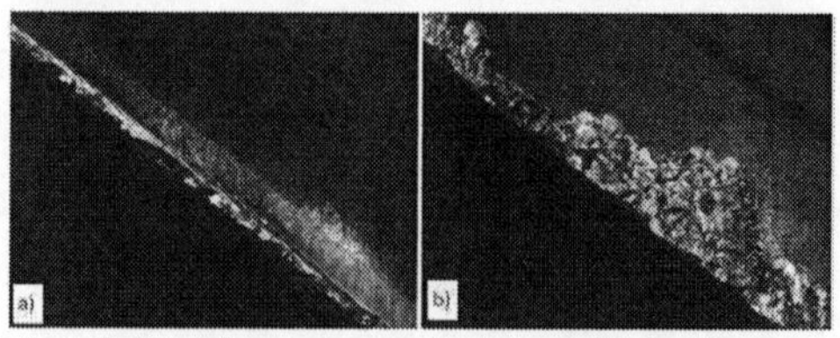

Fig. 5. Tool wear on a) A5 and b) C2

The results, shown in the graph Fig. 4 and which are written in the Tab.2 is brightly seen that external cooling is not suitable cooling method for rough milling S235JR steel. There was always reached more than half tool life in comparison with tool which were cooled by air, at the same cutting conditions. The lower tool life was caused by thermal cracks, which were perpendicularly on the cutting edge. Substrate of the tool started to crumble away or was broken out. Thermal cracks are close-knit with milling technology and cyclic thermal loading of the cutting edge. A cooling liquid always has negative effect and deteriorate the tool life. This was mentioned in theoretical part of this article.

3.2 Comparison of each cutting tools in terms of economic

For mutual comparison of each cutting tools it is necessary to introduce their purchase price (see in the Table. 3). The purchase price of a cutting tool is depending on many factors, which are: a supplier, a region, a country, a discount for faithful customer, etc. For requirements of this article, there are written real prices in the Table 3. The prices include VAT.

Tool producer	Purchase price [€]
A	89
B	48
C	141
D	113

Table 3. The purchase price of each cutting tool

The first step is expression the value of the stock removal (Q):

$$Q = v_f \times a_p \times a_e \quad \left[cm^3 \cdot \min^{-1}\right] \qquad (1)$$

where: v_f ... feed speed [mm/min]
a_e ... width of cut [mm]
a_p ... depth of cut [mm]

The tools A3 ÷ D2 removed 80.634$cm^3.min^{-1}$. Volume of removed material from 1piece was 201.059cm^3. The next step, it is necessary to count for each tool following:

• volume of removed material of each cutting tool:

$$V_{CTi} = T_i \times Q \quad \left[cm^3\right] \qquad (2)$$

where: T_i ... tool life of i-th tool [min]
Q ... stock removal [$cm^3.min^{-1}$]

• price per removed cubic centimetre:

$$C(1cm^3)_{CTi} = P_{CTi} \div V_{CTi} \quad \left[Euro \cdot cm^{-3}\right] \qquad (3)$$

where: P_{CTi} ... the purchase price of i-th cutting tool [€]

• price per milling of 1piece:

$$CM_{CTi} = C\left(1cm^3\right)_{CTi} \times 201.059 \quad \left[Euro\right] \qquad (4)$$

In the Fig. 6 there are compared the tools in term of price per milling 1piece.

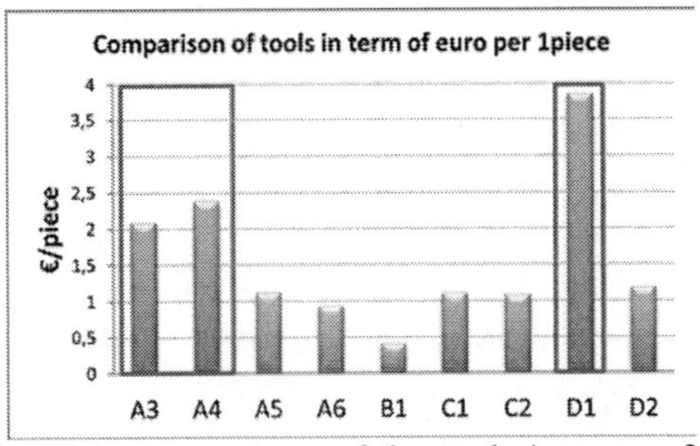

Fig. 6. Comparison of the tools in term of euro per 1piece

Although the tool B does not reach the same tool life as C tools, so the machining of 1piece is the cheapest with this tool. Tools A4, C1, C2 and D2 is possible to machine 1 piece for the same price. This conclusion shows, that the right choice of the cutting tool is possible to cheapen the production.

4. CONCLUSION

This experimental machining brightly demonstrated that external cooling is not suitable for roughing milling S235JR. Removing cooling liquid is possible to increase the tool life more than half and it also have positive effect to cheapen the production. At the same time there were specified a recommendation for the tool producers. For tools, which are extremely force stressed, create the weldon facet. We did it for tools B and D (can be seen in the Fig. 3). Our future steps will be comparison iMachining strategy and classical rough milling without coolant focused on tool life and cutting forces. It should demonstrate concrete benefits of iMachining strategy.

5. ACKNOWLEDGEMENTS

This paper includes results created within the project CZ.1.05/2.1.00/03.0093 - Regional Technological Institute. The project is supported by the European Regional Development Fund and the state budget of the Czech Republic.

6. REFERENCES

1. Zeman, P. Účinek řezného prostředí na trvanlivost břitu. MM *Průmyslové spektrum [online]*. Praha: Vogel Publishing, 2005, roč. 2005, č. 12

2. Hnátík, J., Kutlwašer, J., Sklenička, J. Měření řezných sil při obrábění metodou iMachining. *Strojírenská technologie: časopis kateder obrábění a montáže a kateder příbuzných České a Slovenské republiky*. Ústí nad Labem: ÚJEP, 2014, XIX, č. 2.

3. iMachining: "Simply Amazing". SolidCAM [online]. 2013 [cit. 2015-04-29.http://www.solidcam.com/us/imachining/imachining-overview/

4. Huang, X., Zhang X., Mou, H., Zhang X., Ding, H. The influence of cryogenic cooling on milling stability. *Journal of Materials Processing Technology* [online]. 2014, vol. 214, issue 12, s. 3169-3178. DOI: 10.1016/j.jmatprotec.2014.07.023.

5. Liu, Z., An, Q., Xu, J., Chen, M., Han S. Wear performance of (nc-AlTiN)/(a-Si3N4) coating and (nc-AlCrN)/(a-Si3N4) coating in high-speed machining of titanium alloys under dry and minimum quantity lubrication (MQL) conditions. *Wear* [online]. 2013, vol. 305, 1-2, s. 249-259. DOI: 10.1016/j.wear.2013.02.001.

6. Dry or with fluid. *Sandvik Coromant* [online]. 2012 [cit. 2015-04-29]. http://www.sandvik.coromant.com/en-gb/knowledge/milling/getting_started/general_guidelines/dry_or_with_fluid/Pages/default.aspx

7. Machining Dry Is Worth A Try. *MMS Online* [online]. 2003 [cit. 2015-04-29]. http://www.mmsonline.com/articles/machining-dry-is-worth-a-try

8. Canter, N. The possibilities and limitations of dry machining. *STLE* [online]. 2009. https://www.stle.org/assets/document/40_DryMachining.pdf https://www.stle.org/assets/document/40_DryMachining.pdf

9. Da Silva, et. al. Tool wear analysis in milling of medium carbon steel with coated cemented carbide inserts using different machining lubrication/cooling systems. *Wear* [online]. 2011, vol. 271, 9-10, s. 2459-2465. DOI: 10.1016/j.wear.2010.12.046.

7. ADDITIONAL DATA ABOUT AUTHOR

Ing. Jaroslava Fulemova, Research Worker, University of West Bohemia, Faculty of Mechanical Engineering, Regional Technological Institute, Univerzitni 8, 306 14 Plzen, The Czech Republic, fulemova@rti.zcu.cz,

10th International DAAAM Baltic Conference
"INDUSTRIAL ENGINEERING -
12-13 May 2015, Tallinn, Estonia

MANUFACTURING OF STEERED FIBER COMPOSITE LAMINATE

Haavajõe, A.; Mikola, M.; Herranen, H. & Pohlak, M.

Abstract:
The main goal of the study is to develop a laminating head for an industrial robot that would be competitive from the aspect of the price and versatility, and also to find mathematical models that could describe well enough the relations between the laminating parameters and the properties of the final product.
Key words: *Advanced Fiber Placement Technology, Automated Fiber Placement, Automated Tape Laying, Fiber Reinforced Composites, Laminates*

1. INTRODUCTION

Historically, advanced composites have been manufactured by hand lay-up of prepreg to produce composite parts that are then consolidated and cured in an autoclave. This labor-consuming process results in high fiber volume fraction, void-free, well-consolidated composite structures with excellent mechanical properties. However, the costs of these structures are high due to the labor costs for hand layup processing. Automated Fiber Placement (AFP) and Automated Tape Layup (ATL), as one of the steered fiber composite manufacturing techniques, have gained attention for their cost-effective, flexible, and automatic process.

Application of the AFT and ATL techniques allows to design of composite structures with variable stiffness [1-5]. Pioneering work in this area has been done in Delft University of Technology by research group headed by Z. Gürdal. Their latest results concerning numerical modeling of AFP/ATL techniques are presented in [1-2]. In [3-5] the simulation models and optimization techniques are developed for design of discrete variable stiffness structures. Design and modeling of composite materials and structures has been main topic of the current work group [6-11]. Special attention was paid to optimal material orientation, related tightly with the topic of the current study [12-13].

The main issues today concerning these technologies are improving the heating process controllability and deeper understanding the influence of heating parameters, compaction process, compaction force, positional/speed errors, also possible variations of the robotic machinery and laminating head under a working condition.

In the current study a set of experiments were carried out using a simple testing approach. For measuring of the compaction force, tensile-compression measurement system was used. The temperature of the consolidation area and the heat distribution were screened with the thermal camera. Gas torch heater was used as a heating source. Material used in the experiment was unidirectional carbon fiber reinforced polyamide tape.

Findings show that in addition to the main parameters – the compaction force and the temperature, there are many minor parameters such as the diameter of the compaction wheel, the pretension of the laminating tape, etc, all influence the quality of the final product.

Thus designing a low price machine for manufacturing of steered fiber products, yet making no trade-offs from the aspect of the controllability of the laminating parameters, proves to be a real challenge and needs further investigation.

2. WORKING PRINCIPLES OF AFP

One of the main technologies for producing advanced composites is AFP which belongs to the robotized fiber placement method (RFP). A typical configuration of the RFP system with its main modules can be seen in the Fig. 1.

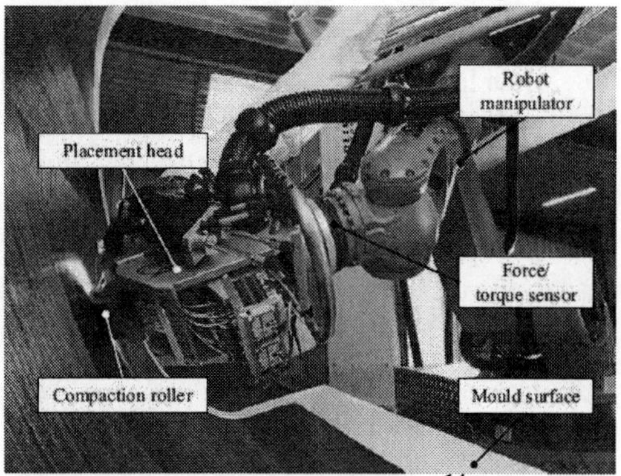

Fig. 1. A typical AFP system [14].

The main parts of the system are industrial robot with its manipulator and force/torque sensor, a mold and the heart of the system- the laminating head.
The basic working principle is shown in the Fig. 2.

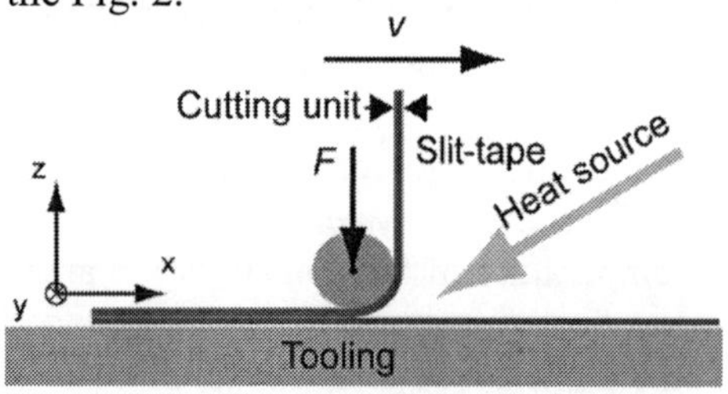

Fig. 2.Working principle of the AFP.

The main parts of the laminating head are the compaction roller that presses the material stripe against the mold or previous ply. Heat source, that heats the material stripe above the melting point to make it sticky and a cutting mechanism that cuts the stripe when necessary.
The major factor that determines the quality of the final product is the compaction force. The experiments carried out by Pitchumani [15] showed that the quality and performance of the final product is directly influenced by the presence of the voids. So the consolidation process plays an important role in the manufacture of high quality composite parts. Moreover, good consolidation is important for eliminating residual stresses and warpage in the product [16].
In response to the demand for high quality of composite structures, the compaction force for a consolidation process needs to be reinvestigated for possible variations of the force that lead to the variations of the consolidation process.
The roller itself is usually made of aluminum but can also be made of different metals with different coatings such as teflon and polyurethane depending on the size and curvature of the mold being laminated. In industrial laminating heads, cooling is used to prevent the composite tape from sticking to the roller.
As a heat source, laser, infrared heater or gas torch can be used.
Laser heating was first introduced by Beyeler and Güçeri [17]. He used CO_2 laser for melting the incoming tape and substrate for thermoplastic filament winding. Laser heating has many advantages over the other methods, like fast response time, excellent energy efficiency and uniform heat distribution and good integration with the overall controlling system. However, the size and weight of a laser requires a large fiber placement head in order to carry it. This restriction limits the application of laser to large machines only not to mention the fact that it is expensive, which is contrary to the main goal of the work group behind this paper.
Infrared heating was investigated by Endres [18] and Calhoun [19] from the

aspect of applicability for filament winding of thermoplastic composite tapes for its energy-efficient characteristics and good response behavior. Although infrared heating is not as good as laser heating, it is, overall, a good technique and available with reasonable price on the market. Therefore, today it is the most widespread solution for heating the towpreg. However, it is inferior to the laser from the aspect of nip-point heating which needs very high intensity heating on a small area.

Hot gas heating for on-line consolidation of thermoplastic composites was first used by Werdermann [20]. In spite of the disadvantages of very low energy efficiency and slow response time, a hot gas heating is the most widely used method in low tech laminating applications for its cost-effectiveness and design flexibility.

AFP as the most common RFP methods for manufacturing advanced composites has successfully been used for manufacturing parts for aeronautics and aircrafts for many years already, yet there is plenty of room for further investigation and research. The main areas that need to be improved are controlling the laminating parameters, especially laminating temperature and compaction force but also laminating speed and tow/tape tension. Also, CIM systems specially designed for path generation of composite laminating. Improving the overall accuracy by using different types of feedback systems. In addition to the main goal - lowering the manufacturing costs, it should be considered the possibility of lowering and simplifying the laminating system itself to be more attainable for small- and medium sized companies dealing with other fields than aerospace and automotive industry. Advanced composites can be successfully used in fields like renewable energy, small ship building, medicine and etc.

3. PROCESS PARAMETERS

The main operational parameters are:

- Compaction force
- Temperature
- Time

And other important parameters are:

- The radius of the roller
- Material and the coating of the roller
- Curvature of the mold, etc.

The following part of this paper concentrates on testing the two main parameters – the temperature and compaction force on carbon fiber reinforced polyamide material PA12-CF60 tape with 0,1 mm thickness, 25,4 mm width and with melting point of 180°C.

4. TESTING

Firstly, to find an optimal laminating temperature to the material, simple tests were carried out. The tests comprised cutting the material into right length and positioning it to the simple fixture made of cardboard to ensure approximately the same overlay surface area during every test attempt. The device can be seen in the Fig. 3., where the red rectangle in the middle indicates the location of the contact area that is going to be heated and compressed.

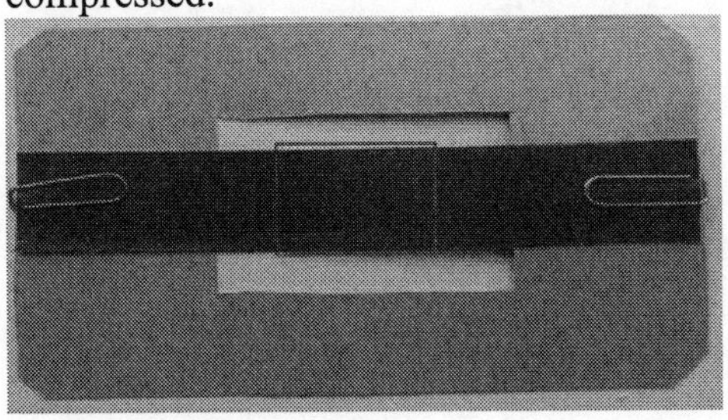

Fig. 3. The test piece.

The estimated contact area is 645 mm^2. The opening in the cardboard is for heating and for pressing, so that during the process all the force is applied to the contact area of the material stripes and

the thickness of the cardboard does not influence the result. The two test pieces are fixed with simple paperclips.
The test setup can be seen in the Fig. 4.

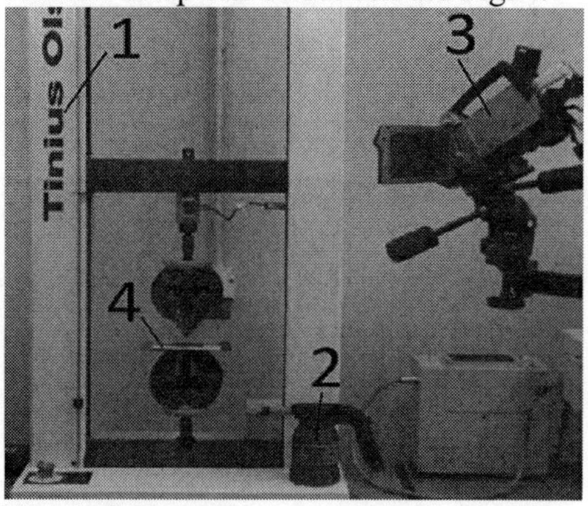

Fig. 4. Test setup.

For applying compressive force, a testing machine Tinius Olsen H10KT (nr 1.) and some simple jigs were used. As a heating source, a gas torch (nr 2.) was used. The temperature was observed with a thermal camera Flir ThermaCAM SC640 (nr. 3). In the Fig. 5, there is a close-up picture of the testing jigs and testing parts.

Fig. 5.Jigs.

The tests were carried out as follows: 1. Firstly the lower steel plate was heated above the melting point of the material to approximately 200 -220°C. 2. The test pieces fixed in the cardboard jig were placed on the lower steel plate imitating the mold. Extra heat was added to the testing pieces to ensure the uniformity of the temperature in mold and material. Temperature was observed with thermal camera seen in Fig. 6. 3. An upper pressure plate was positioned on the material and constant force 400 N was applied during 30 seconds. 4. The test piece was removed from the machine.

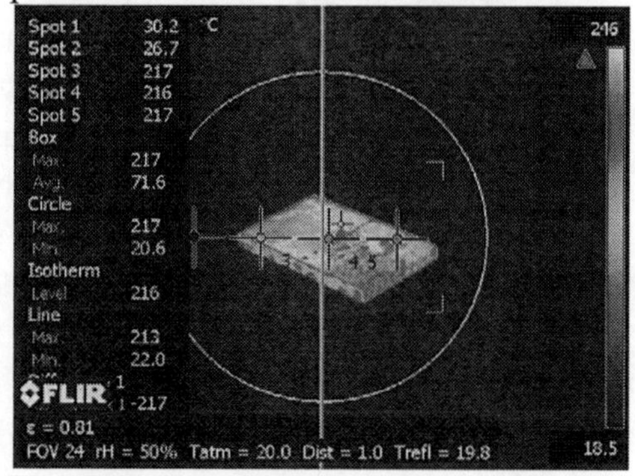

Fig. 6. Thermal image of the mold.

The test was repeated several times varying the heating temperature up to 300° C. Test pieces were torn apart in the tensile machine and investigated from the aspect of adhesion. Findings from the temperature tests show that: 1.The optimal temperature to get the perfect seam is between 200 – 220° C.
2. Using lower temperature can cause improper adhesion. 3. Using higher temperature is not recommended because high temperature can cause the matrix material (PA) to flow out of the seam or to evaporate leaving gaps or cracks in the seam zone (Fig. 7 a)). 4. Even heat distribution is crucial to get high-quality seam. Uneven heating causes improper adhesion (Fig. 7 b)). 5. Excessive heating can cause flashing of the carbon fiber. 6. Considering point 4 and point 5, gas torch as a heat source is bad solution for the current material used.

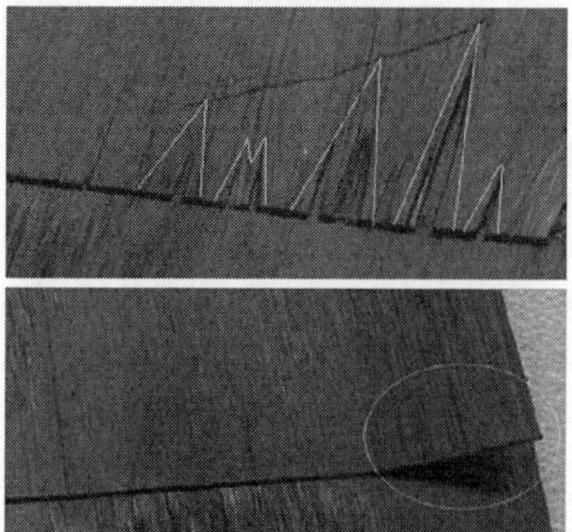
a)
b)
Fig. 7. Heating defects.

The next stage of the testing PA12-CF60 material comprised the investigation of the compaction force. The test was carried out as follows: 1. Testing pieces were positioned on the mold, in that case a planar sheet of steel. The same contact surface area (645 mm^2) was used. 2. Material was heated to the optimal temperature (200 – 220° C). 3. Using compaction roller, the material was pressed together using three different pressures: 0,62 MPa, 1,09 MPa and 1,55 MPa. The manufactured test pieces were then torn apart with a tensile machine to find out the force when the seam breaks. The results are shown in the Fig. 8.

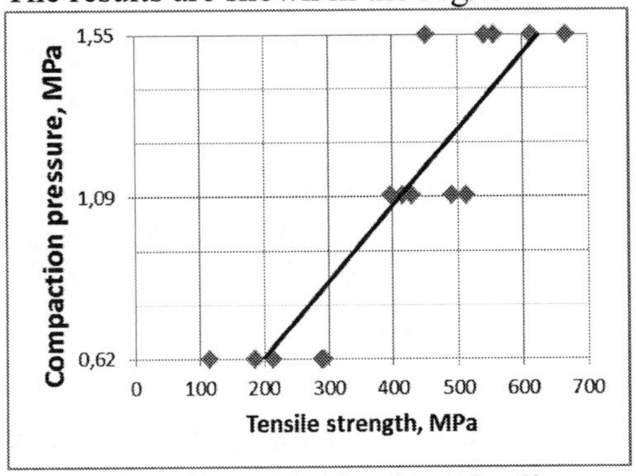

Fig. 8. Tensile strength of test bodies.

It can be seen that there is a positive correlation between the compaction force and the tensile strength of the seamed testing body. Increasing the compaction force has positive effect on tensile strength.

5. CONCLUSION

Automated Fiber Placement as one of the Robotized Fiber Placement method is one way to manufacture high tech composite parts. Traditionally, the method has been used in aeronautics, automotive industry and other large scale industries that have the necessity and means for adapting such technologies. If the costs of the equipment were lower, the technology could be easily harnessed in smaller industries. The purpose of this research was to make a preliminary study concerning AFP process in order to design later affordable AFP equipment. For that we have chosen a material and tested it to find the relations between the main operational parameters and a way to control them. As a result of the experimental part, it appeared that gas torch is not suitable for heating the material PA12-CF60 because of the uneven heat distribution. Proper heating of the material is one of the key factors when high quality is the goal. This is one of the reasons why laser heating has found its place in ATF technology. Secondly, maintaining the constant consolidation force during the lamination process is crucial because it greatly affects the mechanical properties of the final product.

6. ACKNOWLEDGMENT

This research was supported by the EU structural funds project „Smart Composites - Design and Manufacturing“ 3.2.1101.12-0012, Baseline Funding Grant B44 and Estonian Science Foundation grant ETF9441.

7. REFERENCES

[1] Y. Guo, M. Ruess, Z. Gürdal, A contact extended isogeometric layerwise approach for the buckling analysis of delaminated composites. Composite Structures. 2014, 16, 55-66.

[2] A. Khani, M. M. Abdalla, Z. Gürdal, Optimum tailoring of fibre-steered longitudinally stiffened cylinders, Composite Structures, 2015, 122, 343-351.

[3] J. Sliseris, K. Rocens, Optimal design of composite plates with discrete variable stiffness. Composite Structures, 2013, 98, 15-23.

[4] J. Sliseris, G. Frolovs, K. Rocens, V. Goremikins, Optimal Design of GFRP-Plywood Variable Stiffness Plate. Procedia Engineering, 2013, 57, 1060-1069.

[5] J. Sliseris, H. Andrä, M. Kabel, B. Dix, B. Plinke, O. Wirjadi, G. Frolovs, Numerical prediction of the stiffness and strength of medium density fiberboards. Mechanics of Materials, 2014, 79, 73-84.

[6] J. Kers, J. Majak. Modelling a new composite from a recycled GFRP. In: Mechanics of Composite Materials, 2008, 44(6), p 623-632.
[7] H. Herranen, O. Pabut, M. Eerme, J. Majak, M. Pohlak, J. Kers, M. Saarna, G. Allikas, A. Aruniit. Design and Testing of Sandwich Structures with Different Core Materials. Materials Science-Medziagotyra, 2012, 18, 1, 45-50.
[8] Majak, J.; Shvartsman, B.; Kirs, M.; Pohlak, M.; Herranen, H. (2015). Convergence theorem for the Haar wavelet based discretization method. Composite Structures, 126, 227 - 232.
[9] Majak, J.; Pohlak, M.; Eerme, M. (2009). Application of the Haar Wavelet based discretization technique to orthotropic plate and shell problems. Mechanics of Composite Materials, 45(6), 631 - 642.
[10] Karjust, K.; Pohlak, M.; Majak, J. (2010). Technology Route Planning of Large Composite Parts. International Journal of Material Forming, 3(Suppl:1), 631 - 634.
[11] Kers, J.; Majak, J.; Goljandin, D.; Gregor, A.; Malmstein, M.; Vilsaar, K. (2010). Extremes of apparent and tap densities of recovered GFRP filler materials. Composite Structures, 92(9), 2097 - 2101.
[12] Majak, J.; Pohlak, M. (2010). Decomposition method for solving optimal material orientation problems. Composite Structures, 92(8), 1839 - 1845.
[13] Majak, J.; Pohlak, M. (2010). Optimal material orientation of linear and non-linear elastic 3D anisotropic materials. Meccanica, 45(5), 671 - 680.
[14] http://www.coriolis-composites.com/ (11.05.2015)
[15] Pitchumani, R., Don, R. C., Gillespie, J. W., Ranganathan, S., Analysis of On-Line Consolidation during Thermoplastic Tow-Placement Process, ASME Heat Transfer Division, New York, 1994, **289**, 223-239.
[16] Lee, W. I., Springer, G. S., A Model for the Manufacturing Process of Thermoplastic Matrix Composites, Journal of Composite Materials, 1987, **26**, 2348-2410.
[17] Beyeler, E. P., Güçeri, S. I., Thermal Analysis of Laser-Assisted Thermoplastic-Matrix Composite Tape Consolidation, ASME Journal of Heat Transfer, 1988, 110:424-430.
[18] Endres M., Developments in Thermoplastic Filament Winding, 22th International SAMPE Technical Conference, Boston, 1990.
[19] Calhoun D. R., Modeling the On-line Consolidation Processing of Thermoplastic Filament Winding, SME Fabricating Composites, Arlington, 1990.
[20] Werdermann, C., Friedrich, K., Cirino, M., Pipes, R. B., Design and Fabrication of an On-Line Consolidation Facility for Thermoplastic Composites, Journal of Thermoplastic Composite Materials, 2, 293-306, 1989.

8. ADDITIONAL DATA ABOUT THE AUTHORS

Anti Haavajõe, Tallinn University of Technology, Faculty of Mechanical Engineering, Department of Machinery, Ehitajate tee 5, e-mail: anti.haavajoe@gmail.com
Madis Mikola, Tallinn University of Technology, Faculty of Mechanical Engineering, Ehitajate tee 5, e-mail: madismikola@gmail.com
Henrik Herranen, Tallinn University of Technology, Faculty of Mechanical Engineering, Department of Machinery, Ehitajate tee 5, e-mail: henrik.herranen@ttu.ee
Meelis Pohlak, Tallinn University of Technology, Faculty of Mechanical Engineering, Department of Machinery, Ehitajate tee 5, e-mail: meelis.pohlak@ttu.ee.

10th International DAAAM Baltic Conference
"INDUSTRIAL ENGINEERING"
12-13 May 2015, Tallinn, Estonia

MULTI-POLE MODELING AND INTELLIGENT SIMULATION OF A HYDRAULIC DRIVE WITH THREE-DIRECTIONAL FLOW REGULATING VALVE

Harf, M. & Grossschmidt, G.

Abstract: *Composing of multi-pole model and simulation of a hydraulic drive with three-directional flow regulating valve is considered in the paper. Multi-pole mathematical model of a hydraulic drive is presented. An intelligent simulation environment CoCoViLa supporting declarative programming in a high-level language and automatic program synthesis is used as a tool. Simulation examples of a hydraulic drive are presented and discussed.*
Key words: *hydraulic drive, flow regulating valve, multi-pole model, intelligent programming environment, simulation.*

1. INTRODUCTION

Using and solving large differential equations systems in simulation of fluid power system dynamics is not wide spread. It is difficult to compose, guarantee the adequacy and solvability of such systems. In analysis and system synthesis frequently simplified, 3rd...5th order differential equation systems are used [1].
In the current paper an approach is proposed, which is based on using multi-pole models with different oriented causalities [1] for describing components of different levels. Components of the lowest level are hydraulic resistors, tubes, hydraulic interface elements, directional valves, etc [2]. Hydraulic control valves of different types [3] are used as components of the higher level. In such a way models of complex systems can be built up hierarchically.
A special iteration technique is used that allows avoid solving large equation systems during simulations. Therefore, multi-pole models of large systems do not need considerable simplification.
Modeling and simulation of a hydraulic drive including a three-directional flow regulating valve is considered as an example of applying proposed methodology.

2. MULTI-POLE MODELS

In general a multi-pole model represents mathematical relations between several input and output variables (poles). The nearest to physical nature of various technical systems is using multi-pole mathematical models of their components and subsystems [1].
The multi-pole models of the components describe the ports, which have oriented input and oriented output variables in pair, as it is in most real physical systems. Multi-pole models enable to express both direct actions and feedbacks.
The multi-pole model concept enables us to describe mathematical models graphically which facilitates the model developing.

3. SIMULATION ENVIRONMENT

CoCoViLa is a flexible Java-based intelligent simulation environment that includes different simulation engines and is intended for creating and performing simulations in various engineering domains [4]. It provides visual tools and supports full automatic program construction from specifications that are given visually [5].

4. THREE-DIRECTIONAL FLOW REGULATING VALVE

Flow regulating valves [6, 7] are used when the working speed of hydraulic drive should remain almost constant in case of different loads at the user.

Three-directional flow regulating valve (Fig.1) contains adjustable throttle and connected in parallel pressure compensator ensuring constant pressure drop in the throttle. In addition to keeping output volumetric flow constant it keeps pressure at the pump proportional to the load force.

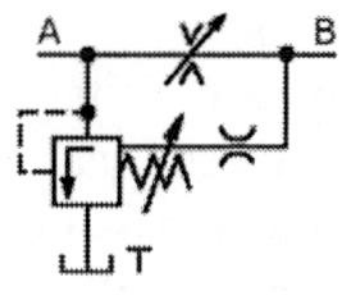

Fig.1. Functional scheme of a three-directional flow regulating valve

In Fig.2 three-directional flow regulating valve of Mannesmann Rexroth is shown.

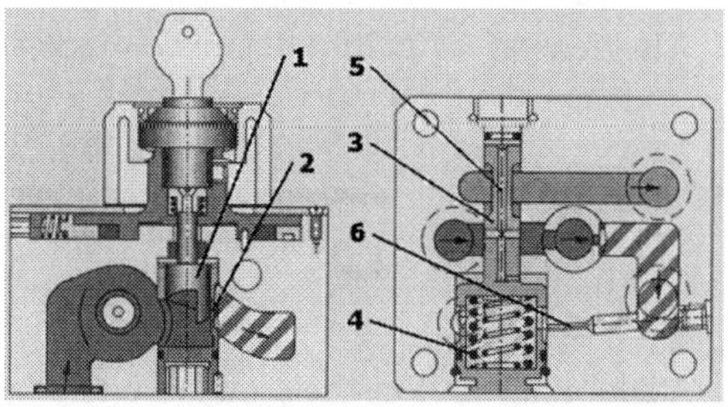

Fig.2. Three-directional flow regulating valve

The valve consists of the throttle pin 1 with orifice 2, normally closed regulating spool 3 with two springs 4, bores 5 and 6 to the spool surfaces.

Multi-pole model of a three-directional flow regulating valve is shown in Fig.3.

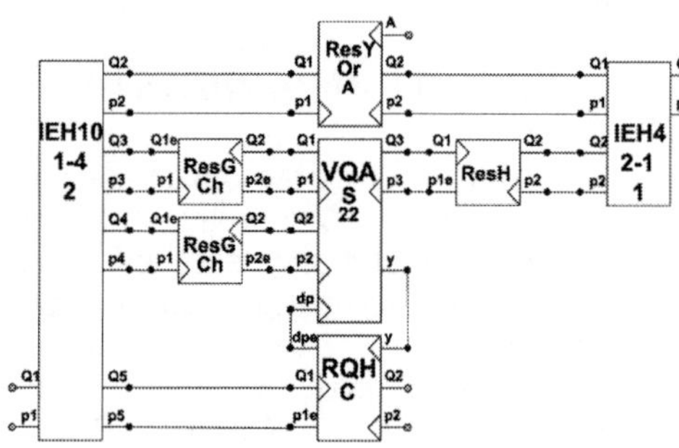

Fig.3. Multi-pole model of a three-directional flow regulating valve

Multi-pole models of components:
ResYOrA – regulating orifice, VQAS22 – pressure compensator spool, RQHC – pressure compensator spool slots, ResGCh, ResH – cushioning resistors, IEH10-1-4-2, IEH4-2-1-1 – interface elements.

Multi-pole mathematical models of control valves of fluid power systems are considered in [2, 3]. Exceptions concerning components used here are described below.

VQAS22 differs from **VQAS21** [3] as follows. **VQAS21** is normally open pressure compensator spool but **VQAS22** is normally closed pressure compensator spool.

In **VQAS21** displacement of the pressure compensator spool

$$y1 = 1 / (1 - B) * (F / c - fV0),$$

pressure compensator spool slot width

$$y = y0 - y1.$$

In **VQAS22**

$$y1 = 1 / (1 + B) * (F / c - fV0), \quad y = y1.$$

Pressure compensator spool slot RQHC
Inputs: pressure $p2$, displacement y of the pressure compensator spool, volumetric flow $Q1$.

Outputs: pressure $p1e$, pressure drop dpe in poppet-valve slot, volumetric flow $Q2$.

Through-flow area of the pressure compensator spool slot

$$A = Pi * d1 * y * \sin(\beta * Pi / 180),$$

where
- d1 diameter of the spool sleeve,
- β half of spool cone angle.

Turbulent flow resistance

$$RT = \rho / (2 * \mu^2 * A^2),$$

where
- ρ fluid density,
- μ discharge coefficient.

Output pressure

$$p1e = p2 + (RT * abs(Q1)) * Q1.$$

Difference of pressures:

$$dpe = p1e - p2.$$

5. HYDRAULIC DRIVE WITH THREE -DIRECTIONAL FLOW REGULATING VALVE

Functional scheme of a hydraulic drive with three-directional flow regulating valve is shown in Fig.4.

The pump **PV** is driven by electric motor **ME** through clutch **CJh**. The outlet of the pump is provided with three-directional flow regulating valve **FCV** and safety valve **SV1** (spool VS and throttle edge RV in Fig.5).

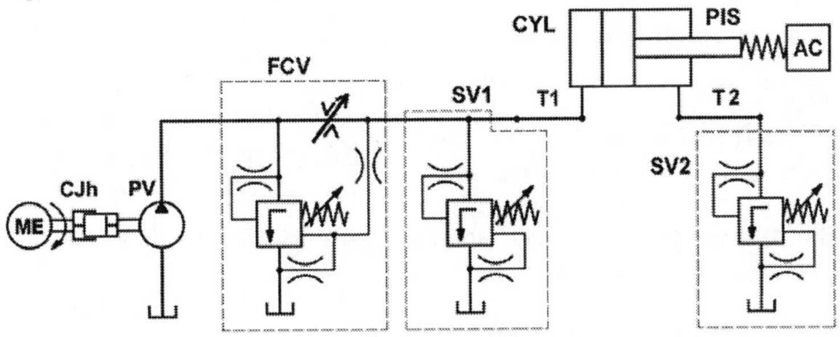

Fig.4. Functional scheme of a hydraulic drive with three-directional flow regulating valve

Tubes **T1** and **T2** are located in inlet and outlet of hydraulic cylinder **CYL**. Piston and actuator are denoted respectively as **PIS** and **AC**. Constant pressure in outlet of the cylinder is ensured by pressure valve **SV2**.

6. SIMULATION OF STEADY STATE CONDITIONS

Simulation task of steady state conditions of a hydraulic drive with three-directional flow regulating valve is shown in Fig.5.

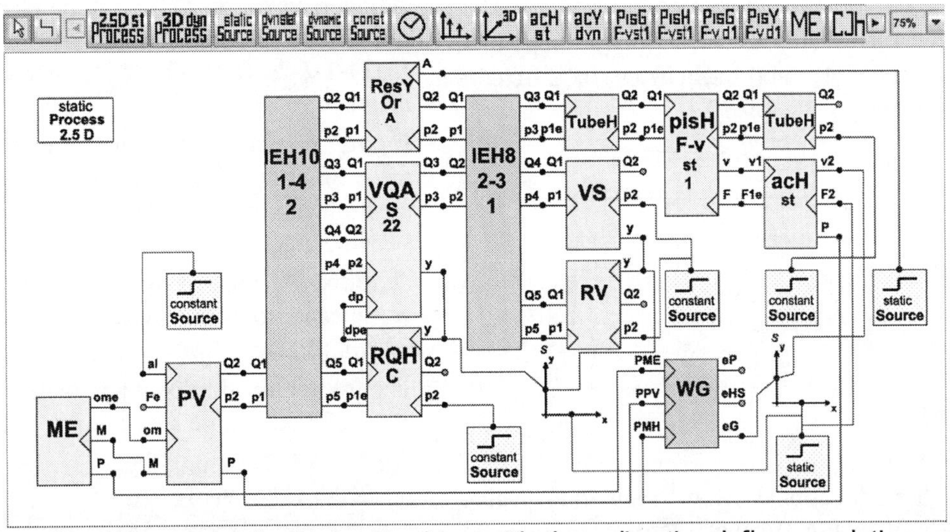

Fig.5. Simulation task of a hydraulic drive with three-directional flow regulating valve for steady state conditions

Multi-pole models: **ME-** electric motor, **PV** - axial-piston pump, **ResYOrA** – regulating throttle orifice, **VQAS22** – pressure compensator spool, **RQHC** – pressure compensator slot, **VS** – safety valve spool, **RV** – throttle edge of safety valve spool, **pisH_F-v_st1** – piston, **acH** – actuator, **TubeH** – tube, **IEH10-1-4-2**, **IEH8-2-3-1** – interface elements, **WG** – efficiency coefficient calculator [2, 3, 8, 9].

Inputs: outlet pressures *p*2, regulating orifice area *A*, constant position angle *al* of the pump regulating swash plate.

Outputs: actuator velocity *v*2, efficiency coefficient *eG* of the entire hydraulic drive.

Simulation manager: static Process 2.5D.

The following parameter values are used for steady state simulations.

For **VQAS22:** d1=0.008 m, d2=0.03 m, μ=0.8, β=30 deg, ds1=0.003 m, Ds1=0.022 m, n1=5, ds2=0.0025 m, Ds2=0.014 m, n2=4, G=8e11 N/m, m=0.04 kg, kfr=2e-9 N/Pa, Ff0=3 N, h=5 Ns/m.

For **RQHC:** d1=0.015 m, d2=0.012 m, μ=0.7, β=30 deg.

For **ResYOrA:** μ=0.7.

For **TubeH:** d= 0.019 m, l= 2 m.

For **pisH_F-v_st1:** piston diameter dpi= 0.10 m, diameters of rods dr1= 0 m, dr2= 0.056 m, piston friction force Ffpi= 100 N, rod friction force Ffr= 50 N.

For **acHst:** Ffr= 100 N, h= 100 Ns/m.

For **VS:** d= 0.008 m, spring stiffness c= 8950 N/m, preliminary deformation of spring fV0= 0.00722 m, h= 20 Ns/m.

For **RV:** d=0.008 m, μ=0.8, β=45 deg.

Results of simulation of steady state conditions depending on the load force for three different values of the regulating orifice area A = (18, 10, 2) e-6 m^2 are shown in Fig.6 and Fig.7.

In Fig.6 graphs of actuator velocities (graphs 1) and efficiency coefficients (graphs 2) are shown.

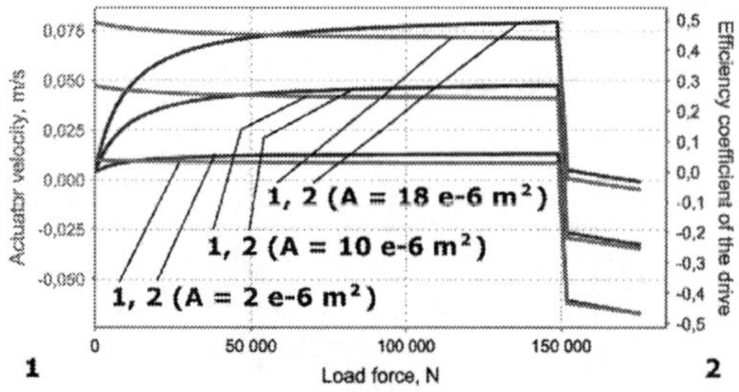

Fig.6. Graphs of simulations of steady state conditions

Three-directional flow regulating valve (FCV in Fig.4) causes actuator velocity to remain almost constant. After opening the safety valve (SV1 in Fig.4), actuator moves backward (actuator velocity becomes negative). Increasing the load force causes efficiency coefficient to rise.

In Fig.7 graphs of pressure compensator spool displacements (graphs 1) and safety valve spool displacement (graph 2) are

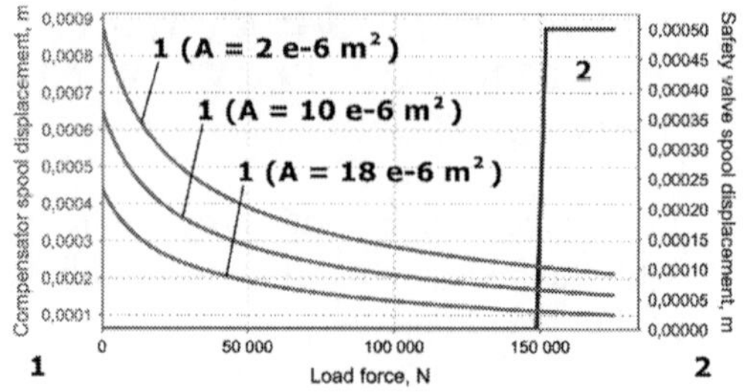

Fig.7. Graphs of spool displacements

presented. Pressure compensator spool displacement decreases by increasing of the load force. The safety valve (SV1 in Fig.4) opens at load force of ~150 000 N.

6. SIMULATION OF DYNAMICS

Simulation task of a hydraulic drive with three-directional flow regulating valve for dynamics is shown in Fig.8.

Additional and different multi-pole models from steady state conditions: **CJh** – clutch, **TubeH**, **TubeY** – inlet and outlet tube, **pisY_F-v_dyn1** – piston, **cylY** – cylinder, **veZ1**, **veZ2** – volume elasticities of cylinder chambers, **acYdyn** – actuator, **IEH4-1-2-2**, **IEH4-2-1-2** – interface elements [2, 3, 8, 9].

Inputs: constant outlet pressures *p*2, load force *F*2, regulating orifice area *A,* constant position angle *al* of the pump regulating swash plate.

Outputs: actuator velocity *v*2, outlet volumetric flows *Q*2, cylinder position *xfi*.

Simulation manager: dynamic Process3D.

The following additional parameter values are used in dynamic simulations.

For **VQAS22:** m=0.04 kg.

For **ResYOrA:** A=1e-5 m^2.

For **ResGCh:** d= 0.0005 m, l= 0.02 m.

For **ResH:** d= 0.001 m, l= 0.02 m.

For **VS:** m= 0.02 kg, h= 20 Ns/m.

For **TubeH**, **TubeY:** d= 0.019 m, l= 2 m.

For **pisY:** elasticity of piston rod er2= 1e-10 m/N.

For **veZ1**, **veZ2:** lengths of cylinder chambers l1= l2= 0.2 m,

For **acYdyn:** m= 20 kg, h= 3e3 Ns/m.

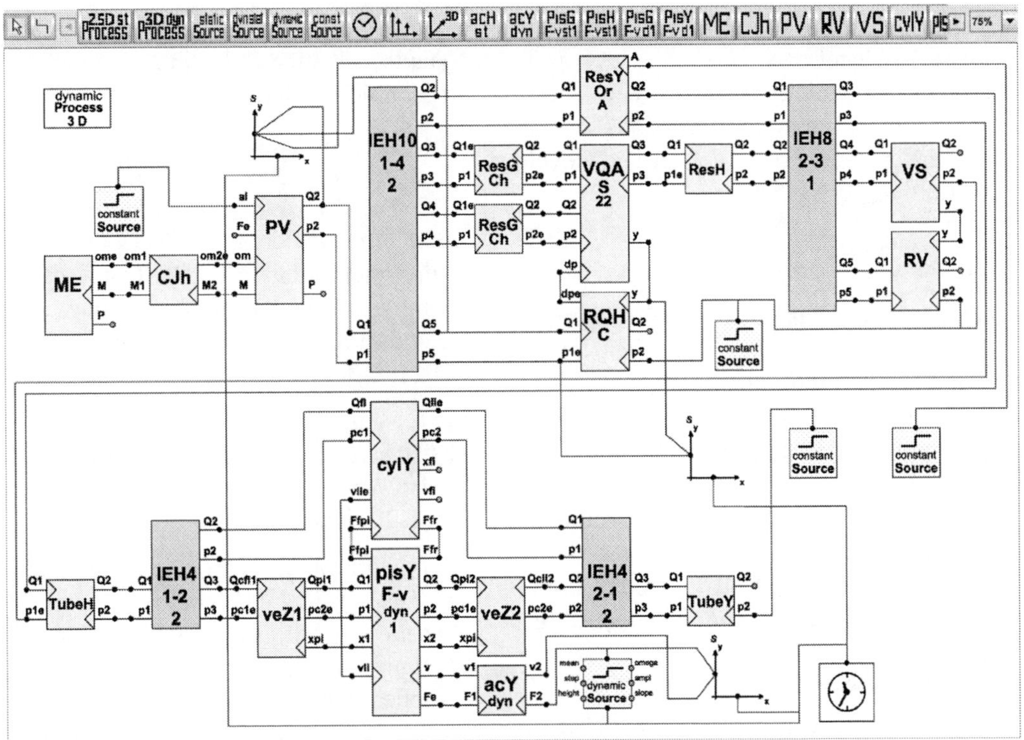

Fig.8. Simulation task of dynamics of a hydraulic drive

Results of simulation of dynamic responses caused by applying the hydraulic drive with three-directional flow regulating valve actuator step load force *F2* = 5E3 N (step time 0.05 s) as input disturbance are shown in Fig.9...Fig.11.

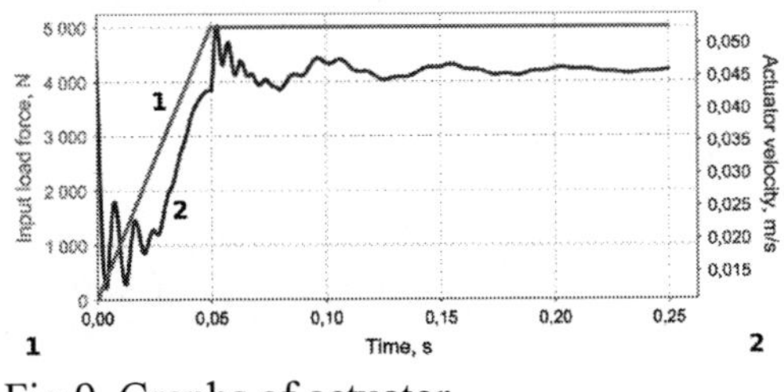

Fig.9. Graphs of actuator

Input load force step change (graph 1) initially causes actuator velocity (graph 2) to drop down. After load force rises to a new level, actuator velocity stabilizes.

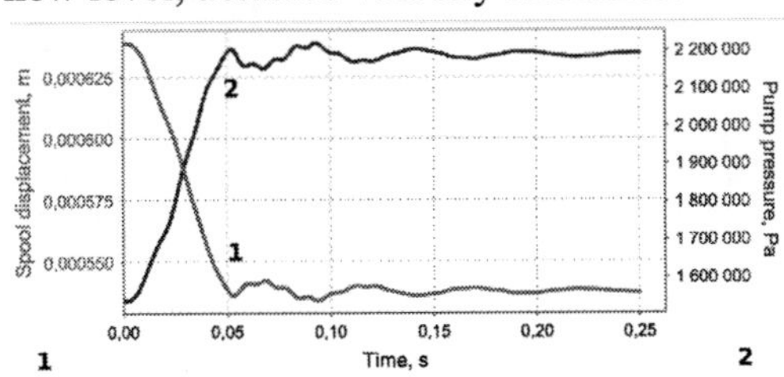

Fig.10. Graphs of flow regulating valve

Flow regulating valve spool (graph 1) takes a new position and causes pump pressure (graph 2) to rise.

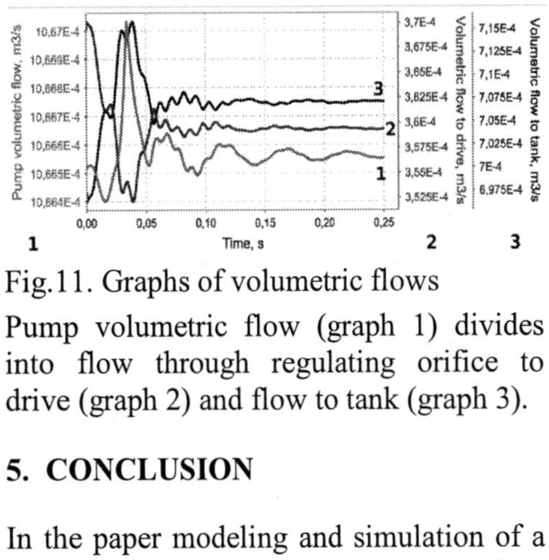

Fig.11. Graphs of volumetric flows

Pump volumetric flow (graph 1) divides into flow through regulating orifice to drive (graph 2) and flow to tank (graph 3).

5. CONCLUSION

In the paper modeling and simulation of a hydraulic drive with three-directional flow regulating valve is considered.

As experiments show, control valve parameters are to be adjusted for each particular case to attain the best performance of the hydraulic drive.

Control valve models e.g. those we described and used in the paper can be used when composing models of fluid power systems whatever type.

Using methodology described here enables to try out different configurations and find optimal parameters in design and develop of fluid power systems.

6. ACKNOWLEDGEMENTS

This research has been partially supported by the *European Regional Development Fund* (ERDF) through:

- Estonian Centre of Excellence in Computer Science (EXCS).
- The project no 3.2.1201.13-0026 „Model-based Java software development technology".
- The project no 3.2.1101.12-0012 „Smart composites - design and manufacturing".

7. REFERENCES

[1] Grossschmidt, G. and Harf, M. *"COCO-SIM - Object-oriented Multi-pole Modeling and Simulation Environment for Fluid Power Systems, Part 1: Fundamentals".* International Journal of Fluid Power, 10(2), 2009, 91 - 100.

[2] Grossschmidt, G.; Harf, M. *Simulation of hydraulic circuits in an intelligent programming environment (Part 1, Part 2).* 7th International DAAAM Baltic Conference "Industrial engineering", 22-24 April 2010, Tallinn, Estonia, 148 -161.

[3] Harf, M. ; Grossschmidt, G. *Multi-pole Modeling and Intelligent Simulation of Control Valves of Fluid Power Systems (Part 1, Part 2).* In: Proceedings of the 9th International Conference of DAAAM Baltic "Industrial Engineering"; 24-26 April 2014, Tallinn, Estonia; (Ed.) Otto, Tauno, Tallinn; Tallinn University of Technology, 2014, 17-28.

[4] Kotkas, V., Ojamaa A., Grigorenko P., Maigre R., Harf M. and Tyugu E. *"CoCoViLa as a multifunctional simulation platvorm",* In: SIMUTOOLS 2011 - 4th International ICST Conference on Simulation Tools and Techniques, March 21-25, Barcelona, Spain: Brussels, ICST, 2011, 1-8.

[5] Matskin, M. and Tyugu, E.. *Strategies of structural synthesis of programs and its extensions,* Computing and Informatics, Vol. 20, 2001, 1-25.

[6] Murrenhoff, H. *Grundlagen der Fluidtechnik,* Teil 1: Hydraulik, 4. neu überarbeitete Auflage. Institut für fluidtechnische Antriebe und Steuerungen, Aachen, 2005.

[7] Gebhardt, N., Will, D. and Nollau, R. *Hydraulik: Grundlagen, Komponenten, Schaltungen,* 5. Neubearb. Auflage, Springer Verlag Berlin Heidelberg, 2011.

[8] Grossschmidt, G.; Harf, M. *COCO-SIM - Object-oriented multi-pole modelling and simulation environment for fluid power systems. Part 2: Modelling and simulation of hydraulic-mechanical load-sensing system.* International Journal of Fluid Power, 10(3), 71 - 85.

[9] Grossschmidt, G.; Harf, M. *Effective Modeling and Simulation of Complicated Fluid Power Systems.* The 9th International Fluid Power Conference, 9. Ifk, March 24-26, 2014, Aachen, Germany, Proceedings Vol 2, 374-385.

10th International DAAAM Baltic Conference
"INDUSTRIAL ENGINEERING
12-13 May 2015, Tallinn, Estonia

DIFFICULTIES IN SMES AND KPI SELECTION MODEL AS A SOLVER

Kaganski, S.; Paavel, M.; Karjust, K.; Majak, J.; Snatkin, A.

Abstract: *Small and medium enterprises are seen as „backbone of the European economy"[1] Nowadays, in „harsh environment" and difficult economic situation, to be able to survive, SMEs should not only optimize production, find new investors, rise effectiveness and productivity, but also change the way of thinking and try to understand all processes in production and company itself instead of just „driving forward with closed eyes". The measurement of processes and company's condition at present moment should be one of the keys to success. Continuous development and analyse would provide management team with plan of actions. The main objective of this paper is to introduce difficulties and problems, with which enterprises are faced in their life. Furthermore, to show concept of KPI selection model for enterprises, which would help to understand, what KPIs should be taking into account and studied by management and how those metrics can change the situation and solve all difficulties. In addition, during the applying of model, the amount of data and data flow in enterprise would be optimized.*

Key words: *Key performance indicators (KPI), Small and medium enterprises (SME), KPI selection model.*

1. INTRODUCTION

The last twenty years have seen profound change in the private sector's relationship with society. Globalization, deregulation, privatization and a reconsideration of the links between state and market have changed the basic principles on which private companies are expected to contribute to the public sector [2, 3]. Additionally, the economic conditions in 2011/2012 in the World and in the European Union with new difficulties nowadays, due political issues and food embargo as a counter from Russian Federation, have crucial impacts on SMEs but still, this form of business is remaining the most important and widely spread. They are showing a better performance in compering to big enterprises and corporations [4,5], additionally, the dynamic role of SMEs-as a chine of the European economy-seems to have been played important role in the recovery from the global crisis since 2008 [1]. On the one hand, SME cannot effort high cost researches and developments (R&D) like large enterprises, due the financial and economic aspects, on the other hand, the speed of implementation of new technologies and methods (Lehtimaki considered the importance of new ideas for product innovations in SMEs of Finland to top of management [6]), comparing with large enterprises, are high and not so much time and money consuming. Additionally, SME can be innovative in other ways-modernization of products and processes to win new markets (LE are not interested in small markets, they are trying to get a really wide spreading markets and big clients).

2. DIFFICULTIES THAT ARE APPEARING IN SME

Watt has distinguished following steps in the risk management process, which should be taken into account by managers [7]:

• Establishing the SMEs risk strategy;

• Determining the SMEs risk appetite;
• Identification and assessment of risk;
• Prioritizing and managing risk.

Also, if we try to divide problems, with which enterprises are facing, then there will be two main groups:

• Financial or economic problems (SMEs success is tied in with the local economy as the SME sectors market growth is usually at the same rate as the macro economy as a whole, therefore, if there is an economic downturn, SMEs will usually also experience difficulty [8].

• Enterprise based problems (human resource problems, multi-functional management, high employee turnover rate [one of the common problems nowadays], lack of skills and experience, low productivity and difficulties of finding quality staff [9].

Considering financial problems, SMEs have very limited bank finance, which is only around 10 per cent, while self-finance remains the major source of finance contributing 76.5 per cent of fixed capital and 51.8 of working capital [10]. In critical situations SMEs don't have the buffer for not only investments in new technologies but also for covering additional costs during prices growth or projects recalculation. For example, according World Bank survey (2002) the lack of money for the majority of Bangladesh's SMEs (55%) was the main issue, during their operation.

To reduce the impact of economic/finance issues on SMEs entrepreneurs should to [11]:

• Definite market opportunities;
• Pay more attention to team working;
• Choose or develop suitable marketing entry strategy;
• Operate the profitable ventures.

It is important to know, who the clients are and of course do not forget about competitors and comparing to them, what would be the main strengths and weaknesses. Additionally, entrepreneurs should think about the logistics and the prices should be also competitive. To open new firm and to start business is quite easy, but to stay afloat and continue to grow is very difficult.

When we are talking about measurements of economic aspects, then nowadays, the majority of SMEs has not established strict financial accounting system, including real-check, card-check and account-check. It's is difficult to carry out the financial accounting procedure [12]. Although most SMEs apply basic financial accounting system but it does not match smoothly with the logistics, manufacturing, sales and cannot provide enterprises with complete information [13].

Considering the enterprise based problems, Employee Turnover is one of the common problems. "It is the ratio of the number of workers that had to be replaced in a given time period to the average number of workers" (Agnes, 1999) [14]. It is often utilized as an indicator of company performance and can easily be observed negatively towards the organization's efficiency and effectiveness (Glebbeek & Bax, 2004) [15]. Due to limited growth of SME most of the skilled employees leave SMEs. According Levy, SMEs are knowledge creators but poor at knowledge retention [16]. Employee job satisfaction has influence on employee turnover in organizations. The extent to which an organization is able to retain its employees' depends on the level of job satisfaction that is made available to these workers. [17]. However, taken into account nowadays situation, the main reasons of high turnover is salary. Young specialists are searching the best place, which could include a good salary, an interesting job and good additional opportunities for further rise. According to the European Statistic 2010, In a recent OECD (Organisation for Economic Co-operation and Development study (2009)) covering eleven countries for job turnover and twenty-two for labour turnover and using harmonized data, job turnover rates were estimated at 22% (of total employment) over the period 1997-2004, and annual average labour turnover

rates at 33% (of total employment) between 2000 and 2005 [18].
The high turnover rate is not only problem, that companies in Europe and in other countries should face. The Boston Consulting Group in their research is mentioning that in the nearest future, companies will face five critical HR challenges [19]:

• Managing talent;
• Managing demographics;
• Becoming a learning organization;
• Managing work-life balance;
• Managing change and cultural transformation.

Considering the nowadays situation, the lack of qualified workers is becoming more problematic with every year. HRM should be on the same level of importance, like the economic issues (turnover, consumer leverage ratio, retail sales and etc.). Managers should not forget that the main job is done by workers and the beneficial HRM is directly influencing the financial indexes of company. KPI of HR for management can help to make right decisions to changes the situation in enterprise.

3. KPI SELECTION MODEL

To understand the main purpose of bottlenecks and difficulties at company, the necessary measurements should be performed by management to be able to evaluate the impact of each factor. KPI selection model [20] should be seen as tool, which is able to identify the critical spots in enterprise and solve them in nearest future perspective. The idea is to use traditional methodical questionnaires to make management understandable and to bring out the main bottlenecks and weaknesses in the production and general enterprise processes.
On Fig 1 the main first phase of KPI selection model is been illustrated. The draft version was described in article Kaganski 2014 [20], however, during practical study, it would get final review and optimized. The first phase is questionnaire, which could be called also as a preparation phase. Survey for KPI has been built concerning the productivity KPIs. In this study, questions are constructed in this way, that by responding on them, the critical spots and problems would be identified. The questions are linked with KPIs that would be investigated after data collection. In turn, KPIs are divided by 3 groups:

• Direct KPI – indicators, which are in explicit relation to the responses;
• Indirect KPI – indicators, which are in connection to more than one question;
• Suggested KPI – indicators, which are proposed to the management for further studying.

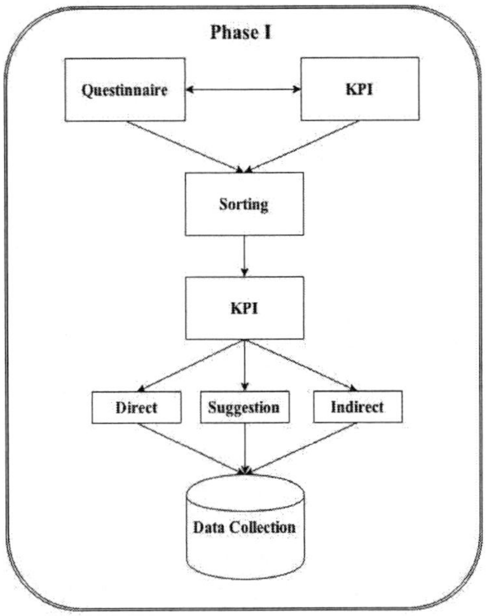

Fig. 1. The first phase of KPI selection model

In table 1, the example of distribution for KPIs from first phase has been shown. This kind of approach provides possibility to show management, if right KPIs have been followed (direct KPIs). Furthermore, depending of the problems, which would appear after analyse of the answers, suggestions can be provided to company, what could be measured.

Question	KPI		
	Direct	Indirect	Suggestion
What's the average age of white collars (Admin person) in your company?	% of employee over age 55		
Your age range	Average age range of employee		
What's the average age of Blue collars (factory floor persons) in your company?			
How long have you been employed in the current enterprise?	staff turnover ratio	Satisfaction	
How many white collar employees have left your company during last year?	staff turnover ratio		
How many Blue collar employees have left your company during last year?	staff turnover ratio		
Company is offering appropriate training for the job, as well as specific Occupational Health's Safety & Welfare training?			Employee Training Index

Table 1. Example of question vs KPI

Indirect KPIs can be specified only by answering at least on two questions, which are connected to it. As for example, KPI-satisfaction can be followed in company only by knowing the situation with staff turnover, trainings and benefits availability and etc.

To eliminate misunderstanding and check the relationship between pairs "question-KPI" the sorting was performed in WEB resource "Optimal workshop" [21] by case study group. The case study group is group of researches, who has no connection to the study, but in spite of that has necessary knowledge and experience in the field.

Survey should be filled by workers from all areas, including "blue collars", "white collars" and CEO. The independent point of view of each employee would help to construct the right representation.

4. CASE STUDY

The selection model would be tested on different manufacturing companies, which are dealing within divergent fields to evaluate the compatibility of it. Taking into account, that prevailing language at companies in Estonia can be divided to 3 main groups: Estonian, English, Russian, -

the survey should be providing the opportunity to be used in any condition. After acquiring mandatory amount of data, analyse should be done. Each studied company would receive the methodology of improvement steps for the future and also, the whole picture of all enterprises by detection standard bottlenecks, would be analysed. Furthermore, the questionnaire is only one step of data collection and can be considered as manual way. The other opportunity is to use advantage of PMS (product monitoring system) and get data directly from machines, production lines to the database for further study. Wireless sensors would be attached to machine park and provide data of parameters: vibration, temperature, voltage consuming, which in turn could be used to prognoses the condition of tool. [22] The usage of wireless sensors would reduce costs (not need for cables and wires) and simplify the assembly and installation. [23] The advantage of this approach is the possibility to study KPI: OEE (overall equipment efficiency) and all related with it metrics. Online data flow gives the opportunity to make right decisions in instance and drive the production in right direction, which would save time, resources and nervous.

5. FURTHER RESEARCH

The KPI selection model would be tested in practice and acquired data analysed. Taking into account, that testing model on one enterprise isn't enough for optimization, next points should be done/analysed:

• Optimization of KPI selection model (questionnaire & linked KPIs);
• Data from PMS as first step for further study;
• Different SMEs for collecting right amount of data.

6. CONCLUSION

Considering the efficiency of production in SMEs, HR issues, material flow and other critical subjects at companies, the measurement with further improvements and optimization remains general task for management. The difficulties, which were described in article, cannot be eliminated in an instant; however, the necessary measures could be done in order to avoid further difficulties. The KPI model structure and difficulties, which are occurring at companies, were described in this paper. Later steps were defined for next researches.

The KPI selection model could become the main fundament on which decisions and improvements would relay. Furthermore, it should simplify the work of management and make production more transparent. Optimization, data collection and analysing are foreseen as next tasks.

7. ACKNOWLEDGEMENTS

This research was supported by ETF grant 9441 and Innovative Manufacturing Engineering Systems Competence Centre IMECC (supported by Enterprise Estonia and co-financed by the European Union Regional Development Fund, project EU30006).

8. REFERENCES

1. Wymenga, P.; Spanikova V.; Barker A.; Konings J.; Canton E. EU SMEs in 2012: at the crossroads. *Annual report, Rotterdam,* September 2012.
2. Raynard P.; Forstater M.; Implications for small and medium enterprises in developing countries. *United Nations Industrial Development Organization*, 2002.
3. Karjust, K.; Pohlak, M.; Majak, J. Technology Route Planning of Large Composite Parts. *International Journal of Material Forming, vol 3, 631 – 634*, 2010
4. Snatkin, A.; Karjust, K.; Eiskop, T. Real time production monitoring system in SME. *In: Proceedings of the 8th International Conference of DAAAM Baltic Industrial Engineering 19-21st April 2012.* (Toim.) Otto, T.. Tallinn: Tallinna Tehnikaülikooli Kirjastus, 2012, 573 – 578, 2012
5. Siringoringo H.; Tintri D.; Kowanda A.; Problems faced by small and medium business in exporting products. *Delhi Bussiness Review, Vol.10, No.2,* 2009
6. Lehtimaki, A.; Management of the Innovation Process in Small Companies in Finland. *IEEE Transactions on Engineering Management, 38 (2): 120–6,* 1991
7. Watt J.; Strategic risk management for small businesses. In:Reuvid, J. (ed.). *Managing Business Risk 2nd Edition – a practical guide to*

protecting your business. London – Philadelphia. Kogan Page, 2007

8. Berry A.;The role of the small and medium enterprise sector in Latin America: implications for South Africa. *Unisa Latin American Report, 18(1): 4-14,* 2002
9. Smit Y.; Watkings J.A.; A literature review of SME risk management practice in South Africa. *African Journal of Business Management, 30 May,* 2012
10. Akterujjaman S.M.; Problems and prospects of SMEs Loan Management: A study on Mercantile Bank Limited, Khulna Branch. *Journal of Business and Technology, Volume V, Issue 02, July-December,* 2010
11. Barrow C.; Brown.R. Barrow. C; Principles of Small Business, August 1997
12. Zhengjin; Small and medium-sized enterprises financial management problems and countermeasures. *Business Administration (06),* 2011
13. Xuhui Y..; Ruoxi Z.; Discussion on SME financial management problems and countermeasures, *International Converence on Artificial Intelligence and Software Engineering (ICAISE 2013)*
14. Agnes, M. Webster"s New World College Dictionary (4th Edition).New York, 1999
15. Glebbeek, A.C & Bax, E.H.; Is high employee turnover really harmful: An empirical test using company records. *Academy of Management Journal, 47 277-286,* 2004
16. Levy, M.;Loebbecke, C. and Powell, P. SMEs, Cooperation and Knowledge Sharing:the Role of InformationSystem. *European Journal of Information System, Vol. 12,* 2003
17. Mbah, S.E.; Ikemefuna C.O. Job Satisfaction and Employees' turnover intentions in total Nigeria plc. In Lagos State. *International Journal of Humanities and Social Science, Vol. 2, no.14,* 2012
18. Employment in Europe 2010, Youth and segmentation in EU labour markets [WWW] http://ec.europa.eu/employment_social/eie/chap3-2_en.html (17.03.2015)
19. Caye, J.M.; Strack, R.; Leicht, M.; Villis U. The Future of HR in Europe, Key challenges through 2015, *Boston Consulting Group,* 2007
20. Kaganski, S.; Paavel, M.; Lavin, J. Selecting key performance indicators with support of enterprise analyze model. *9th International DAAAM Baltic Conference,* Tallinn, 2014
21. Optimal workshop [WWW] https://www.optimalworkshop.com (12.03.2015)
22. Aruväli, T.; Serg R.; Preden, J.; Otto, T. Inprocess determining of the working mode in CNC turning, *Estonian Journal of Engineering,* 2011
23. Aruväli, T.; Preden, J.; Otto, T. Modern monitoring opportunities in shopfloor. *Annals of DAAAM for 2010 & Proceedings of the 21st International DAAAM Symposium, Volume 21, No. 1*

9. ADDITIONAL DATA ABOUT AUTHORS

PhD student Sergei Kaganski; PhD student Marko Paavel; Ass.Prof. Kristo Karjust, Res.Prof. Jüri Majak; PhD studend Aleksei Snatkin; Department of Machinery, Tallinn University of Technology, Ehitajate tee 5, Tallinn, 19086, Estonia, E-mail:
sergei.kaganski@outlook.com
marko.paavel@ttu.ee
kristo.karjust@ttu.ee
juri.majak@ttu.ee

10th International DAAAM Baltic Conference
"INDUSTRIAL ENGINEERING" -
12-13 May 2015, Tallinn, Estonia

STUDY OF COMBINED MACHINING PARAMETERS ON 3D ROUGHNESS BEHAVIOUR IN MOULDS AND DIES

A. Logins, T. Torims, S. C. Gutiérrez Rubert, P. Rosado Castellano, R. Torres Carot, F. Sergeyev

ABSTRACT

In order to process materials with special characteristics (e.g. brittleness, hardness and stiffness), high-speed machining (HSM) has gradually become one of the most popular metal cutting and machining methods. In the die-mould manufacturing industry, HSM is mainly used for workpiece machining. By applying HSM technology[1], it is possible not only to obtain specific surface roughness parameters, but also to improve surface quality generally. This is because the technological parameters of the machining have an impact not only on surface quality parameters such as surface texture and roughness, but also on surface micro-hardness.[2] This paper contains detailed results on how high-speed milling impacts on 3D surface texture and roughness parameters. The ISO25178 standard of geometric product specification was used to characterize surface roughness. This paper also offers a comprehensive review of how HSM impacts on industry and the surface micro-hardness with regard to particular materials. Carefully designed experiments were conducted, varying technological parameters, which were recorded along with the relevant 3D roughness parameter measurements. Significant conclusions are based on the compilation of statistical models, to find differences between groups of means. Analysis of variance was applied in this research, using R-commander software for ANOVA analysis. This paper also proposes initial recommendations for mould and die manufacturers to deploy HSM to improve the machining process and surface quality.

1. INTRODUCTION

Surface topography measurements are an important area in characterization of machined steel surfaces. If in the past 2D surface roughness measurements were sufficient to describe the machined surface, there is nowadays greater potential to improve machining processes with information obtained from 3D surface topography measurements. Wide ranging factors of influence, which it is necessary to take into account, make it very difficult to select the most appropriate operating conditions.[3] Several studies have already been conducted, to identify how technological parameters affect surface topography parameters.[3],[4] For the most part, such research only considers what has been generally established in science and uses general topography parameters such as S_a – average roughness S_q – root mean square roughness and S_t – max. peak-to-valley roughness.[5] These parameters are inadequate to describe machined mould surfaces. Several other, more appropriate surface texture parameters can be used for mould and die manufacturing.

Our research focuses solely on the most correlative parameters, which are specific to this industry and mathematically proven. S_a – arithmetic mean surface height, S_{tr} – texture aspect ratio, S_{ku} – kurtosis of the scale-limited surface and S_{vi} – valley fluid retention index.

Machined material surface quality parameters involve not only surface texture, but also the micro-hardness of the machined surface and geometrical surface and shape errors. Therefore the research also had to include mathematical analysis of how the high-speed machining technological parameters influence surface micro-hardness, measured before and after sample processing. Geometrical surface deviations are dependent on machining technological parameters and thus also have to be taken into account.[6]

2. EXPERIMENTAL PROCEDURE

Initially, several samples were machined using high-speed milling on three different die-mould materials (C45 - 1.1730, 40CrMnMo - 1.2312 and BT1-0). In these samples, such parameters as cutting depth, strategy, overlap and tool-nose radius were modified.

After confirming the influence of these parameters (Fig. 1), both in isolation and combined, the relevant effect of other variables not taken into account so far was observed. The variables for sample No. 3 and 5 are represented in Table 1.

Table 1

Technological parameters for sample Nos. 3 and 5

Sample Nr.	*Strategy*	*Overlap, mm*	*Feed, mm/min*	*Mode*
3	*LP*	*0.1*	*12566*	*Up*
5	*CP*	*0.1*	*12566*	*Down*

Inter alia, the relation between up-and-down machining, overlap and tool-nose radius variables were analyzed. An appropriate number of samples were machined to demonstrate the influence of these new variables. A new type of material was then added (Toolox 33).

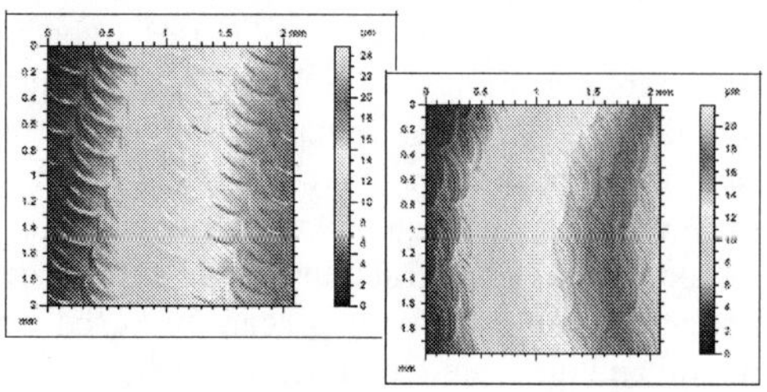

a) b)

Fig. 1. *Marks and traces in samples n° 3 (a) and n°5 (b), top view. Material: Toolox 33*

3. APPROACH

To provide useful recommendations for die and mould manufacturers, it is important to use relevant cutting technological parameters in accordance with the die mould manufacturers' practices. In the experiment however, the materials used were chosen for their mechanical properties from suggestions from the plastic part moulding industry and die-mould manufacturers. Therefore information will be required from further investigation before it is applied in manufacturing. In the first stage of this research, two different types of materials were chosen, each with different mechanical and chemical properties. Both materials are commonly-used industrial die-mould steels. The first is 1.1730 moulding steel, a surface-hardened steel with a tough core, suitable for die-mould manufacturing. This steel has lower tensile strength and lower hardness than other materials used in die-mould manufacturing. The second material is a widely used 1.2312 die-mould steel, with higher tensile strength and hardness, but with lower elongation properties. It has higher resistance and is appropriate for mechanical treatment and machining.

To conduct extensive and more accurate research into the influence of the materials' mechanical properties on 3D surface topography parameters, it was decided to introduce an additional material with a different chemical composition and mechanical properties – unalloyed titanium.

In the second stage of this research, the authors enhanced the number of machined samples with another new material, Toolox 33. This is a die-mould steel with similar mechanical properties to 1.2312 steel, but with a different chemical composition.

All of the material mechanical properties and chemical compositions are shown in Table 2.

Table 2

Chemical and mechanical properties of the materials

Material	***Chemical composition***	***Tensile strength - Yield strength - Elongation – Hardness***
Mould Steel 1.1730	0.45C – 0.27Si – 0.7Mn	640N/mm^2 – 340N/mm^2 – 20% - 190HB
Mould Steel 1.2312	0.4C – 0.4Si – 1.5Mn – 0.03P – 0.08S – 1.9Cr – 0.2Mo	990N/mm^2 – 860N/mm^2 – 15% - 280-325HB
BT1-0 Grade 2	0.18Fe – 0.07C – 0.1Si – 0.04N – 98.61/99.7Ti – 0.12° – 0.01H – 0.3Mo	400-450N/mm^2 - 300-420N/mm^2 – 30% - 210HB
Toolox 33	0.24C – 0.11Si – 0.8Mn – 0.01P – 0.02S – 1.2Cr – 0.3Mo	980N/mm^2 – 850N/mm^2 – 16% - 300 HB

The GENTIGER GT-66V-T16B HSM high-speed milling (HSM) machine was used to machine all the samples in this research. It is equipped with a Mitsubishi type VC2ESB spherical ball-end milling tool. The 4mm radius ball nose is coated with Al, Ti and N. The milling tool was adjusted at 90° to the work surface.

To analyze the influence of the technological parameter it is necessary to apply different combinations of cutting conditions on each of the chosen material samples. Therefore, each material sample was divided into 16 subsamples, which were machined with different cutting parameters. The chosen cutting parameters are:

Cutting strategy:

1) Linear Path (LP) - the tool movement on the material is straight and one-way.
2) Circular Path (CP) - the tool moves on the surface in a spiral path outwards from the centre.
3) Two Linear Paths (TLP) – combination of two linear paths, one along the X axis and the other along the Y axis.

Cutting overlap: Two different overlap values were chosen according to the peak heights. The chosen values are 0.05 and 0.1mm.

Feed speed: Feed speed is potentially one of the most important technological parameters, as in conventional milling. The feed rates were chosen in accordance with the tool manufacturer, who recommended using 3 different linear feeds:

a) 2,513mm/min (0.08mm/tooth);
b) 6,283mm/min (0.2mm/tooth);
c) 12,566mm/min (0.4mm/tooth).

To ensure the correct machining of the titanium, the feed speeds were altered:

a) 587mm/min (0.08mm/tooth);
b) 1,466mm/min (0.2mm/tooth);
c) 2,933mm/min (0.4mm/tooth).

Milling mode: As in conventional milling, there are two possible milling modes – up-milling and down-milling. For up-milling, the tool rotation direction is opposite to the feed direction. In down-milling, the tool rotation is the same as the feed direction. Both milling modes were used for all selected machining strategies and feed types.

All other technological parameters were kept constant, in accordance with the die-mould manufacturers' recommendations and HSM technology.

Cutting depth: The cutting depth was kept constant at 0.3mm, to ensure part finishing conditions.

Spindle speed: The spindle speed was maintained at 16,000 rpm as recommended by the tool manufacturer and HSM technology, to ensure the correct chip formation and appropriate machining processes.

Surface texture measurements

The first 3D surface topography measurement stage was conducted at Riga Technical University, Latvia. The results were obtained with the Taylor Hobson Form Talysurf Intra measuring device. It provides simultaneous measurement of surface texture parameters, dimensions, form and surface roughness. The device is equipped with Talymap expert analysis software. Some 4 samples of the Toolox 33 material were also measured on this device to compare the results, check the standard deviations and absolute errors. The second measurement stage took place in the Faculty of Mechanical Engineering at Tallinn University of Technology, in cooperation with associate professor Fyodor Sergeyev. All the Toolox33 material samples were measured on the Bruker Contour GT3 optical microscope (Fig. 2.), equipped with Vision64 software.

Fig. 2. *Measurements taken using the Bruker Contour GT3 optical microscope*

This measuring method is quicker than the contact method using the Taylor Hobson device and provides more precise results, excluding errors caused by needle shape defects (Fig. 3.).

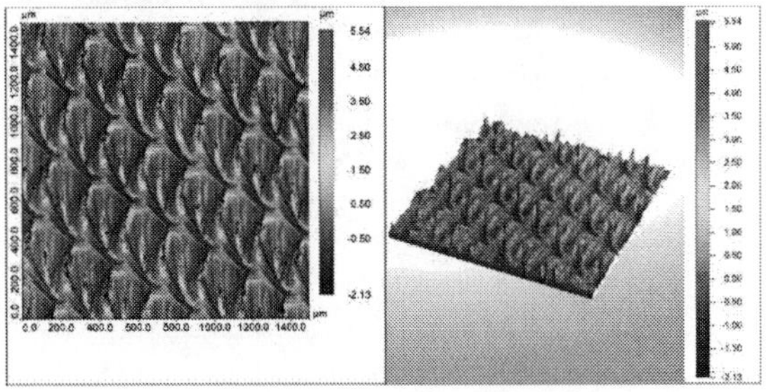

Fig. 3. *Marks and traces in sample n° 3 Material: Toolox 33 measured by the Bruker Contour GT3*

4. DATA ANALYSIS

In the first stage, the measurements were taken for all 16 subsamples of the 3 material types. For the additional material, several samples were measured, to conduct additional analysis on the influence of the material's mechanical properties on surface topography. The obtained results were sorted by group, according to the surface parameter they describe. These groups were then subjected to an analysis of mathematical correlation. The correlation matrix of *n* random variables was then drawn up. The matrix included a set of 3D roughness parameters from each group, to determine the most relevant HSM parameter for each group.

To conduct the analysis and determine the technological parameters' influence on 3D surface roughness parameters, a multifactorial analysis of variance (ANOVA) was undertaken between all types of materials. New values were introduced into the analysis, with technological parameters being replaced by factorial values, as required by the method of analysis. These values were replaced by two or three factors. In this analysis, the surface parameters were used as a response function of technological parameter's interaction with the analysis function. R-commander mathematical analysis software was used to perform the mathematical verification.

In the first stage of this research, three types of materials were compared in terms of the dependence of their surface topography behaviour on the cutting parameters. A graphical spreadsheet analysis was prepared using the ANOVA results. The graphical spreadsheet shows the trend line coefficients of the polynomial regression equation with argument x. The analysis showed that some technological parameters, such as the cutting tool strategy type, have a significant influence on surface texture parameters, although this is not a major contributor (Fig. 4.). Almost all other chosen impact factors had a higher influence rate in HSM.

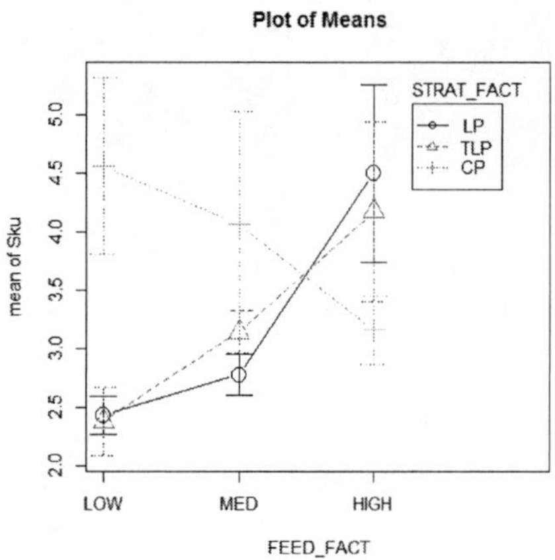

Fig. 4. *Plot of means for the parameter S_{ku} of strategy and feed influence*

This parameter is affected by strategy type up to 2.2 times, depending on the strategy type, comparing the strategy and feed factor influence. Only the parameter S_a value is lower on average than the other texture parameter values, which was generated by applying a circular path strategy type. All other chosen 3D surface texture parameter values are higher with a lower feed factor, even for Titanium samples. It can be seen in Fig.4. that by using the circular path strategy, the influence of the feed speed is absolutely different – increasing the feed speed decreases the S_{ku} value. This may be caused by the chaotic distribution of the surface texture (peaks) over the measured surface area, which is generated by the circular strategy type.

Also the overlap influence was detected as an insignificant cutting parameter. Of course, this parameter is predicative, in that increasing the overlap value increases the peak height (Fig. 5). Roughness parameter distribution is similar for both overlap values, but they are dependent on material type. In the case of previously machined materials, the average roughness difference was higher than with the material Toolox33.

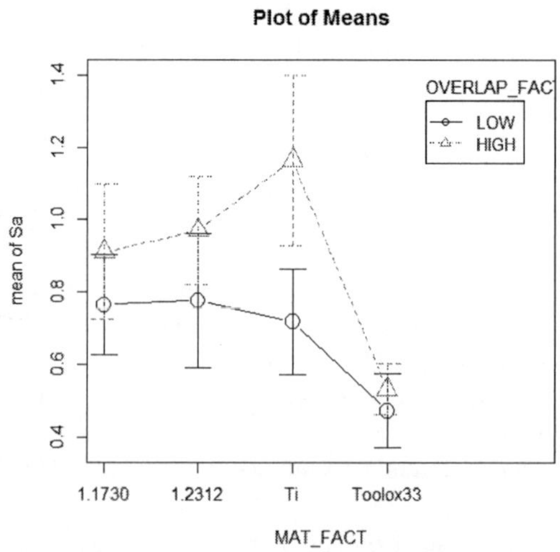

Fig. 5. *Plot of means for parameter S_a of overlap and material influence*

Mathematical analysis confirms that the material factor is a more important influence than overlap (see Fig. 6.).

```
> Anova(AnovaModel.5)
Anova Table (Type II tests)

Response: Sa
                       Sum Sq Df F value  Pr(>F)
MAT_FACT               1.6854  3  2.8537 0.04527 *
OVERLAP_FACT           0.7044  1  3.5780 0.06373 .
MAT_FACT:OVERLAP_FACT  0.3280  3  0.5553 0.64673
Residuals             11.0248 56
---
Signif. codes:  0 '***' 0.001 '**' 0.01 '*' 0.05 '.'
```

Fig. 6. *Mathematical analysis of material factor and overlap influence*

Relative to the other parameters, we can see in Fig. 6 the plot of means for tool feed. In the case of overlap influence, increasing the feed rate should increase the topography parameters. This is true for parameters S_a, S_{ku} and S_{vi}, S_{tp}, but not for S_{tr}, since with this parameter, at the highest speeds, the cutting marks and topography texture aspect are better preserved by the direction and behaviour, but at lower speeds, the machined surface is more isotropic. The S_a parameter has a different distribution according to types of materials and feed rates. Materials with lower tensile strength at lowest feed rate have higher texture parameters, whereas with increased material mechanical properties, the surface texture parameters diminish. (Fig. 7)

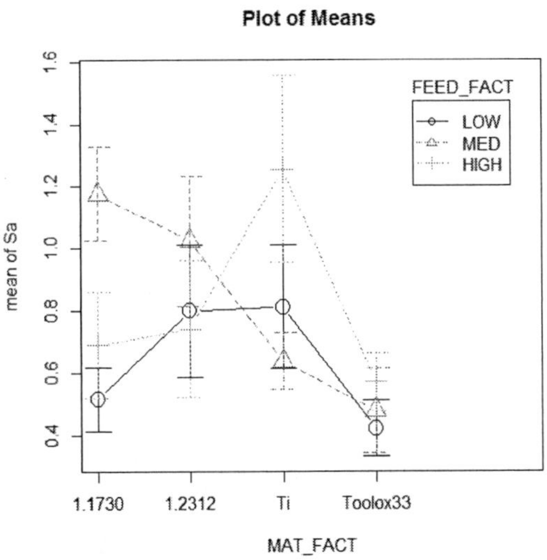

Fig. 7. *Plot of means for parameter S_a for feed rate and material influence*

This conclusion provides real data for manufacturers.

As mentioned above, the number of samples was enhanced to prove the results achieved in first stage of research. As the figures show, the results for the different materials sometimes diverge up to 6 times, depending on the influence of the material's mechanical properties on cutting forces.

The influence of the milling mode (up/down) can be seen in Fig. 8. As shown, the milling mode significantly affects the surface parameters and the surface texture parameters are distributed differently for each material.

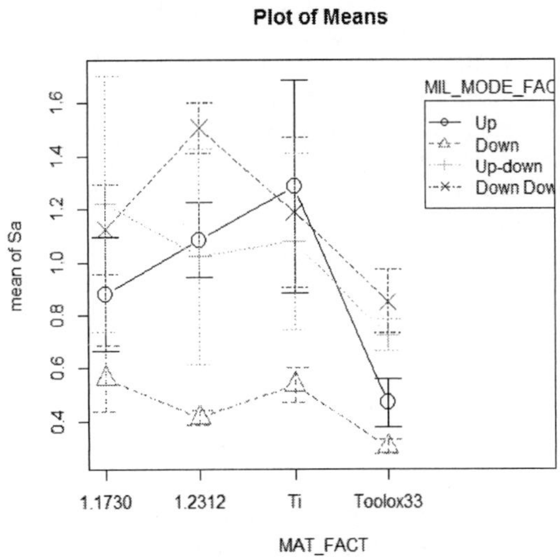

Fig. 8. *Plot of means for parameter S_a of milling mode and material influence*

The down-milling mode generally has the best influence on cutting processes, the best chip formation and lowest vibrations, as shown in Fig. 8., surface texture parameters are growing together with material mechanical properties In this case, material Toolox33 is different, because mechanical properties are similar to material 1.2312. In most cases, samples of material Toolox33 have the lowest texture and the best surface roughness parameters. Thus the material is a major influence, although this may not only be related to material mechanical properties, but also to the material's chemical composition and manufacturing nuances.

5. CONCLUSIONS

After analyzing the results, the authors identified which factors had the most impact on machining and surface texture parameters. The following conclusions were obtained:

1. The up/down milling mode is the most relevant parameter for 3D surface roughness. The correct chip formation and acceptable volume of removed material reduce the process forces and vibrations. The down milling mode provides the lowest surface texture parameters S_a and S_{ku} – for mould and die materials. Parameter S_{vi} and S_{tr} reaches the lowest values

by using TLP strategy with the appropriate milling mode.

2. Based on the mathematical ANOVA multifactorial analysis results, the second most significant influence on texture parameters is the material (or the material's mechanical properties). Toolox 33, which has different mechanical properties and chemical composition, confirmed the arguments of the material's influence on surface texture parameters. On observing the results, Toolox33 is the most appropriate material to use for mould and die manufacturing with HSM.

3. Analytically, the machining strategy is the third major factor of influence in HSM. The differences between texture parameters can be up to 2.2 times greater. The lowest texture parameter values were obtained by using circular path (CP) and linear path (LP). Both were used in combination to produce mould and die workpieces. Two linear paths (TLP) milling has a negative effect on the material surface roughness and probably also on mechanical properties.

4. Small overlaps and small cut feeds lead to the concentration of the cutting zone around zero radiuses. The negative effect of these phenomena masks the results.

To prove these phenomena, as well as the TLP effect on surface texture and mechanical parameters, surface micro-hardness measurements and analysis are required. These surface micro-hardness measurements are still in progress.

6. REFERENCES

[1] D. Hioki, A. E. Diniz, A. Sinatora, 2013, "Influence of HSM cutting parameters on the surface integrity characteristics of hardened AISI H13 steel", Journal of the Brazilian Society of Mechanical Sciences and Engineering, 35 (4), pp. 537-553

[2] M. Kumermanis, J. Rudzitis, A. Avisane, Investigations of 3D Surface Roughness Characteristic's Accuracy, Proceedings of the 11th International Conference of the European Society for Precision Engineering and Nanotechnology, Lake Como, Italy, 23-27 May, 2011, pp. 145-149

[3] J. Vivancos, C.J. Luis, L. Costa, J.A. Ortíz, Optimal Machining Parameters Selection in High Speed Milling of Hardened Steels for Injection Moulds", Elsevier, Journal of Materials Processing Technology, 2004, pp. 1505-1512

[4] D. Begic-Hajdarevic, A. Cekic, M. Kulenovic, Experimental Study on the High Speed Machining of Hardened Steel, 24th DAAAM International Symposium on Intelligent Manufacturing and Automation, 2013, pp. 291-295

[5] I. Buj-Corral, J. Vivancos-Calvet, H. González-Rojas, "Influence of Feed, Eccentricity and Helix Angle on Topography obtained in Side Milling Processes", Department of Mechanical Engineering, Universitat Politecnica de Catalunya, International Journal of Machine Tools & Manufacture, vol. 51, 2011, pp. 889-897

[6] G. Bunga, Ē., Geriņš, Apstrādes ar Atdalīšanu Tehnoloģijas, Rīgas Tehniskā Universitāte, 2007, 85 pages, ISBN: 978-9984-39-157-1

10th International DAAAM Baltic Conference
INDUSTRIAL ENGINEERING
12-13 May 2015, Tallinn, Estonia

PRODUCTIVITY IMPROVEMENT BY IMPLEMENTING LEAN PRODUCTION APPROACH

Mahmood, K., Shevtshenko, E.

Abstract: *This paper aims to provide a better understanding of lean production approach in order to enhance productivity, reduce cost and maximize customer value while minimizing waste during the production processes. Lean tools enabling a company to differentiate value from waste and facilitate to maximize customer value while minimize waste. Although there are many key factors for this methodology but here authors would be focusing on the Value Stream Mapping (VSM), Pull system (Kanban) and Dedicated Flow that are contribute to change the process by eliminating different kind of wastes (such as inventory) which slows down the process. A case from metal manufacturing company will be taken into account that focus on lowering down the inventory (waste) levels with the help of lean tools.*
Key words: Productivity, Lean, wastes, VSM, Kanban (pull system), Dedicated Flow

1. INTRODUCTION

In today's advance world the companies are striving to get better and better, the competiveness in the world market has always existed and is a key element for improvements, because it forces people and companies to drive themselves in order to bring up new tools or methods and take stance in front of innovative gadget, lean approach is the way to achieve improvement either in manufacturing and service sector.

Problem – Nowadays, the problem facing by medium sized manufacturing companies is how to increase productivity and reducing cost in order to stay competitive in the market and satisfies customer. Specially, in Baltic region medium size manufacturers have a fear to losing competitiveness against other low cost manufacturing countries. Hence, they must be more effective to survive and in order to do this they are seeking for process improvement methods.

Purpose – The solution for above stated problem is to adopt lean production. The aim of this paper is to create awareness about lean production with the help of a simple model of manufacturing system alongwith recognization of deadly seven wastes of lean and lean tools.

Methodology/Approach – A thorough search of literature and the qualitative case study research methodology [[1]] are used in order to find the solution of the stated problem.

Findings – Lean Production has been one of the most well-known methodologies to eliminate waste in the manufacturing and service industry. The emphasis of lean is to eliminating non-value-added activities and the goal has been a goal of industrial engineering i.e., to improve the efficiency of all processes. Many firms have been practicing lean production to improving productivity and as a process improvement approach, small-medium enterprise (SME) manufacturers also looking forward to get under the umbrella of lean production. In Baltic SMEs willing to learn Lean and it can be achieved through Lean Lego Games, hence, there is also a need to build such kind of Lean Labs where companies can practice Lean.

2. REVIEW OF LITERATURE

In this section the contribution of lean concept and its background with principle will be discussed. A brief explanation of productivity as a competitive tool will also be a part of this section.

2.1 Brief History of Lean

In the beginning of 20th century, Henry Ford developed a new way of manufacturing, mass production, which has been a major evolution and first applied to the automotive industry, and then his philosophy has been applied in every kind of production plants. A second major evolution, which is nowadays considered as the best of way manufacturing has been developed by Toyota, during the second half of the 20th century is Lean manufacturing [[2]]. It describes the opposite way of mass production, where quality is preferred to quantity [[3]].

The approach of lean was first initiated by Toyota, lean production is orginated from Janpanes manufacturing method known as TPS - Toyota production system.

The term "lean process" in the literature has many definitions. Lean production uses half the human effort in the factory, half the manufacturing space, half the investment in tools, half the engineering hours to develop a new product in half the time. It requires keeping half the needed inventory, results in many fewer defects, and produces a greater and ever growing variety of products [[4]].

"Shah and Ward [[5]]" defined lean process as "an integrated socio-technical system whose main objective is to eliminate waste by concurrently reducing or minimizing supplier, customer, and internal variability".

"Calborg et al., [[6]]" suggests in their study of "A lean approach for service productivity improvements" that standardizing services and increasing reliability in service processes through lean principles can increase efficiency. Further, Lean principles can be beneficial in order to improve productivity in services with an appropriate approach in which customer satisfaction must be considered, otherwise the positive long term effects of a lean approach in services will be absent.

2.2 Productivity as a Competitive Tool

"Productivity is a relationship (usually a ratio or an index) between output (goods and/or services) produced by a given organizational system and quantities of input (resources) utilized by the system to produce that output" [[7]]. The input are usually classified as labor, capital, material (inventory) and energy. In this paper inventory as an input will be under consideration, during the case study the inventory reduction with the help of lean tools leads to improvement in productivity of a case company.

Productivity enhancement is a common factor in fulfilling producer demands and customer demands as well. In this area the potential for improvement is always high to certain extent. Some researchers and company managers in western production industries argue that there is a potential to increase productivity by 50 per cent in many western companies. The more obvious fact is that when productivity is too low, operations are transferred to low-cost countries. Nowadays, this threat is also a reality in operations outside manufacturing. The most probable reasons for moving operations to low-cost are usually due to lower wages that will lead to increased profitability and/or enhanced competitiveness. Even some customers asking their supplier to move their operations to low-cost countries in order to expected reduction in prices. Unfortunately, this kind of discussion disregards the fact that the probable gains in profit due to lower wage level are often eliminated by costs related to longer lead times, increased tied-up capital, poor control, increased transportation costs, etc.

Lean is an approach to operations that specifically aims at increasing productivity in order to attract customers as well as investors. Nevertheless, it is not a short term approach to cure the wound; it is a long term approach that eventually focuses on efficient resources utilization by eliminating waste, increasing workforce commitment and keeps an eye on customer expectations [8].

2.3 Lean Principles

Lean methodology has five principles according to"Womack and Jones [9]" and they are defined as:

- Identify value from customer point of view
- Value stream mapping (process map)
- Create flow – redesigning processess to minimize waste and optimize customer service
- Establish Pull – produce when it is needed to fulfill customer demand
- Pursue perfection – Zero Defects

3. METHODOLOGY

After reviewing the literature it is somehow obvious that a proper lean implementation is a basic need in current manufacturing industries and in this research focusing on more to the identification of indicators, practices or tools or techniques for the implementation of lean in manufacturing for productivity improvement. The purpose is to get the right product at the right time and with right quality in order to gain profitability and stay competitive by continuing the sales growth [10].

In this paper a simple model of input, process and output of manufacturing system is developed with the identification of wastes known as seven deadly wastes of lean in order to create awareness of lean in manufacturing system.

The figure 2 is presented a system model shows the relation between lean dimensions and wastes. 'Muda' is a Japanese word for waste and "Ohno [11]" has identified seven kinds of waste that are: *overproduction, waiting, transportation, unnecessary motion, poor processing, defects and inventory.* Productivity can be improved by eliminating these wastes from the system by implementing lean tools effectively.

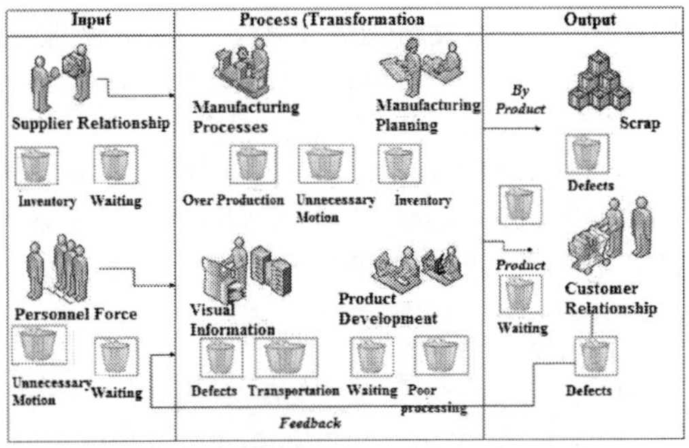

Fig. 1. Lean dimensions in a manufacturing system and its relation to wastes.

4. DISCUSSION

VSM, Kanban and Dedicated flow are the three tools of lean production that were under investigation in this research, they can be used to clean the system especially all kind of inventory waste and focus on productivity enhancement. This section discusses the above stated tools that how these lean tools play their role in effectiveness and influence on the improvment of a company.

4.1 Value Stream Mapping (VSM)

It is a map of process flow for better visualization, VSM is a mapping tool that is used to map a productive process or an entire supply-chain of a product. It maps not only material flows but also the information flows that signal and control production. Moreover, a value stream is a collection of all actions (value added as well as non-value-added) that are required to bring a product through the main flows, starting with raw material and ending with finish product to reached customer. The ultimate goal of VSM is to identify all types of waste in the value stream and take steps for trying to eliminate the wastes [12].

4.2 Kanban (Pull)

Kanban provides the same idea as pull system which means a 'signal', 'signboard' or 'ticket' that is used to activate demand of a product or service. It specifies when and where more material is needed for sake of better flow internally and externally both. These cards are used to control production flow and inventory. This system facilitates high production volume and high capacity utilization with reduced production time and work-in-process inventory [13].

4.3 Dedicated Flow

Dedicated flow is based on product families, the similar function products are divided into the same product category and this family of product proceeds according to their respective process machines or equipment that results in optimal routing of a product family.

5. CASE STUDY

A metal manufacturing company was selected for the case study and to implement the tools for process improvement as discussed in above section. This company was producing copper strips by casting process followed by slitting and rolling mill processes, more precisely the company produces copper strips for automotive and electrical applications

Company had some internal challenges such as lengthy transportation, low quality, a lot of scrap and excessive stock as well as high work-in-progress, and long manufacturing and delivery lead times. That makes the management to initiate the lean production improvement operation with the focus to increase the productivity by minimizing wastes (inventory levels) within the production facility, to concentrate on quality and reduce costs. The main problems in the production flow were:

- High Work-In-Process (WIP) Inventory
- Unpredictable process and time
- Long changeover time

The *current (as-is)* state of stream was mapped from fetching of raw material to dispatch of finish goods. The current state was a mess and creates chaos due to bad handling of information and lack of communication with some wrong practices. As there was imbalance between casting and rolling processes with respect to production speed, a great amount of WIP inventory was present that creating queues and occupying extra space on the production floor. The work load on rolling and slitting machines was uneven that also creates buffer in terms of inventory (finish goods- FG) as shown in fig.3. It was because of variation in product mix and which was not handled properly, moreover, no grouping of product and dedicated flow of product with respect to machines (RM – raw material, inventory).

Fig. 2. Process Flow with marked inventories

Steps that were followed during the analysis and improvement process were: first to create overall current production flow (mapping) through VSM, next step was to clean the process by implementing lean tools – Kanban and dedicated flow to reduce inventory levels and lead times. Third was to standardize the improved process by creating standard operating procedures with better developed process map (VSM).

Inventory type	Quantity (ton)	Time (days)
Raw Material	349	4.8
WIP	1026	14.3
Finish Goods	470	6.6
Total	**1845**	**25.8**

Table 1. Levels of Inventory (Before Lean)

The current inventory levels of all kinds of inventory are shown in the table 1. Lean tools implementation and their results along with future inventory levels will be discussed in the following section.

5.1 VSM Deployment

Figure 4 represents the To-Be process map as an improved values stream map where the information and material flow was clearly identified. The awareness of employees about this suggested improved process map

also created. The VSM helps to develop a new alloy combination also to avoid certain quality issues and to cope up with scrap problem especially in foundry process.
Moreover, new flow of products leads to avoid waiting time and extra transportation along with equal distribution of workload, while the concept of freezing window for product groups is taken into the account to minimize re-planning issues. In order to avoid and mitigate the problems (delaying in finish goods deliveries) that occurred due to unpredictable process, extra delivery day is taken into consideration to fulfill the commitment of customer and to abstain rescheduling deliveries.

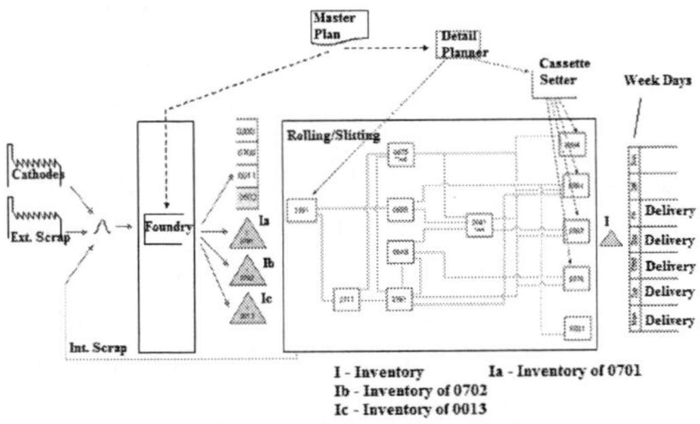

Fig. 3. Value Stream Map (To-Be)

5.2 Kanban Deployment

The Kanban replenishment system was successfully implemented that reduced the WIP remarkably and enhance the process performance. A kanban sticker/card was introduced that contains a product information – when, where and how much is needed to produce for the subsequent process as the signal from upstream has triggered. This means information flow and the material flow run parallel, this kanban (pull system) effectively control the buffer between casting mill and rolling mill processes.
The kanban method start from production machine in rolling area that generate a signal by means of card (contains information what they required), this card goes to the planner who plans casting production accordingly, casting produced that specific product (replenishment) and in the end demanding machine get the desired product (withdrawal). The process is shown in the figure 5. Although, there were some challenges to keep this system work properly and understand by everyone in the production area. This kanban approach is postponed twice during the transformation to lean implementation due to some issues in foundry management. But in the end this system start working and the WIP level has reduced from 1026 tons to 800 tons.

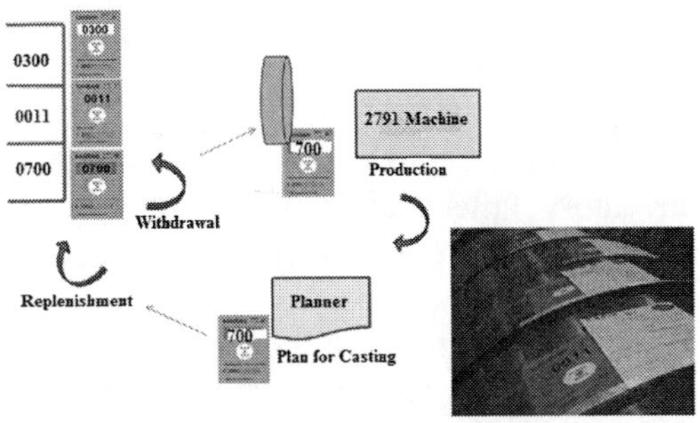

Fig. 4. Kanban Replenishment System

5.3 Dedicated Flow Deployment

In order to keep the flow continuous that leads to reduction in buffer time, a dedicated flow has been implemented – in this a group of similar products would be process through their designated machines. This result in clear and smooth flow, also leads to waiting time (buffer) gets smaller and hence, transportation has also optimized with better utilization of machines.
The products were divided into different families according to the processes that would be performed on them. Product family 'A' belongs to those products that have annealing point greater than 100 my and thick in dimension, likewise the products passes through annealing between 70 to less than 100 my with single pre-rolling correspond to family B and family C products are those that have annealing point less than 70 with double pre-rolling. The process can be seen in the figure 6. Furthermore, the challenges have been faced during this implementation was the mix of product (A, B, C) with no agreement with sales to fit customer demand to capacity.

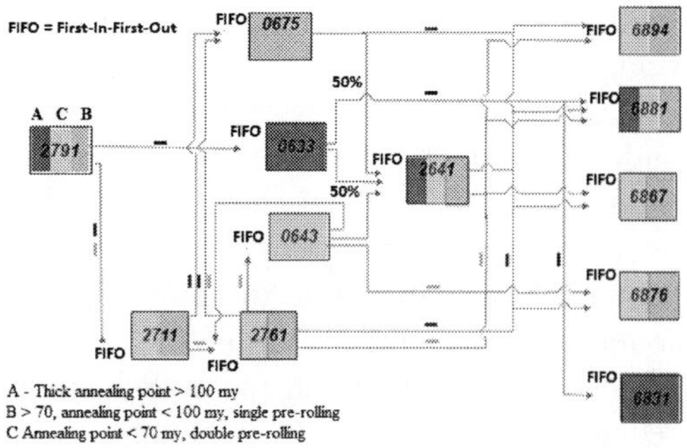

Fig. 5. Dedicated flows of product families

6. RESULTS

The future inventory levels of all kinds of inventory after lean implementation are shown in table 2. The total inventory level is reduced to 1600 tons from 1845 tons that leads to productivity improvement. Aslo the number of days for inventory are reduced to 22 days as compare to 26 days that speed up the production process.

Moreover, the graph in figure 7 shows a comparison between as-is and to-be state of the production process in terms of all kind of inventory levels along with percentage change. The improvement achieved in overall inventory level is 13.3% after the implementation of lean approach. To sum-up this case study it has been obeserved that the lean approach is not a short term approach it is a long term philosophy that required continuous practice to get bigger achievements. Although, we have seen improvements in the metal manufactuirng company's production process by means of reduction in inventory levels and smooth flow after lean implementation in a short period of time but it can be enhanced by doing more and more efforts in order to strive for perfection and lean area to see huge improvements.

Inventory type	Quantity (ton)	Time (days)
Raw Material	400	5.6
WIP	800	11.2
Finish Goods	400	5.6
Total	**1600**	**22.4**

Table 2: Inventory Level (After Lean)

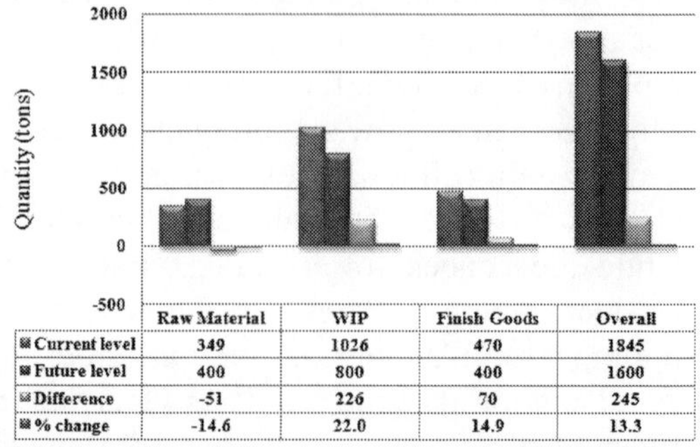

Fig. 6. Comparison graph As-Is (Before) Vs To-Be (After)

7. CONCLUSION

Lean tools and activities support stability. The tools – VSM, Kanban, Dedicated Flow improves safety and productivity. The impact of lean tools and techniques was quite noticeable as the case company had a big problem of high work-in-process inventory and it was consider to a fat that has been reduced by lean approach and lean implementation have shown improvement by reducing WIP inventory from 1026 tons to 800 tons while overall inventory level has fallen down from 1845 tons to 1600 tons. The consequence of this study is that Lean is a very global and vast system, and it's not easy to implement just a small portion, it is moreover, all or nothing. People must be ready to change the whole organization, not only production lines. Sales, Logistics, Marketing, Product development departments will be affected by this change and if one of them does not follow, Lean will not sustain.

One of the future researches could be the effectiveness and barrier of lean approach in SMEs; on the other hand research on lean effectiveness in service sector for service productivity improvement can also be a future research.

8. REFERENCES

1. Yin, R.K. *Case Study Research – Design and Methods*. Sage publications, Thousand Oaks, CA, 1994.

2. Liker, J.K. *The Toyota Way*. McGraw-Hill, New York, 2004.

3. Juran, J M., De Feo, J A. *Juran's Quality Handbook: The complete guide to performance excellence.* McGraw-Hill, USA, 2010.

4. Womack, J. P., Jones D. T., Roos, D. *The Machine that changed the world: How Lean Production Revolutionized the Global Car Wars.*, (New ed. of 1990 edition), Simon & Schuster UK Ltd, Great Britain.

5. Shah and Peter, Ward, T. Defining and developing measures of lean production. *Journal of OM.*, 2007, **25**, 785–805.

6. Carlborg,P. Kindstrom, D. Kowalkowski, C.A. Lean approach for service productivity improvements: synergy or oxy.?. *Managing Service Quality.*, 2013, **23**, 291-304.

7. Hannula, M. Total productivity measurement based on partial productivity ratios. *Int. J. of PE.*, 2002, **78**, 57–67.

8. Petersson, P. Johansson, O. Broman, M. Blucher, D. Alsterman, H. *LEAN – turn deviation into success!.* Part development AB, Bromma, Sweden, 2010.

9. Womack, J.P. Jones, D.T. *Lean Thinking: Banish waste and create wealth in your corporation.* Free p. business, London, 2003.

10. Natasya, A. Wahab, A. Mukhtar, M. Sulaiman, R. A Conceptual Model of Lean Manufacturing Dimensions., *Procedia Technology.*, 2013, **11**, 1292–1298

11. Ohno, T. *Toyota Production System: Beyond large-scale production.* Productivity press, Portland, 1988.

12. Abdulmalek, F.A. Rajgopal, J. Analyzing the benefits of lean manufacturing and value stream mapping via simulation: A process sector case study. *Int. J. of PE*, 2007, **107**, 223-236.

13. Eaton, M. *The lean practitioner's handbook.* Kogan page limited, London, Philadelphia, New Delhi, 2014.

DATA ABOUT AUTHOR

Kashif Mahmood
Tallinn University of Technology, Faculty of Mechanical Engineering, Department of Machinery,
E-mail: kashif.mahmood@ttu.ee

10th International DAAAM Baltic Conference
"INDUSTRIAL ENGINEERING -
12-13 May 2015, Tallinn, Estonia

INFLUENCE OF INITIAL SETUP OF PARTS BEFORE ROUGHNESS EVALUATION

Melichar Martin, Kutlwašer Jan, Kubátová Dana

Abstract: *This article deals with issue of initial setup of the tested part before roughness measuring process. In cooperation with industry partners it has been tested the influence of accuracy of the positioning of parts for measuring parameters inside high precision holes with small diameter. During the experimental phase it was evaluated by high-end stationary laboratory roughness device thousands of samples of one type of high precision hole to get statistical relevant data. The other part of the article deals with results and current research about possibility of solving the problem with noncontact unit.*

Key words: Surface Metrology, Roughness, Initial

1. INTRODUCTION

The meaning of machining quality of the individual components increases recently. It has the major impact on the functionality of the entire device. Especially by the contact surfaces (eg. bearings) quality is roughness one of the decisive factors. It can influence the operation and use of the device as a whole. The surface quality has a significant effect on the life and reliability of components, such as machine parts running accuracy, their noise, friction losses, heat transfer, or wear resistance. Therefore, it is preferable to monitor surface roughness of functional surfaces and evaluate the measured parameters. [[6]]
The surface structure has a decisive weight on how the component will behave in practice and what will have functional properties. General surface contains a number of unevenness (roughness, waviness, the surface shape) forming the surface structure, especially its differing pitches having a different effect on the function surface. Therefore, the development of measurement equipment and metrology surface structure needs to pay more and more attention.
Since the surface structure control for single-use measuring machines, it is clear that their manufacturers sharing the metrology enhancing the focus. The system of assessment and evaluation of the structure of the surface is defined by a plurality of standards that describe the evaluation, measurement, marking texture or calibration of measuring instruments etc. These are called. Standard GPS (Geometrical product), which specify the geometric requirements for the product [[2]]. These standards are continuously gradual development of measurement equipment and general metrology developed, refined and supplemented.

2. ISSUE OF INITIAL SETUP

When analyzing the surface integrity, especially roughness, one of the key factors in achieving relevant results is the correct setting of the mutual position between the measuring device and controlled products [[4]]. The ideal position in terms of measurement seems (when application of contact evaluation methods) theoretically so that the stylus moves perpendicular to the tracks roughness e.g. ideally etalon roughness perpendicular to the lines cuts. How much can a potential difference of perpendicularity affect the relevance of the results of the measurements was investigated in experimental metrology conditions.

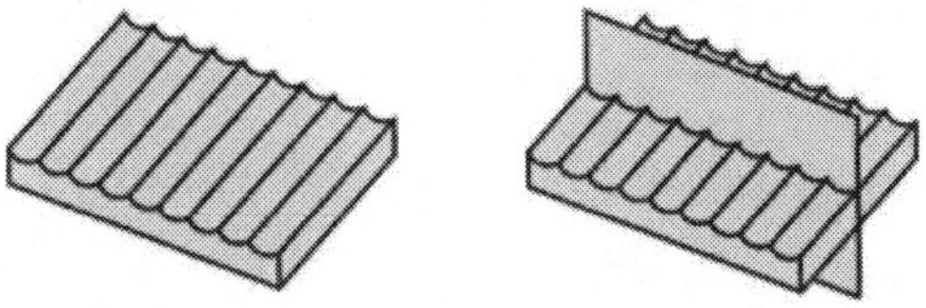

Fig. 1

During the test phase has been under controlled entry conditions experimentally measured several thousand precision holes (this is specific and often solved metrology operation [[5]]) machined with a reamer. The quantity of the obtained results was necessary for reliable statistical evaluation experiment. The actual parameters detection was performed on Roughness/Contour Hommel-ETAMIC T8000 (see Fig. 2), which is capable of in a controlled environment to detect and measure the parameters in the accuracy of fully meeting the requirements of aerospace and automotive industries. The device Hommel Etamic T8000 has an optimal combination for surface roughness measurement and after replacing the sensor head can be used for profile control components (control contours). This dual measurement is made possible thanks to the intelligent arrangement of both probing systems.

Roughness and contour are measured using two different imaging systems. Outline of the probe can be easily interchanged and, in some cases it is possible to parallel roughness and profile part.

An experiment was carried out using a measuring system HOMMEL-ETAMIC T8000. This is a measuring system that allows measurement of surface roughness and waviness profile and topography of different part types. The design of the system has a modular design that allows the interconnection of different shift units, sensor types, columns and granite slabs. This allows relatively high usability measurements for different sizes, types of surfaces etc.

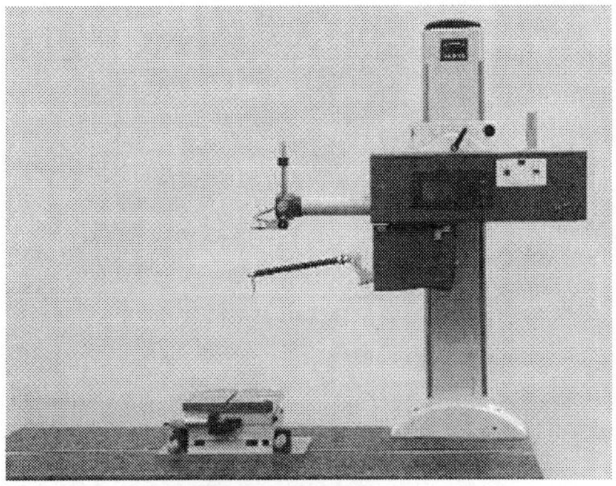

Fig. 2 Hommel-Etamic T8000

This equipment as well as other manufacturers' equipment provides measurements in only two axes and a possible shift the third axis is solved by manually or CNC controlled table, which can be obtained from the manufacturer in the supplementary equipment of machine.

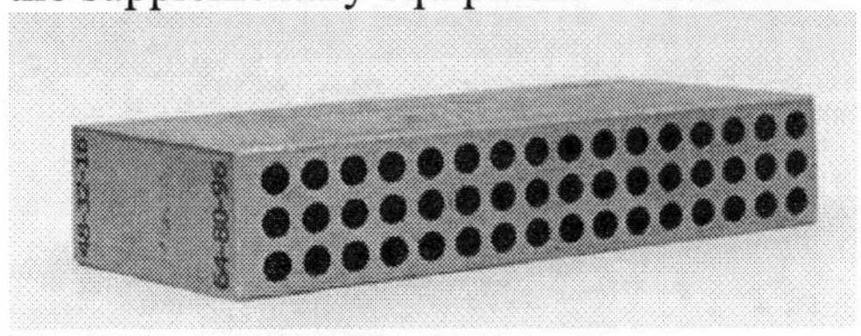

Fig. 3 Measured sample [1]

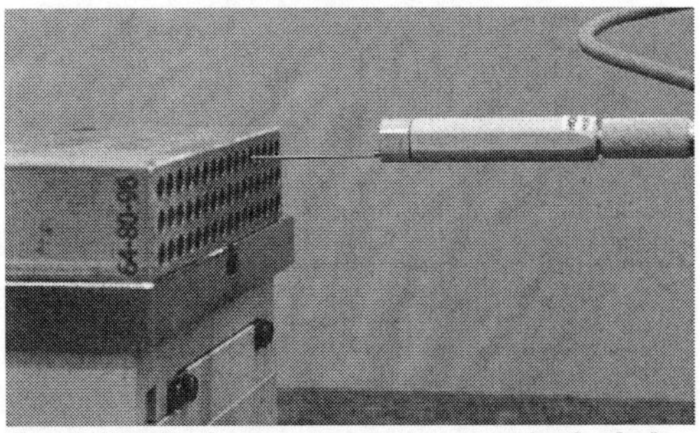

Fig. 4 Detail of contact measurement in the hole [1]

With an accurate manual micrometer stand holes were always tested in an ideal position to measuring machine and installed as incorrect assumption. The angular extent of "false" positioning was limited diameter holes (12mm) in relation to input roughness parameters given in ISO 4288 [[3]]. The output is clear that the initial phase of testing was clearly distinct influence of the initial system setup machine - controlled piece.

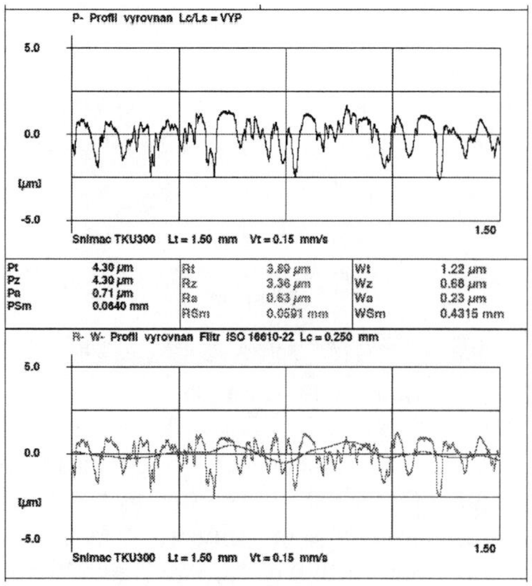

Fig. 5 Correct position of the sensor

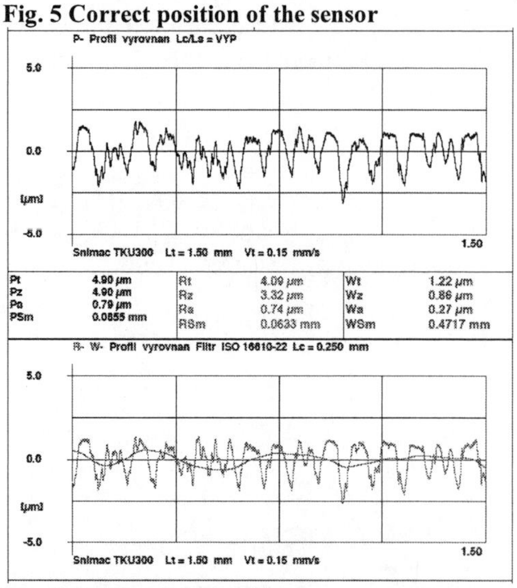

Fig. 6 Position 2,5mm off axis

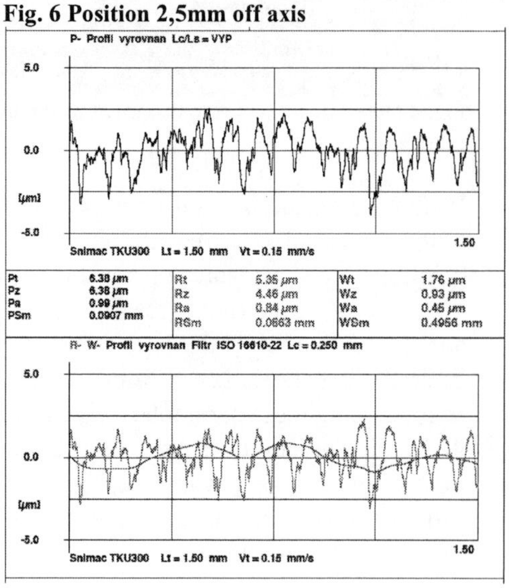

Fig. 7 Position 5mm off axis

To eliminate potential errors were chosen methodology selected holes further evaluated by in terms of shape and parameters on the precise roundness Taylor Hobson Talyrond 585 LT. This device is equipped with automatic levelling and centering piece measured up in the order of tenths of a micrometer. Among its other features include the ability to scan the entire surface of contact both internal and external cylindrical profile and the profile mathematically evaluated.

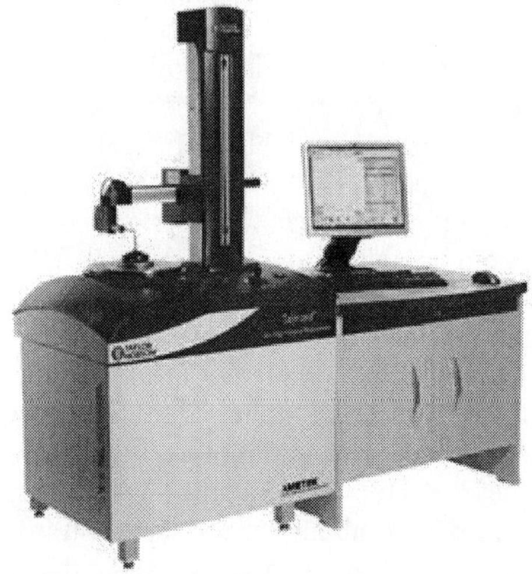

Fig. 8 Taylor Hobson Talyrond 585 LT

Output value of this evaluation corresponds to the initial assumptions and again clearly demonstrates the effect of measurement methodologies and especially the initial position gauge - piece on relevant result.

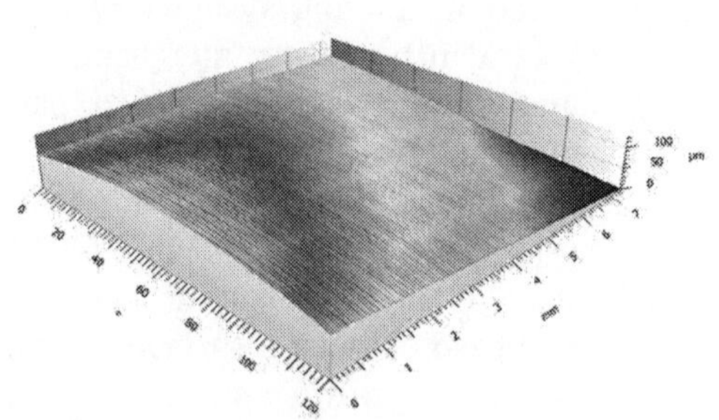

Fig. 9 3D data of evaluated hole

3. FIGURES

Based on the experimental simulations it was found that the described device provides an effective way to initial setup of the third axis of the roundness instrument. Because exerted contactless nature of the methodology and the resolution of HD camera, we obtain high accuracy while keeping a significant increase in the success of initial setup cycle. This is achieved substantial productivity gains the control processes. Also, the value of standard uncertainty remains at the same level as touch setup. Advantage, the decreased cost of the non-contact components compared to standard method based on the manual or CNC table, savings in labor time and expense-consuming machines, to which the contactless setup realized, as there is no need to operator surveillance over the process because of the anti-collision procedure. The proposed device is therefore a significant improvement and streamlining the setup process.

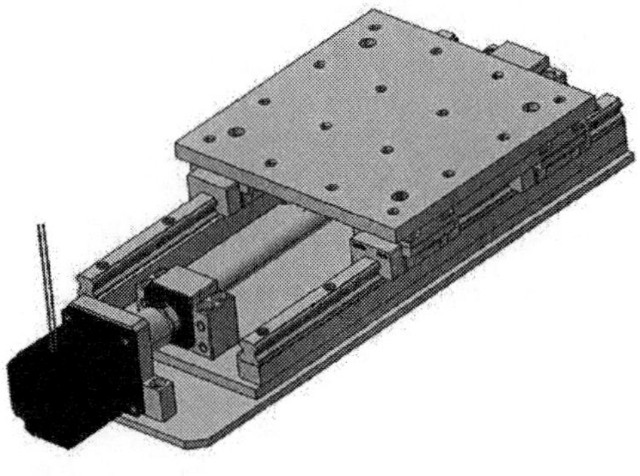

Fig. 10 Prototype of experimentary table

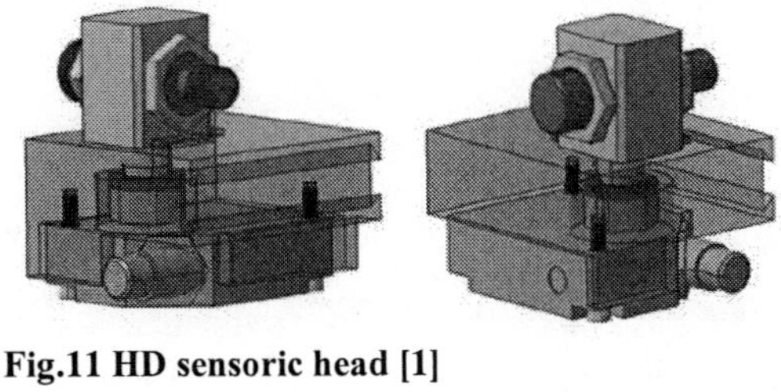

Fig.11 HD sensoric head [1]

4. CONCLUSION

The model devices provides statistical stable results for the applications described above so the next step will be preparation next generation of these models and their applications by using high precision machines by support of Regional Technological Institute (RTI-UWB). RTI under research project cooperate with many companies, which allows the usage of the devices and procedures in the reality of industrial sector.

5. ACKNOWLEDGMENT
This paper includes results created within the project SGS031-2013 and project CZ.1.05/2.1.00/03.0093 - Regional Technological Institute. The project is supported by the European Regional Development Fund and the state budget of the Czech Republic.

6. CORRESPONDING ADDRESS

Ing.. Martin Melichar, Ph.D.
University of West Bohemia
Faculty of Mechanical Engineering
RTI
Univerzitni 8, 30614 Plzen, Czech Republic
Phone: +420 377638534
E-mail: mech@rti.zcu.cz
http://www.zcu.cz

6. ADDITIONAL DATA ABOUT AUTHORS

1) Authors:
Ing. Martin Melichar, Ph.D. – Senior researcher
Ing. Jan Kutlwašer – Ph.D. student
Ing. Dana Kubátová – Ph.D. student
2) Title:
Influence of initial setup of parts before roughness evaluation
3) Address:
University of West Bohemia
Univerzitní 8
Plzeň 30614
Czech Republic

7. REFERENCES

[1] MELICHAR, M., KUTLWAŠER, J. The Issue of Contactless Setup before Measuring Process. Procedia Engineering. 2014, vol. 69, s. 1088-1093. DOI: 10.1016/j.proeng.2014.03.095.

[2] TICHÁ, Š., Engineering Metrology - Part 1 Ostrava: VSB-TU Ostrava, 2004. 112 S., ISBN 80-248-0672-X

[3] ISO 4288:1996. Geometrical Product Specifications (GPS) - Surface texture: Profile method - Rules and procedures for the assessment of surface texture.

[4] MELICHAR, M., Replacement of initial contact setup of Roundness machine by contactless method, PhD thesis, ZČU FST, 2011

[5] AMRANA, M.A., S. SALMAHA, N.I.S. HUSSEINA, R. IZAMSHAHB, M. HADZLEYB, SIVARAOSB, M.S. KASIMB,M.A. SULAIMANB, Effects of Machine Parameters on Surface Roughness Using Response Surface Method in Drilling Process, Procedia Engineering
Vol. 68, 2013, Pages 24–29. DOI: 10.1016/j.proeng.2013.12.142

[6] M.J. KLOPFSTEIN, D.A. LUCCA, Recent Assessment of Surface Integrity Resulting from Fine Finishing Processes, Procedia Engineering, Vol. 19, 2011, Pages 209–221, DOI: 10.1016/j.proeng.2011.11.103

10th International DAAAM Baltic Conference
"INDUSTRIAL ENGINEERING"
12-13 May 2015, Tallinn, Estonia

SHARPENING HOBBING WORM MILLING CUTTERS ISSUE

Milsimerova, A.

Abstract: *The complex shape and kinematics of grinding teeth faces of hobbing worm milling cutters makes it difficult to achieve the necessary accuracy and resultant shape geometry. Grinding wheel shape, size and helix angle influence accuracy during grinding. Inappropriate combination of these parameters causes undercutting of face teeth and reduces accuracy. The main concept of this research is the graphic design of a wheel corresponding to the geometry of the ground groove. Real testing is performed based on data from grinding simulation with the designed wheel shape and from comparison of curves of both wheel shapes. The aim is to find a parametric system which, when variables are changed, recalculates and modifies the solution for specific tools and their parameters.*

Key words: hobbing worm milling cutters, grinding, groove undercut, grinding wheel shape

1. PROBLEM STATEMENTS

Hobbing worm milling cutters are special tools – each one is designed to manufacture a gearing with specific parameters. The accuracy of the hobbing worm milling cutter has an impact on the accuracy of the rolling gearing. This aspect has high demands on the geometric accuracy of these tools, especially on the cutting area (area which is made by teeth faces). The tool is worn out during the manufacturing process by the cutting conditions and due to the required preservation of parameters grinding wheels are used for grinding the face teeth. Inappropriate combination of the geometric parameters of both input factors into the grinding process causes increased inaccuracy of the helix groove. This process has an impact on the undercut of the groove according to the required tolerance.

2. APPLICATION AREA

The ongoing research is concentrated on the elimination of the helix groove undercut and attaining the required accuracy of the ground surface (e.g. [2]). In the current status the solution is focused on hobbing worm milling cutters with zero face teeth angle, indicated as γ, according to the axis of the tool (see Fig. 1. Face teeth angle).

Fig. 1 Face teeth angle

The insertion of the grinding wheel cutting parts into the face teeth of the hobbing worm is caused with consideration of the ground tools geometry and with the inclusion of the influence of the grinding wheel geometry. The helix pitch angle control of the hobbing worm and the shape and the diameter of the grinding wheel are the main influences on this negative effect. The higher helix curling and bigger wheel diameter, thus there is larger insertion and clearly higher inaccuracy (= undercut). If the wheel diameter is theoretically small, as it were be made by point, no undercut will occur. There would only be point contact. However, with the increasing wheel diameter, surface contact occurs. But there is the requirement of achieving the tangent contact through the whole grinding process for the maintaining accuracy. Taking the helix curvature into account, a wheel with a bigger diameter cannot be placed in every position of its running with tangent contact with the ground face teeth but with the surface contact.

3. RESEARCH COURSE

In the current phase we are designing and adjusting a wheel shape, which should correspond to the ground groove and copy its curvature, based on analysis of the input factors, which negatively affect the resulting accuracy of the ground surface (e.g. [[1]]), and taking the complexity of the issue into account. This effect should eliminate the helix groove undercut with no respect to the grinding wheel diameter. The graphic method is exploited for this design solution. The benefit of this solution in contrast with an analytic is the description and display of facts and the possibility to look into the process at every current point and plane of its running. These methods are priceless in terms of describing the issue and in terms of more easily understanding the processes involved.

4. METHOD USED

For the investigation of default tool surfaces a graphic method of cutting planes is used as standard. This method is primarily used for investigation of default tool surfaces for profiling – for helix groove machining with the grinding wheel in this case. The method is based on orthogonal projection on two projection planes. Intersections of particular circles of the tool and the surface of the helix groove in each cutting plane are obtained by creating a cutting plane system perpendicular to the tool axis (including consideration of the tool rotation according to the helix groove angle). These intersections create involute curve point system. The intersection system, which creates particular points of the default tool surfaces equal to a given solution type, is made by projection of intersections in every cutting plane into the tool axis.

4.1 The wheel shape design

Based on the cutting planes definition auxiliary geometry was composed from default selected point of guide helix, auxiliary planes, surfaces and lines which define the wheel geometry necessary for the correct compilation of particular wheel cutting planes. The circles (define wheel diameter geometry in every cutting plane) tangent to the teeth surface were created in every plane (see Fig. 2. Circle system).

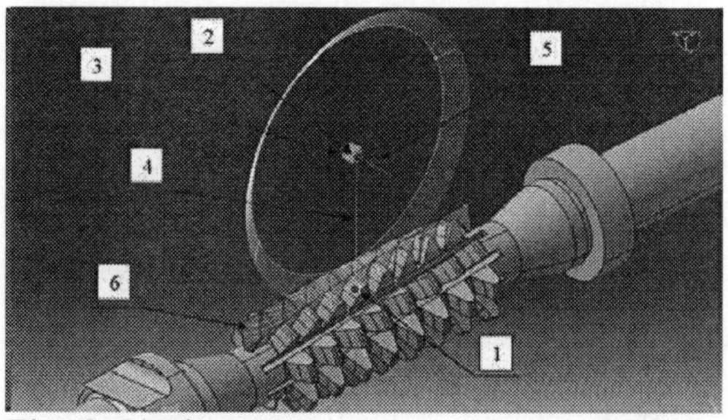

Fig. 2 Circle system

Explanations:

1. The intersection of perpendicular plane to the pitch angle and guide helix
2. Middle wheel point
3. Cutting planes system
4. The line defines big wheel diameter
5. The line defines the wheel axis
6. Cutting planes intersections into the teeth face

This effect enabled creation of the wheel involute curve into the material of the hobbing worm (see Fig. 3. Involute curve).

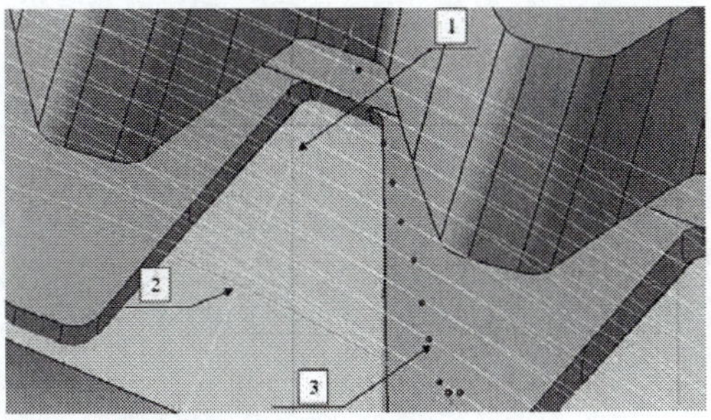

Fig. 3 Involute curve

Explanations:
1. The line defines cutt
2. The intersection of line and circle in a current cutting plane
3. The point of contact – involute curve point in a current cutting plane

New wheel shape points were investigated by successive creation of intersections of particular cutting planes circles and the line defines the wheel diameter. Linking these points created the new wheel profile sketch (white highlited) and the suitable wheel shape was made by its rotation (see Fig. 4. New wheel shape).

Fig. 4 New wheel shape

The width of this wheel depends on the depth of the groove; i.e. the result corresponds only to the teeth face and this shape will be considerably narrower than the real wheel shape. The original wheel shape was created in the same default position in the same way (see Fig. 5. The original wheel shape).

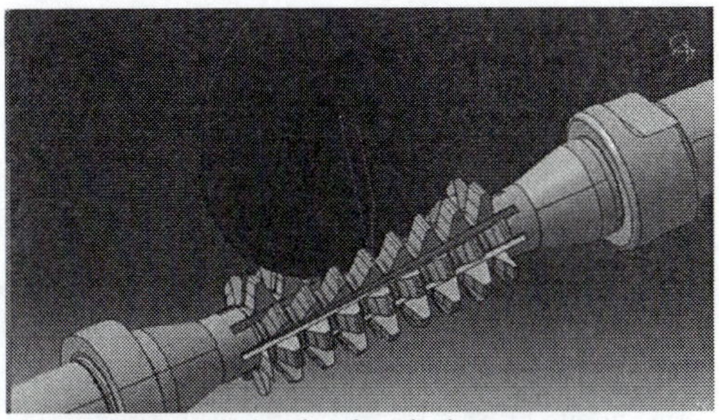
Fig. 5 The original wheel shape

All these constructions were made in the perpendicular plane to the helix at a selected point. Equal wheel shape and investigation of current involute curve of the tool into the hobbing worm material is the result as for real grinding. The accuracy of the solution depends on the quantity of created planes and their distance. Thus the more intersections and found points we have, the closer we are to an ideal solution. The shape of the disc for solving the issue is created by the connection of the intersection points of individual curves and lines defining the diameter of the disc. This connection must be created in the perpendicular plane on the helix because in this plane the adjusting position of the grinding wheel in the groove was defined according to the helix angle. This place is the current contact of the grinding wheel and hobbing worm. When rotating the disc in the groove as in real grinding, this place will be in other sections of the face, thus we reach a different plane perpendicular to the helix.

5. STATUS

The grinding wheel shape is designed using graphic method for solution of helix groove undercut. Currently the wheel shape is designed based on the construction method, which should correspond to ground groove geometry. From the comparison of the view of the groove, with the original wheel shape and the newly designed wheel shape, it is clear that the original wheel inserts its cutting part into the teeth faces during the grinding process. The newly designed wheel shape traces the teeth curvature (see Fig. 6. Comparison of views of the groove).

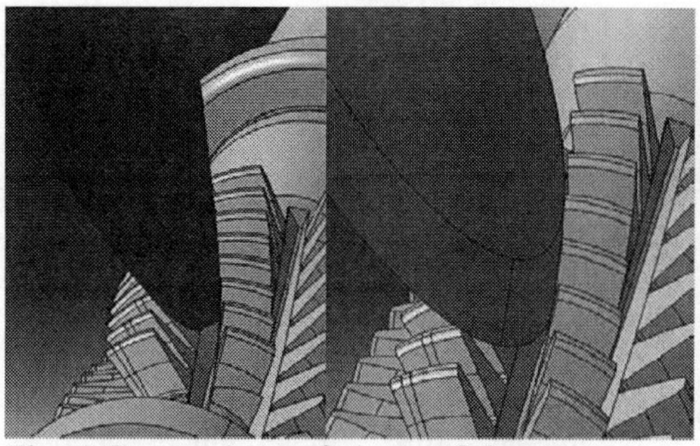

Fig. 6 Comparison of views of the groove

5.1 The curve of the grinding wheel profile

The newly designed wheel shape is created by a curve (red highlighted), which was made by connecting particular points obtained using the graphic method (see Fig. 7. The curve of the grinding wheel profile).

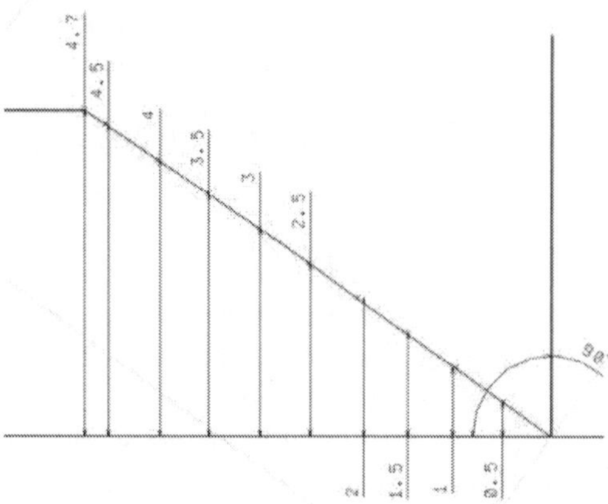

Fig. 7 The curve of the grinding wheel profile

These points each have the distance of 0.5mm according to the distance of the construction cutting planes. Points were connected through spline curves. At first sight it seemed that the curve is rectilinear and there is no curvature. Thus the generated curve was projected on the model with original wheel geometry and from the comparison it is clear that there is real divergence of both curves (see Fig. 8. Comparison of wheel shape curves).

Fig. 8 Comparison of wheel shape curves

Distances of particular points of the new designed curve were taken based on this comparison (see Tab. 1. Distances of particular points).

Curve point	X[mm]	Y[mm]	X′[mm]
0	0	0	0
1	0.71	0.5	0.7
2	1.42	1	1.4
3	2.14	1.5	2.1
4	2.87	2	2.8
5	3.61	2.5	3.5
6	4.36	3	4.2
7	5.12	3.5	4.9
8	5.88	4	5.6
9	6.65	4.5	6.3
10	6.96	4.7	6.6

Tab. 1 Distances of particular points

A graph was created from these points. The values of the original rectilinear shape were also plotted on this graph (see Graph 1. Wheel shape curves).

Explanations:
Y – Graph axis Y
X – Graph axis X of the designed curve
X′ – Graph axis X of the original straight curve

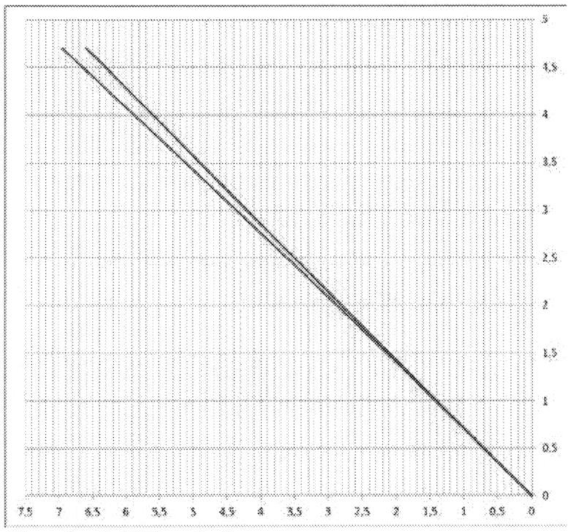

Graph 1 Wheel shape curves

6. RESULTS

The grinding simulation was created with the original wheel shape and with the designed wheel shape for verification of obtained results. The verification was calculated using the analysis of the remaining material and checked in several random points. The toleration was set on the value up to 0.01mm.

6.1 Grinding simulation - verification

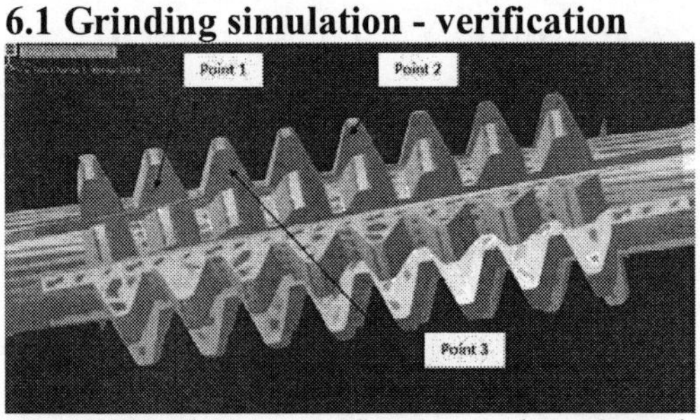

Fig. 9 Simulation verification – the original wheel shape

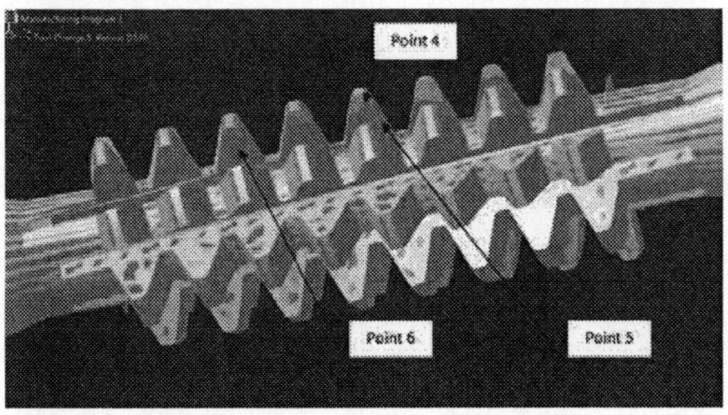

Fig. 10 Simulation verification – the designed wheel shape

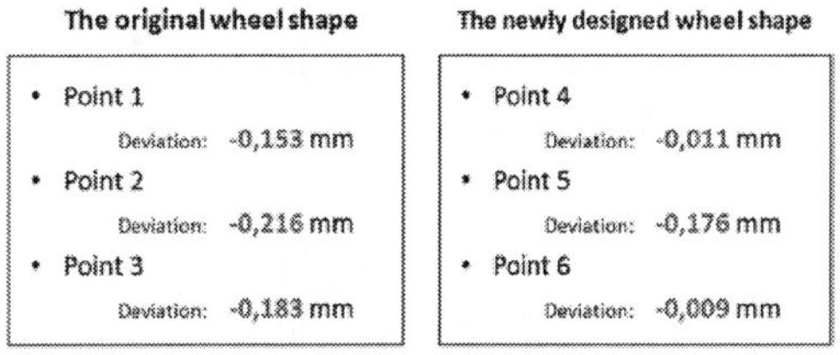

Fig. 11 Simulation results

The simulation confirmed accuracy increasing in hundredths and thousandths orders. It´s not a fixed result because we must consider some inaccuracy and graphic error. But it is definitely a degree of innovation in this issue. The undercut occurred in point 5, is still close to the -0,2mm value. This effect occurs at the teeth heel. There was a probability this would happen due to curve shapes, because there is no, or too small divergence, between the original and the designed curve. Based on this effect was designed construction for the wheel shape using above mentioned method. But in places with no divergence of both curves with 0.1mm cutting planes distance because due to a verification of the curve running, whether it would be the same or different.

6.2 Construction refinement

A graph was created from particular points of designed curve (see Tab. 2 Distances of particular points – refinement). The values of the original rectilinear shape were also plotted on this graph.

Curve point	X[mm]	Z[mm]
0	0	0
1	0.14	0.1
2	0.28	0.2
3	0.42	0.3
4	0.56	0.4
5	0.71	0.5
6	0.85	0.6
7	0.99	0.7
8	1.13	0.8
9	1.28	0.9
10	1.42	1
11	1.56	1.1
12	1.71	1.2

13	1.85	1.3
14	2	1.4
15	2.14	1.5
16	2.29	1.6
17	2.43	1.7
18	2.58	1.8
19	2.73	1.9
20	2.87	2
21	3.61	2.5
22	4.36	3
23	5.12	3.5
24	5.88	4
25	6.65	4.5
26	6.96	4.7

Tab. 2 Distances of particular points – refinement

It is fully clear that both designed curves overlap each other precisely (see Graph 2 Designed curves - overlapping). Based on this effect there is an assumption of curve shape close to the ideal.

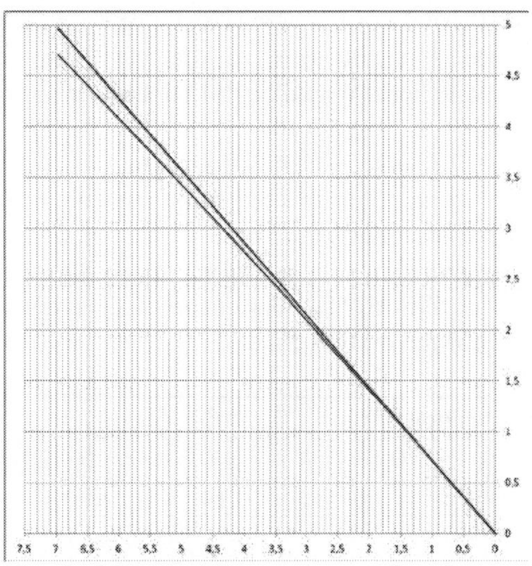

Graph 2 Designed curves - overlapping

7. FURTHER RESEARCH

Currently we are testing the grinding with the disc, which profile is dressed according to designed curve. Based on results there is a possibility to consider the accordance of real results with result of used method. The next step is the creation of series hobs with different parameters and the design of the wheel shape curve corresponding to each ground groove. If there would be a possibility to create a parametric hobing worm model, it would be possible to find a parameter, which would be directly linked with construction method for a grinding wheel. There is an assumption that it should be possible to find analytic description of the solution in this phase. Then there should be a possibility to recalculate and modify the solution, if a variable would be found, for a specific tool.

8. REFERENCES

[1] Karpuschewski, B., Jandecka, K., Mourek, D., Automatic search for wheel position in flute grinding of cutting tools. *CIRP Annals – Manufacturing technology. Elsevier. Netherlands.* 2011, **60**, 347-350.

[2] Ehmann, K., F., DeVries, M., F., Grinding Wheel Profile Definition for the Manufacture of Drill Flutes. *CIRP Annals – Manufacturing technology. Elsevier. Netherlands.* 1990, **39,** 153-156.

9. ADDITIONAL DATA ABOUT AUTHORS

Ing. Aneta Milsimerova
University of West Bohemia,
Faculty of Mechanical Engineering,
Department of Machining Technology,
Univerzitni 8, 30614 Plzen,
Czech Republic
Phone: +420721951094,
E-mail: anetam@kto.zcu.cz

10th International DAAAM Baltic Conference
"INDUSTRIAL ENGINEERING"
12-13 May 2015, Tallinn, Estonia

MATERIAL RECYCLING AND IMPROVEMENT ISSUES IN ADDITIVE MANUFACTURING

Mägi, P.; Krumme, A. & Pohlak, M.

Abstract: *The objective of this study is to find ways of reusing polymeric material left over in Selective Laser Sintering (SLS) in other Additive Manufacturing (AM) processes, i.e. Fused Deposition Modelling (FDM). Furthermore, attempts are made to improve the material's mechanical properties. This paper provides information about the changes of tensile properties taking place in the un-sintered polymeric material after it has been exposed to SLS process and the tensile properties that the recycled material obtains when additives are added into its matrix.*
Key words: Additive Manufacturing, Selective Laser Sintering, Fused Deposition Modelling, material recycling, Polyamide-12

1. INTRODUCTION

Additive Manufacturing is being falsely advertised as 'the technology that creates no waste'. Additive and subtractive manufacturing are both widely used for fabricating customized products. Subtractive manufacturing begins with a block of material from which the desired object is cut out by removing material producing waste as a by-product. In contrast, AM adds raw material layer by layer in order to form the desired object; this doesn't mean absolutely no waste is created. There are a number of different types of AM processes; most of them create waste in the form of supporting material used in the manufacturing process.

In Selective Laser Sintering a part's three-dimensional design is created in a computer aided design program. That 3-D design is virtually sliced into layers with a thickness 0.1 mm [1]. A recoater in the SLS machine applies a layer of polymer powder onto the print bed and a laser beam traces the model's cross section fusing the polymer particles into a solid object layer after layer, leaving the non-traced powder loose [2].

One of the most widely used polymer powders used in SLS is polyamide. All powder present in SLS process is subjected to high heat, which deteriorates polymer particles [3]. The heat and radiation initiated degradation of polyamides may involve the following: change in chemical structure of the polymer molecule, change in crystallinity (molecular orientation), dipole rearrangement, hydrogen bridging and change in the amount of molecularly associated materials that have plasticizing effect, such as water. The polymer molecule breaks at the C-N bond; moreover, breaking of a very small proportion of the C-N bonds can already result in an appreciable decrease in molar mass. Increase in crystallinity is proposed accompanying the decrease in molar mass. All these degradation effects may result in appreciable change in physical characteristics of polyamide [4]. Therefore, only certain amount of processed (presumable degraded) material can be used in mixture with non-degraded, new material to preserve characteristic mechanical properties of the material. This results in excess of polymer powder with changed properties that cannot provide sintered parts with as high quality as would be achieved by using virgin powder. Therefore a considerable amount of raw material goes to waste; only a certain percentage, 30 - 70% (depending on the material) [1], of leftover can be blended with virgin material and re-introduced into the process assuring production of high quality parts. Since the powder deteriorates differently at different parts of the build chamber the quality of the recycled powder can vary greatly. If too high percentage of heavily deteriorated recycled powder is blended with virgin powder, the produced parts will have surface defects such as 'orange peel'. In order to avoid the risk of producing damaged parts, producers frequently decide to blend just a small percentage of recycled powder into virgin powder [1]. By producing more polymer powder waste than is recycled there will always be leftover powder that will be permanently scrapped.

This research encourages environmentally friendly thinking and acting by studying the waste generated in SLS process and aiming to develop new materials of this waste that would be suitable for using in FDM process. This way all leftover polymer powder which otherwise would be scrapped, could be made into new raw material.

The ultimate goal of this study is to develop a set of polymeric materials that could function as the raw material for FDM printing hand prostheses. Currently a USA based non-profit organization e-NABLE and its global network of volunteers FDM print prosthetic hands at their own expense for people in need across the world [5]. Plastic filament makes the bulk of the cost of such prostheses. If material left over from SLS could be made into filament that would classify as acceptable prostheses material, it would be possible to make printing hands economically more viable. In addition to tackling the economical factor, opportunities to make a version of that recycled material that is stiff and a version that is flexible are explored. Another goal is to make an electrically conductive polymer that is suitable for using in mechatronic equipment.

In this research, polyamide-12 (PA-12) that is left over from SLS process in Tallinn University of Technology has been studied with a focus on giving the material new mechanical properties. PA's mechanical properties were manipulated by adding thermoplastic polyurethane (TPU) ground pellets and aramid fibres into recycled PA-12's matrix.

TPU is a linear block copolymer that is made up of hard and soft segments, which are separated during extrusion or injection. This separation of microphase segments gives the material unique properties including high tensile strength, high elongation at break, good wear and tear resistance and low temperature elasticity. The shortcoming of TPU is expressed in low temperature resistance, which may become a problem during processing the material. Blending TPU with other (heat resistant) materials, such as polyamide, helps to overcome this issue [6]. Aramid is considered to be a high performance fibre [7]. High levels of molecular alignment result in high levels of stiffness and strength of aramid fibres [8].

Polymers are reinforced with short fibres to increase their strength and load carrying capacity of the components. Among the fibre reinforcements as glass and carbon, aramid fibres are widely used. Polymer composites reinforced with these fibres are usually one to four times stronger and stiffer than their unfilled equivalents. The reinforcement reduces also coefficient of friction and hence allows the material to be used for higher duties without exceeding the softening point of the matrix [9]. Good adhesion (which is necessary for reinforcing effect) between PA matrix and aramid fibres is expected due to similar chemical nature.

TPU and polyamide mixture provide material with toughness, low temperature flexibility, tensile strength, tear strength, abrasion resistance, transparency, resistance to oils, resistance to hydrolysis, resistance to kinking and hardness without brittleness [10].

In addition to perfecting current materials' composition and structure, further studies will tackle PA's electrical conductivity and visual properties. Depending on the achieved properties of the material some possible applications are considered, e.g. limb prostheses, mechatronic equipment, etc.

2. IMPROVEMENT OF MATERIALS

This study investigated the change in mechanical properties of un-sintered PA 2200, which is a modified version of polyamide 12 supplied by EOS GmbH [11], after the polyamide powder was processed in Formiga P100 (EOS GmbH) and after that reused powder was added TPU and aramid.

All raw materials were dried in a convection oven at 105°C for four hours; after drying the raw materials were stored in airproof containers until processing them in the injection moulding machine. Four different sets of test specimens were prepared at 215°C using micro-injection moulding machine Babyplast 6/10P [12]. Specimens were prepared of the following materials 1) 100% virgin PA-12, 2) 100% reused PA-12, 3) 90% reused PA-12 and 10% mechanically milled TPU pellets, 4) 95% reused PA-12 and 5% aramid fibres.

The brand name of TPU that was used is Elastollan® which is made by BASF. This material has excellent low temperature flexibility [13]. Short fibre reinforced PA-12 was prepared using aramid fibres with a length of 2 mm.

3. TESTING OF MATERIALS

All specimens were prepared according to the standards of ISO 527-2 type 5B. A small specimen size was chosen in order to preserve unnecessary waste of materials and to shorten the testing time.

Tensile testing was carried out on all four different sets of prepared specimens using Instron 5866 machine. The crosshead speed was set at 50 mm/min. Each specimen was individually measured. The results of 10 successful tensile tests of each set of specimens were recorded.

A sample filament was produced of recycled PA-12 using a laboratory compounder manufactured by Brabender.

Following properties were measured: maximum tensile stress, tensile strain at break, modulus and energy at break. Recycled PA-12 filament was used as raw material in Vellman K8200 FDM printer [14] to 3-D print an object in order to be able to analyse FDM-printability of produced material.

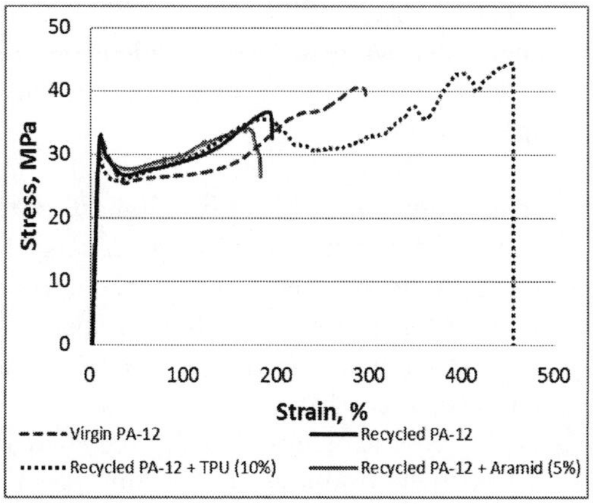

Fig. 1. Comparison of tensile stress against tensile strain of specimens made of different materials.

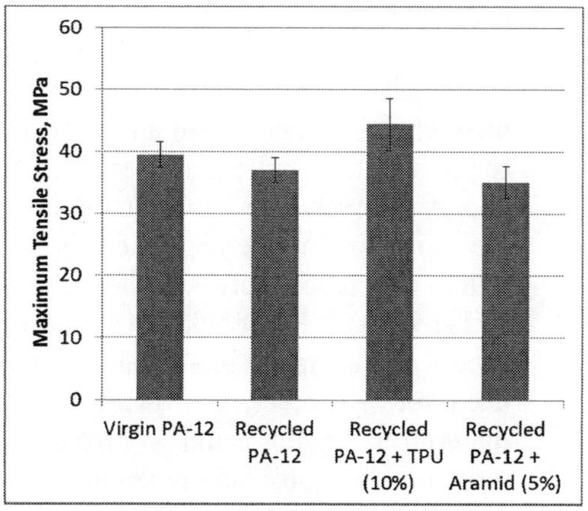

Fig. 2. Tensile strength of different compositions.

Material	Maximum Tensile Stress, MPa		Tensile Strain at Break, %		Modulus, GPa		Energy at Break, J	
	Value	Standard Deviation	Value	Standard Deviation	Value	Standard Deviation	Value	Standard Deviation
Virgin PA-12	39,5	2,1	296,8	21,3	0,60	0,10	5,90	0,60
Recycled PA-12	37,0	2,0	198,5	34,3	0,60	0,10	3,90	0,70
Recycled PA-12 + TPU (10%)	44,4	4,1	421,4	132,1*	0,56	0,05	9,21	3,18
Recycled PA-12 + Aramid (5%)	35,1	2,6	172,6	23,7	0,64	0,07	3,32	0,55

Table 1. Mechanical properties of different PA -based materials (* high variability is caused probably by two defected specimens).

4. RESULTS

Fig. 1 shows typical tensile test diagrams of the four different material sets of specimens studied. Fig. 2 represents the comparison of average tensile strength of each material type. The standard deviation is marked for each set of specimens.

The mechanical properties of all four tested PA -based materials are shown in Table 1. Some preliminary tests were made to analyse the suitability of recycled PA-12 in an FDM machine. Fig. 3 shows a flower printed with recycled PA-12 in an FDM printer.

Fig. 3. FDM printed Recycled PA-12 flower

5. DISCUSSION

The results show that the biggest difference between virgin and recycled PA reflects at tensile strain at break and also in energy at break (see Table 1). Other results are more similar. Perhaps there is not that much difference between these materials and recycled PA could be successfully used for most applications that virgin PA is used for. However, the quality of virgin material is much better controlled; it is more homogenous and predictable.

Adding 10% of TPU to PA may have been too small amount for making it softer since the modulus did not change significantly compared to pure PA. However, the maximum strain increased considerably, and also the tensile strength increased. More studies for identifying the influence of different levels of TPU in PA are needed.

Adding 5% of short aramid fibres made the material only slightly stiffer. Small difference may have been caused by either too low reinforcement content or non-uniform fibre distribution inside the test specimens. Therefore, some further studies are needed.

Studies show that the recycled PA-12 is FDM-printable, but at this point the quality is far from acceptable level (see Fig. 3). The main issues to deal with are achieving control over the thermal expansion of the material and assuring good adhesion of the material to the building platform of the FDM machine.

6. CONCLUSION

The study focused on recycling issues of polyamide powder left over in Selective Laser Sintering process. The attempts were made to improve the material and re-use it in Fused Deposition Modelling system. In the study four polyamide based materials were tested: a) unprocessed virgin PA; b) recycled PA; c) recycled PA + TPU (10%); d) Recycled PA + short Aramid fibres (5%). Mechanical properties of the materials were measured. The best results showed recycled PA + TPU. The addition of TPU increased tensile strength and maximum strain.

Future studies for identifying optimal content on additives should be made. A study for focusing on FDM processing using the created materials is planned.

7. ACKNOWLEDGEMENTS

The study was supported by research grant of Estonian Science Foundation ETF 9441.

8. REFERENCES

[1] Wan Yusoff, K. D., Recycling of polyamide 12 based powders in the laser sintering process, *Rapid Prototyping J.*, 2009, **15**, 192 – 203.

[2] Ion, J., *Laser Processing of Engineering Materials - Principles, Procedure and Industrial Application - 8.5.4 Selective Laser Sintering*, Elsevier, 2005.

[3] Beal, V. E.; Paggi, R.A.; Salmoria, G. V. And Lago, A., Statistical evaluation of laser energy density effect on mechanical properties of polyamide parts manufactured by selective laser sintering, *J. Appl. Polym. Sci.*, 2009. **113**, 2910 – 2919.

[4] Bernard G . Achhammer, Frank W. Reinhart, and Gordon M. Kline Journal of Research of the National Bureau of Standards Vol. 46, No. 5, 1951, 391 - 421

[5] e-NABLE [WWW] http://enablingthefuture.org/ (29.04.2015)

[6] Li, W.; Liu, J.; Hao, C., Jiang, K.; Xu, D. and Wang, D. Interaction of Thermoplastic Polyurethane With Polyamide 1212 and Its Influence on the Thermal and Mechanical Properties of TPU/PA1212 Blends. *Polym. Eng. Sci.,* 2008. **48**, 249-256.

[7] Mittal, V. (ed), *High Performance Polymers and Engineering Plastics*, John Wiley & Sons, Inc., Hoboken, 2011.

[8] Young, R. J.; Lu, D.; Day, R.J.; Knoff, W. F. and Davis, H. A. Relationship between structure and mechanical properties for aramid fibre. *J. Mater. Sci.,* 1992, **27**, 5431-5440.

[9] Kukureka, S.N., Hooke, C.J., Rao, M., Liao, P., Chen, Y.K. Tribology International 32 (1999) 107–116

[10] Prusty, M., Wilms, A., Noon, E. S., Götz, W., Serhatkulu, G., Marten, E. Article comprising thermoplastic polyurethane and polyamide 6/66 copolymer. US Patent application no US20120055576, WO 2010076225 (A1), Priority date 31.12.2008

[11] EOS GmbH Electro Optical Systems [WWW] http://www.eos.info/material-p (28.04.2015)

[12] Babyplast [WWW] http://www.babyplast.com/index.php?page=prodotto&code=BAE (28.04.2015)

[13] BASF [WWW] http://www.polyurethanes.basf.com/pu/KUInternet/KU/en_GB/content/product/TPU (29.04.2015)

[14] Vellman [WWW] http://www.k8200.eu/ (30.04.2015)

9. ADDITIONAL DATA ABOUT THE AUTHORS

Piret Mägi (corresponding author), Tallinn University of Technology, Faculty of Chemical and Materials Technology, Department of Polymer Materials, Ehitajate tee 5, e-mail: piretmagi@gmail.com.
Anders Krumme, Prof., Tallinn University of Technology, Faculty of Chemical and Materials Technology, Department of Polymer Materials, Ehitajate tee 5, e-mail: andres.krumme@ttu.ee.
Meelis Pohlak, Tallinn University of Technology, Faculty of Mechanical Engineering, Department of Machinery, Ehitajate tee 5, e-mail: meelis.pohlak@ttu.ee.

10th International DAAAM Baltic Conference
"INDUSTRIAL ENGINEERING -
12-13 May 2015, Tallinn, Estonia

ANALYSIS MODEL DEVELOPMENT TO SIMPLIFY PLM IMPLEMENTATION

Paavel, M.; Kaganski, S.; Karjust, K.; Lemmik, R. & Eiskop, T.

Abstract: *Each new Development and implementation needs plan what consists of good vision, goals and strategy. There is described Analyse model for giving information to different activities before real implementation. Information is gathered from different levels of company t to get the best results. Involving majority of company's workers gives information of problems of different level of company and presents comprehensive picture of company.*
Article gives brief overview of Enterprise Analyse model development first stage. Enterprise Analyse model was initiated to help companies to implement Product Lifecycle Management (PLM)/ Product Data Management system (PDM) by before PLM activities. Through Analysis model to map and recommend Key Performance Indicators (KPIs) in company.
Analyse model structure, construct and answering options are described.

Key words: Analysis model, PLM, PLM implementation, KPI

1. INTRODUCTION

Competition between enterprises makes the companies to improve their processes and their performance. For reacting quicker to market changes and increasing development speed with fewer errors. One opportunity is to improve data management systems, related to products and production. Advantage can be achieved by thought-out implementation of Enterprise Resource Planning (ERP) or PLM/PDM system. [[1,2,3,4]]

2. PLM IMPLEMENTATION

PLM system should help to meet the goals set by the company's management. PLM is a business strategy of different complex IT tools and applications which support digital design and manufacturing practices in different areas of company. This business concept helps the company to manage products and products related data in whole enterprise. [[5,6,7]]
PLM is implemented to get better results. The results are obtained through making Products Lifecycle more transparent and more organized. By different modules and applications. Information moves across the company and closes the missing loops. From Beginning of Life (BOL) to Middle of Life (MOL) and to End of Life (EOL).Also backward from EOL to MOL and BOL by making the product information a whole.[[8,9]]
The expected benefits are divided into four bigger categories; [[5,10,11]]

- Reduce cost by reducing development, material and prototyping cost. Also by faster and controlled design and wright on time information and products movement.
- Improve quality by better and cleared data quality what ensures less errors, rework customer complaints, product returns. Better document control gives faster document search and revision management.
- Project times reduction by lower overrun times, engineering change

times, usage of already developed details and assemblies.

- Other useful extras by better business decisions and results by clearer product lifecycle.

The need for PLM implementation should come from companys overall vison and PLM should be one tool to achieve set goals. From overall vision should come business vision in which depend the PLM vision and PLM strategy. Vision should include The company: objectives, strategy, success factors, key issues, The PLM Initiative, Description of PLM Vision, Next steps: PLM strategy, PLM implementation Strategy, PLM Roadmap, PLM Plan, schedules, resources, value, cost, Return of Inverstment (ROI).These are company specific and there should be described goals and abjectives before contacting the vendor. The Implementation strategy shows how resources will be organized in future and is the starting point for Implementation Plan. PLM Plan shows the components, objectives, action steps, timing and financial requirements. The main purpose of the chain from business mission and objectives through PLM Vision, PLM Strategy, PLM implementation strategy PLM plan and implementation is that the company meets it objectives. [[10,12,13,14]]

3. ANALYZE MODEL

There are lot of different possibilities how to ease implementation. From hiring company to going to depth by itself. There are several further work options like questionnaires, interviews or surveys. Each have their advantages and disadvantages.
This Analyze model is described in two variants. On is in English/Estonian duplicate in English/Russia. Analyze model developed considering Estonian production field.
Current analyze model is developed for two reasons:

- Help implement PLM/PDM system in company.
- Map and recommend KPIs in company.

In this paper we focus how to ease PLM/PDM Implementation. This model is built up in Lime Survey Environment and information is collected through structured questionnaire. Input to this Survey is in accordance with overall corporate business visions and business objectives. Also controlling how it matches with incision of people. Survey enables us to determine the PLM maturity of an individual organization. After determining the as-is situation and to be situation it is possible to raise goals what can be used describing PLM vision. Information from PLM framework can also be used for development of PLM strategy, PLM Implementation strategy and PLM Plan (Fig 1).

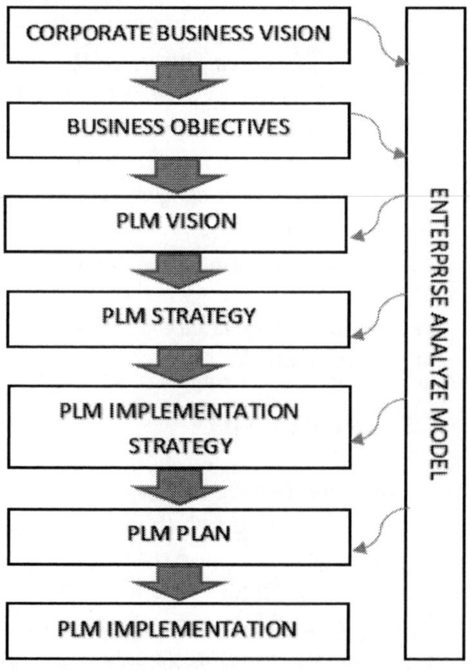

Fig 1. Information movement between different Implementation phases and analyses model.

Also for evaluating companies PLM maturity and .The empirical research will be conducted based on questionnaire.

Questions are covering main business Dimensions: [15]

- Strategy and policy
- Monitoring and control
- Organization and processes
- People and culture
- Information technology

3.1. CONSTRUCTS

Questions in questionnaire are divided into fifteen categories (Table 1). Each category has different number of constructs. Constructs are based on and composed based on research articles, information needed for PLM implementation and constructs connected to KPIs. Information what describes the construct can be based on facts or based on opinion of the worker.

Category	Different constructs	Questions
Customer and Supplier participation	5	9
Electronic Data Interchange	38	43
Enterprise name	1	1
Financial	1	2
General information	5	5
Human Resource	31	42
Logistics	8	14
Mission and goals	7	12
NPI	21	28
Performance management	5	11
PLM implementation	13	14
Production	28	52
Quality management	8	15
Respondent information	6	6
Sales Management	2	5
Sum:	179	259

Table 1. Constructs divination between categories.

Overview of issues asked through questions in categories;

- Enterprise name – Gives opportunity do analyses each enterprise separately.
- General information- describes the location, field where the enterprise is operating and what its research areas.
- Mission and goals – Shows the enterprise mission and goals. How are long and short term goals understood and fallowed by company's workers. Also describes workers awareness about responsibilities and their connection to overall picture.
- Respondent information - brought out respondent information about position, work experience, age and gender.
- Customer and Supplier participation – Gives information about suppliers and customer participation in

enterprise early stage development process.

- Electronic Data Interchange – Shows what type of data and information is managed. How and by whom it's managed. Data movements and times in enterprise. Also gives feedback about data quality and asks for proposed amendments.
- Financial – Shows workers disposition about different financial measures.
- Human Resource – Gives overall background about company's workforce and its distribution between different divisions. Also its skills and satisfaction with his job. This information reflects overall readiness for cooperation.
- Logistics – Describes situation in companies logistic. Are used on transport or outsourcing, information about delivery times and delays.
- New Product Introduction- Information about portfolio and age of portfolio. What is the speed of renewal product portfolio? What metrics are fallowed during New Product Introduction?
- Performance management – Provides information about performant measures what company is fallowing and workers view of current situation.
- PLM implementation – Shows what company has done to date at time. Do they have PLM vision, PLM strategy and how these are matching with overall business strategy and vision? Information about feasibility study, PLM ROI and its payback time. What kind of Advantages Company is hoping to achieve by implementing PLM/PDM system.
- Production – Gives overview of principles enterprise production division is operating. What are the main problems in production field and where to focus attention in future? What parameters are measured and what should be measured according to other company's expectations.
- Quality management – What are problems in field of quality. What are measured and what are expectations for quality measurements by the management. Main reasons of reclamations and errors inside the company.
- Sales Management – Information about winning new clients, and the competence of salesman.

3.2. QUESTIONS

Questionaire consist of 259 different questions what are collecting information for 179 different construct. Not all the questions are asked from everyone. Questions are divided into groups. Based on your work position you can answer only to prescribed questions. There is described 81 different work positions from assembler to Chief Executive Officer (CEO). Job position are divided into larger groups:

- R&D Personnel
- IT Personnel
- business,sales and marketing
- Production
- Quality
- Human Resource
- Purchase and Logistics
- CEO

Questions are divided into two:

- Questions what are fact based.
- Questions what are interested in workers opinion.

For answering the questions what are collecting information about facts there are used more than 40 answering scales. Questions what are collecting information about workers opinions were described through six point discrete scale. This was used because of no middle point (Table 2).

To get correct information these questions are asked twice with similar wording. And answer of the both question is counted when the answers are similar. This guarantees the accuracy of the data and excludes that someone has misunderstood the question.

CT001	Strongly agree
CT002	Agree
CT003	Inclined to agree
CT004	Inclined to disagree
CT005	Disagree
CT006	Strongly Disagree

Table 2.1. Consent Scale (Discrete)

4. QUESTIONS COMPLIANCE TO CONSTRUCTS

During the development there was done questions compliance to constructs. All the survey items were validated by using sorting method to prove that they represent associated constructs. Optimal Workshop software was used to prepare questions and constructs for validation to expert group. Expert group consisted of 6 people from academic and industry.

All constructs and questions were entered to Optimal workshop software by categories. If more than 5/6 workgroup member's opinion matched then the question matched with construct. Little corrections were made based on expert group feedback before finalizing the Survey.

5. SURVEY PUBLISHING

Survey is published and is forwarded for fulfilling into 4 companies what are operating in different field in Estonian production landscape. Including wood industry, electrical installation and metal works.

There's agreement that based on this survey PLM/PDM implementation goals are recorded. Based on written goals PLM vision and PLM strategy is developed. It's agreed to propose KPIs what would be useful for company's further improvement.

6. CONCLUSION

Survey results will be collected from the current project phase. Together with preliminary research model based on constructs will be input for the next project phase where the statistical analysis is done. Information received from answers will be used in implementation early stages. Used for mapping goals what should be set for PLM/PDM implementation. Used for developing companies personal PLM vision, PLM strategy, PLM implementation strategy and plan. According to the results from statistical analysis suggestions can be given for further actions. In the final project phase the research model is validated and project conclusions will be drawn.

7. ACKNOWLEDGEMENTS

This research was supported by ETF grants 9441 and innovative Manufacturing Engineering Systems Competence Centre IMECC (supported by Enterprise Estonia and co-financed by the European Union Regional Development Fund, project EU30006).

8. REFERENCES

[1] Karjust, K.; Küttner, R.; Pääsuke, K. Adaptive web based quotation generic module for SME's. Proceedings of the 7th international conference of DAAAM Baltic industrial engineering, 22-24th april 2010. Küttner, R. (Toim.)., Tallinn, Estonia (375-380). Tallinn: Tallinn University of Technology

[2] Paavel, M.; Kaganski, S.; Lavin, J.; Sonk, K. Optimizing PLM implementation from vision to real implementation in Estonian SME`s. Proceedings of the 9th International Conference of DAAAM Baltic Industrial Engineering 24-26st April 2014 (Toim.) Otto, T. Tallinn: Tallinna Tehnikaülikooli Kirjastus, 2014, 157-162. ISSN 2346-6138

[3] Snatkin, A.; Karjust, K.; Eiskop, T. Real time production monitoring system in SME. In: Proceedings of the 8th

International Conference of DAAAM Baltic Industrial Engineering 19-21st April 2012. (Toim.) Otto, T. Tallinn: Tallinna Tehnikaülikooli Kirjastus, 2012, 573 - 578. ISBN:978-9949-23-265-9
[4] Brandao, R.; Wynn, M. Product Lifecycle Management Systems and Business Process Improvement –A Report on Case Study Research. The Third International Multi-Conference on Computing in the Global Information Technology. ICCGI 2008. July 27 - August 1, 2008 - Athens, Greece.113-118
[5] Saaksvuori, A., and Immonen, A. 2008. Product Lifecycle Management. 3rd Edition. Springer-Verlag, Berlin, Heidelberg.
[6] Z.M. Qiu, Y.S. Wong, Dynamic workflow change in PDM systems, Computers in Industry 58 (2007) 453-463.
[7] A CIMdata White Paper. PLM Market Growth in 2007 A First Look in 2008—Exceeding Expectations. March, 2008
[8] Kiritsis D., Bufardi A., Xirouchakis P. Research issues on product lifecycle management and information tracking using smart embedded systems. Advanced Engineering Informatics 17 (2003) 189-202.
[9] Jun H.-B., Kiritsis D., Xirouchakis P. Research issues on closed-loop PLM. Computers in Industry 58 (2007) 855-868.
[10] Stark, J., Product Lifecycle Management 21st Century Paradigm for Product Realisation. Springer-Verlag, London, 2011.
[11] Lee, Y-C., Sheu L-C., Tsou Y-G. Quality function deployment implementation based on Fuzzy Kano model: An application in PLM system, Computers & Industrial Engineering 55 (2008) 48-63
[12] Stark, J. Global Product: Strategy, Product Lifecycle Management and the Billion Customer Question. Springer-Verlag, London, 2007.
[13] Silventoinen, A; Pels, H.J.; Hannu Kärkkäinen, H.; Lampela, H.;Towards future PLM maturity assessment dimensions. Contents Proceedings PLM11 - 8th International Conference on Product Lifecycle Management 11 13 July 2011. Technische Universiteit Eindhoven, in Eindhoven, The Netherlands. 480-492.
[14] Orčik, A.; Anišić, Z.; Gečevska, V.; Veža, I. Implementation of PLM strategy in the process of the new product development in chemical industry. In 4th International Scientific Conference International Conference Management of Technology – Step to Sustainable Production (MOTSP 2012). 14- 16 June 2012 in Zadar.
[15] Batenburg, R.; Helms, R.W.; Versendaal, J. PLM roadmap: stepwise PLM implementation based on the concepts of maturity and alignment. Int. J. Product Lifecycle Management, Vol. 1, No. 4, 2006. 333-351.

9. DATA ABOUT AUTHORS

PhD student Marko Paavel
PhD student Sergei Kaganski
Assistant Professor Kristo Karjust
PhD student Rivo Lemmik
PhD student Tanel Eiskop

Department of Machinery,
Tallinn University of Technology,
Ehitajate tee 5, Tallinn, 19086, Estonia.
Corresponding Author: Marko Paavel
E-mail: Marko.Paavel@ttu.ee

10th International DAAAM Baltic Conference
INDUSTRIAL ENGINEERING
12-13 May 2015, Tallinn, Estonia

OPTIMAL STRUCTURAL DESIGN OF A SLOTLESS PERMANENT MAGNET GENERATOR

Pabut, O.; Kirs, M;. Lend, H. & Tiirats, T.

Abstract: *Most optimal mechanical layout of an electrical generator in a wind turbine is still under investigation. In current work, a novel lattice passed solution is presented together with corresponding design principles and optimization procedure. Finite element analysis and response surface modelling are used to create the mathematical model of the structure and carry out the design investigation and optimization. Generator layout is modelled at two turbine capacity levels and compared to a similar sheet metal structure. Results prove the suitability of the offered procedure and superiority of the lattice concept in terms of the generator structural mass.*
Key words: Design optimization, lattice structure, response surface modelling, permanent magnet machine.

1. INTRODUCTION

Wind energy sector has gone through a rapid commercial and technical revolution during past decades [1]. For many sub-systems clear technical standardization has taken place, while the concept of the most beneficial electrical generator is still under investigation [2]. Direct-drive generators have been gaining popularity but in terms of machine cost they are still inferior to the established combination of a double-fed induction generator and a gearbox [3]. Researchers have indicated that optimization and component integration of these machines could lead to a considerably lower energy cost, while keeping their known advantages [4]. Most of the cost and heavy mass of these units can be attributed to the mechanical carrier structure [5].

Goal of the current work is to develop design principles and optimization procedure for a new type of mechanical carrier structure for a permanent magnet (PM) generator. Utilization of now available computational power together with various software tools and methods, including finite element modelling (FEM) and response surface modelling (RSM), is used to create the mathematical model of the generator structure. For comparison purposes two alternative mechanical structures are investigated to prove the superiority of the lattice concept against the more common sheet metal solution.

2. GENERATOR MODEL

2.1 Electromagnetic layout

Basic parameters of a direct drive generator are primarily determined by the wind turbine where the generator is used. The power P available in the wind for a wind turbine with given blades can be found according to [6]:

$$P = \frac{1}{2}\rho_{air}\pi r^2 v_{wind}^3 C_p, \tag{1}$$

where ρ_{air} is the air mass density, r is the turbine rotor radius, v_{wind} is the given wind speed and C_p is the power coefficient. Rotational speed of the turbine is given as:

$$\omega = \frac{v_{tip}}{r}, \tag{2}$$

where v_{tip} is the blade tip speed and ω is the rotational speed of the machine. Torque resulting from the power and rotational spccd is dcfincd according to:

$$\tau = \frac{P}{\omega}. \qquad (3)$$

By combining equations (1), (2) and (3), torque of the turbine system can be expressed as:

$$\tau = \frac{\rho_{air}\pi r^3 v_{wind}^3 C_p}{2v_{tip}}. \qquad (4)$$

In order to convert the kinetic energy of the wind into electrical energy, the used generator has to be able to produce equal torque described by [7]:

$$\tau = 2\pi\sigma_{tan}R^2 l, \qquad (5)$$

where σ_{tan} is the tangential component of Maxwell stress, R generator air-gap radius and l generator axial length. Value of σ_{tan} for a generator with defined electromagnetic configuration is a rather constant value over a large range of capacities and is given as [8]:

$$\sigma_{tan} = \frac{1}{2}A_s B_n cos\varphi, \qquad (6)$$

where A_s is the linear current density, B_n normal direction air-gap flux density and $cos\varphi$ is the power factor between active and apparent power. Air-gap flux depends largely on the chosen electromagnetic circuit of the machine. In present work concept developed by Kallaste [9] is utilized with air-gap windings and iron in the stator yoke leading to a magnetic flux density around 0.5 T. Allowed linear current density is limited by the heat dissipation properties of the machine. For passively cooled generators, values leading to high efficiencies range between 40 kA/m and 50 kA/m [10].

For a given generator type, the most economic layouts in terms of active material vs. efficiency vs. acting forces are achieved at a constant electromagnetic aspect ratio k_a determined according to:

$$k_{a} = \frac{l}{R}. \qquad (7)$$

Based on previous research, value 0.125 is chosen for the particular machine. By combining equations (5) and (7) the generator air-gap radius is expressed as:

$$R = \sqrt[3]{\frac{\tau}{2\pi\sigma_{tan}k_a}}. \qquad (8)$$

In order to assess the mechanical designs of machines in equivalent frame of reference, their electrical designs should fulfil certain similarity criteria. Following relative dimensions for the generator active components have been found to be most optimal in terms of cost vs. efficiency:

$$k_p = \frac{w_m}{w_p} = 0.65, \quad k_w = \frac{h_y}{w_p} = 0.32,$$

$$k_{mag} = \frac{h_m}{h_y} = 0.44, \quad k_{coil} = \frac{h_c}{h_y} = 0.29, \qquad (9)$$

where k_p is the magnet, width and pole pitch ratio, w_m is the magnet width, w_p is the pole pitch, k_w is the total air-gap ratio to the pole pitch, h_y is the total air-gap height, k_{mag} is the magnet height ratio to the air-gap, h_m is the magnet height, k_{coil} is the coil height ratio to the air-gap and h_c is the coil height.

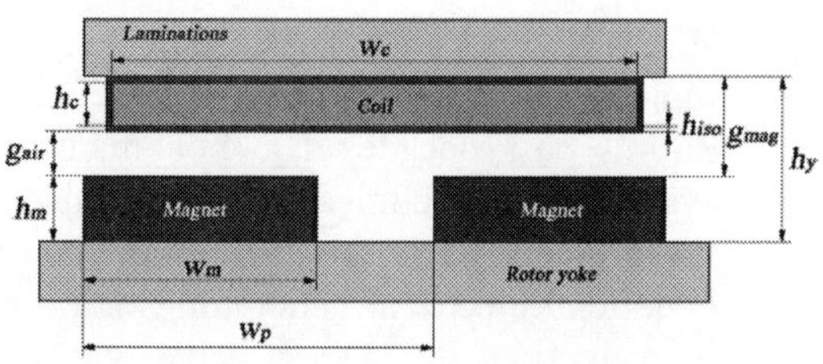

Fig. 1. Air-gap layout and parameters

For simplification purposes magnet length l_m is taken to be equal to the axial length l of the machine. Main parameters of the active components are described in Fig. 1. As common practice, the value of generator mechanical air-gap g_{air} is predefined according to mechanical deformations and assembly precision.

Magnet height h_m as a function of air-gap flux density can be found according to [11]:

$$h_m = \frac{B_n \mu_{rm} g_{eff}}{B_{rm}}, \tag{10}$$

where μ_{rm} is the relative permeability of the PM material, B_{rm} is the remanent flux density of the magnet and g_{eff} is the effective air-gap. As the machine has no slots and coil units are located in the air-gap following simplification can be made:

$$g_{eff} = h_y + 2h_{iso}, \tag{11}$$

where h_{iso} is the isolation layer height around the winding. Combining equations (9), (10) and (11) leads to:

$$h_m = \frac{g_{air} + 2h_{iso}}{\left(\frac{B_{rm}}{B_n \mu_{rm}} - \frac{k_{coil}}{k_{mag}} - 1\right)}. \tag{12}$$

Based on the above described similarity criteria and principles given in [9] two generator layouts with different capacities are defined and presented in Table 1.

Electrical power	P_{el}	MW	3	1
Rated wind speed	v_{wind}	m/s	11	11
Power coefficient	C_p	-	0.48	0.48
Rated tip speed	v_{tip}	m/s	81.5	81.5
Rotational speed	ω	rpm	15.3	26.5
Torque	τ	kNm	1983	385
Air-gap radius	R	m	6.3	3.65
Air-gap width	g_{air}	mm	11.5	8.5
Flux density	B_n	T	0.5	0.5
Current density	A_s	kA/m	40	40
Efficiency	η	%	94.5	93.5
Magnet height	h_m	mm	22	17.5
Magnet width	w_m	mm	100	79.5
Magnet length	l_m	mm	800	460
Magnet flux density	B_{rem}	T	1.34	1.34
Coil height	h_c	mm	14.5	11.5
Coil width	w_c	mm	200	160
Coil length	l_c	mm	1000	620

Table 1. Used generator layouts

2.2 Structural layout

Design of the mechanical carrier structure for a PM generator is in general driven by the allowable air-gap deflection. For machines with high flux densities ($B_n = 1.1$ T) and large aspect ratios ($k_a > 0.4$) well studied mechanical layouts can be found from literature. These structures are made up from thin walled ($0.001R$ to $0.0025R$) discs, structurally profiled spokes, support arms and stiffness ribs [2]. For generator's with ultra large radiuses and small air-gap flux densities, more complicated solutions with pre-stressed tension rods or air-gap bearings have been employed [12]. One option not considered in the available literature is the utilization of a lattice structure. This has been used in bridge and crane structures for centuries and is generally characterized by its light weight, high loading capacity and high stiffness to length ratio, which makes it a good candidate for an optimal structural concept.

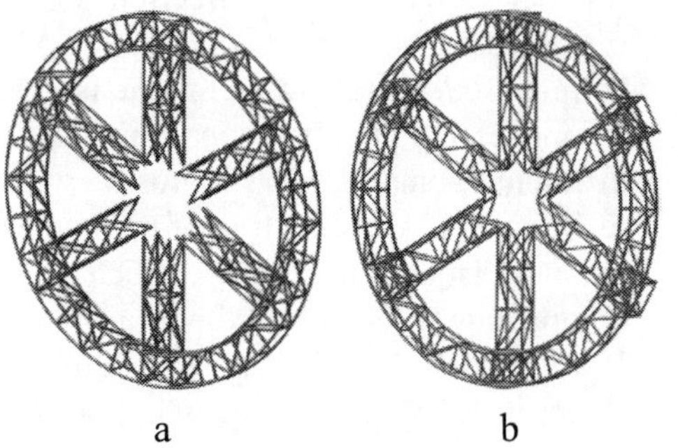

Fig. 2. Layout of the lattice rotor (a) and the lattice stator (b)

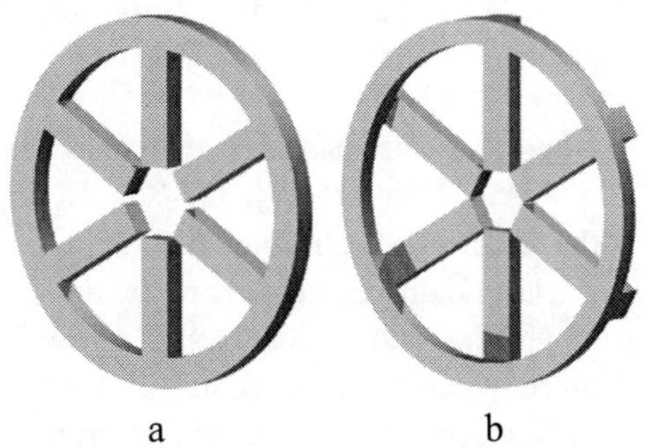

Fig. 3. Layout of the sheet rotor (a) and the sheet stator (b)

Principle layouts of the employed lattice configurations for rotor and stator are depicted in Fig. 2a and Fig. 2b, respectively. In order to have a more direct comparison to a sheet structure with similar size a corresponding sheet model is developed and analysed in parallel. Principle layouts of the rotor with internal support arms and stator with cantilever support beams are shown in Fig. 3a and Fig. 3b, respectively.

2.3 Mathematical model

Direct-drive generators with sheet metal structures are well studied and analytical formulas have been developed to determine the air-gap deflections for regular mechanical layouts [5]. As lattice structure has so far not been utilized for a generator, no ready-made analytical tools are available. It is possible to use known truss calculation methods like Maxwell-Cremona diagram for determining the internal forces and deflections of the members. In current study, FEM in conjunction with RSM technique is used to compose the mathematical models for different support structures. RSM technique is a combination of mathematical and statistical methods, which allows composing an objective function based on learning data [13, 14]. These functions are represented by certain degree polynomial models. In current work quadratic model is used with following general form [15]:

$$Y = \beta_0 + \sum_{i=1}^{k} \beta_i X_i + \sum_{i=1}^{k} \beta_{ii} X_i^2 + + \sum_i \sum_j \beta_{ij} X_i X_j, \quad (13)$$

where Y is the measured response, k is the number of factors, β_0, β_i, β_{ii} and β_{ij} are the regression coefficients for intercept, linearity, square and interaction and X_i and X_j are the independent design variables. In present paper the procedure is carried out in *Design Expert 9* software environment.
Full factorial experimental (FFE) method is employed to create the initial population of designs for RSM. For both structure configurations the number of factors k is selected to be 4 and the number of levels n for each factor is set to be 5. As the sheet and lattice structures are different in nature, the used factors vary between the structure types. Factors and their limit values are depicted in Table 2. General structural variables and variables specific for the sheet structure are described in Fig. 4a and variables used for the lattice structure in Fig. 4b.

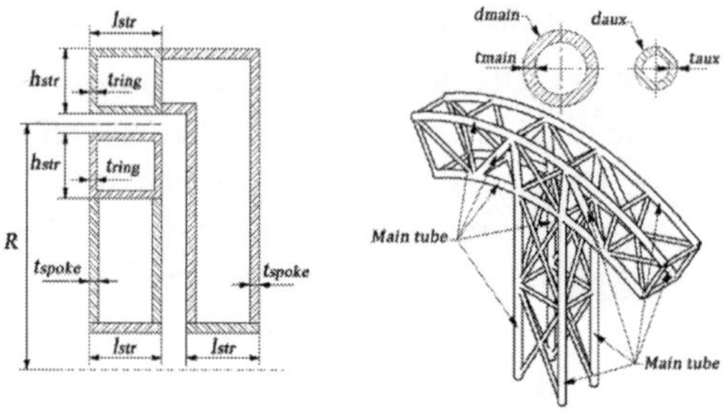

a b

Fig. 4. General and sheet variables (a) and lattice structure variables (b)

For both of the structure types, the height of the rotor and stator rings h_{str} is varied as a state variable defined by:

$$h_{str} = k_{str} l_{str}, \quad (14)$$

where k_{str} is the structure aspect ratio and l_{str} is the structure axial length.
Lattice structure bracing is subjected to following dependencies in order to keep the number of variables at a reasonable limit:

$$d_{aux} = 0.7 d_{main}, \quad t_{aux} = 0.5 t_{main}, \quad (15)$$

where d_{aux} is the bracing tube diameter, d_{main} is the main tube diameter, t_{aux} is the bracing tube wall thickness and t_{main} is the main tube wall thickness.
Calculations to compose the learning data are carried out on a parametric geometrical computer aided design model in *ANSYS Workbench* software. General description of the applied loads and the corresponding beam model can be found in [16].

3. OPTIMIZATION PRINCIPLES

Generator structural mass $F_m(\bar{x})$ is chosen as an objective subjected to minimization while the air-gap deformation ε is defined

Factor	Symbol	Unit	3 MW lattice	3 MW sheet	1 MW lattice	1 MW sheet
Structure axial length	l_{str}	mm	700-1300	700-1100	400-700	385-585
Structure aspect ratio	k_{str}	-	0.7-1.3	0.5-0.9	0.7-1.3	0.5-0.9
Main tube diameter	d_{main}	mm	110-150	-	30-50	-
Main tube wall thick.	t_{main}	mm	10-18	-	4-8	-
Spoke sheet thick.	t_{spoke}	mm	-	6-18	-	4-10
Ring sheet thick.	t_{ring}	mm	-	6-18	-	4-10

Table 2. Factors and levels used in the FFE

as a constraint. Therefore, the optimization problem can be formulated:

$$F_m(\bar{x}) \rightarrow \min, \qquad (16)$$

subjected to linear constraints applied to the design variables vector $\bar{x}$:

$$x_i \leq x_i^*, \quad -x_i \leq x_{i^*}, \quad i = 1, \ldots, n, \qquad (17)$$

and non-linear constraint applied to the sum of rotor and stator air-gap deflections ε^{max} as:

$$\varepsilon^{max}(\bar{x}) \leq \varepsilon^*, \qquad (18)$$

where indexes * in (17) and (18) refer to upper limit value of the variable.

For regular slotted generators the allowable air-gap deflection is in most cases fixed to 10% of the initial value, as larger deflections would result in unreasonable changes to flux densities and generated electromagnetic force. The machine under investigation has air-gap windings and as a result, the magnetic air-gap g_{mag} (see Fig. 1) and mechanical air-gap g_{air} are not equal. Consequently, the generator is electromagnetically less sensitive towards mechanical air-gap deformations. Therefore, a double criterion is applied to the deformations in order to allow for sufficient flexibility and guarantee both mechanical and electrical safety:

$$\varepsilon^{max} \leq 0.12 g_{mag}, \ \varepsilon^{max} \leq 0.3 g_{air}, \qquad (19)$$

As the stator is designed in a cantilever form and rotor is supported from both sides, rigidity of the rotor can be achieved with smaller weight penalty. The allowable deformations for rotor ε_{rot}^{max} and stator ε_{stat}^{max} are found according to:

$$\varepsilon_{rot}^{max} = 0{,}25\varepsilon^{max}, \ \varepsilon_{stat}^{max} = 0{,}75\varepsilon^{max}. \qquad (20)$$

Even though stress is observed as one of the outputs during the FFE, initially no criterion is applied to the resulting stresses. As the optimization is against single criteria with applicable constraints, no normalization of the inputs or outputs is necessary.

4. RESULTS AND DISCUSSION

Optimal configurations of the structures with respect to the criteria described in (19) and (20) together with corresponding values of the coefficient of determination for deformation (R_{def}^2) are listed in Table 3. At 3 MW level, total masses of different generators are almost equal in terms of weight and for the 1 MW level, the lattice concept is about 17% lighter. Selection of the stiffness criteria was based on an assumption that structures designed for high rigidity are also optimal in terms of ultimate strength and fatigue resistance [[17]]. Evaluation of the determined optimal solutions against fatigue, linear buckling and additional tower top accelerations revealed this to be untrue. Found encumbrance can be mostly attributed to the fact that rigidity of such a large structure is a global property, while fatigue and buckling problems depend on local topology. For each structure type not compliant with the applied set of requirements, another optimization run was conducted with additional design criteria for maximum stress range and minimum load multiplier for buckling. The achieved results are presented in Table 4.

Factor	Unit	3 MW lattice		3 MW shell		1 MW lattice		1 MW shell	
		Stator	Rotor	Stator	Rotor	Stator	Rotor	Stator	Rotor
Structure ax. length	mm	700	1234	700	700	400	400	385	385
Aspect ratio	-	1.04	1.30	0.88	1.30	1.00	1.30	0.54	1.01
Deformation	mm	2.5	0.85	2.5	0.85	2.0	0.55	2.0	0.47
Stress	MPa	54	43	83	59	80	54	106	60
Part mass	t	16.85	13.74	15.77	13.64	2.22	2.08	2.79	2.38
Generator mass	t	30.59		29.41		4.30		5.17	
R^2_{def}	-	0.995	0.997	0.980	0.980	0.987	0.991	0.965	0.986

Table 3. Optimization – stiffness criteria

Factor	Unit	3 MW lattice		3 MW shell		1 MW lattice		1 MW shell	
		Stator	Rotor	Stator	Rotor	Stator	Rotor	Stator	Rotor
Structure ax. length	mm	700	1000	700	700	400	400	385	385
Aspect ratio	-	1.04	1.30	0.50	1.30	1.08	1.13	0.50	1.17
Deformation	mm	2.5	0.85	2.4	0.85	1.57	0.55	2	0.47
Stress	MPa	54	43	70	55	60	50	113	50
Mass	t	16.85	13.74	19.32	14.68	2.80	2.38	3.33	2.54
Generator mass	t	30.59		34.00		5.18		5.87	

Table 4. Optimization – stiffness, fatigue and buckling criteria

With the additional criteria the lattice generator is 10% lighter at 3 MW level and 12% lighter at 1 MW level. For lattice stators, smallest allowed axial lengths and aspect ratios around 1 lead to low weight structures. This is mainly caused by the cantilever nature of the stator, where additional overhang due to the increased axial length leads to large deformations. Increasing the aspect ratio beyond 1 does not yield any benefits, as the bending is mostly seen in the stator spokes. For the lattice rotors, higher aspect ratios are preferable as the angles of diagonal bracings are turned into better alignment with the vectors of normal stress. Similar phenomenon can be seen for the shell rotor and stator aspect ratios and axial lengths.

5. CONCLUSIONS

Analytical equations and optimization procedure are proposed for designing a PM generator with a novel type of carrier structure. Similarity criteria for a slotless air-gap winding generator are introduced and used to define two generator electromagnetic configurations at different power levels. The resulting loads and mechanical parameters are used as an input to set up a lattice type carrier structure for the generator and to compare it to a more common sheet metal solution. Due to the complex geometry of the lattice design, RSM approach is used to create the analytical models for both structure types with the aid of FEM. Obtained results are used as an input for RSM with quadratic curve fitting to optimize the structural mass for both concepts. Based on the process and achieved outcome it is concluded that:

- RSM with quadratic fitting can be used to predict the structural behaviour of generator support structures;
- when considering only stiffness criteria, conclusion regarding optimality of a generator structural concept cannot be made;
- if stiffness, fatigue, buckling and tower top acceleration criteria are considered, lower structural weights can be obtained with the lattice concept;
- stiffness criteria can be used for initial definition of the described generator structure, however fatigue resistance criteria has to be applied either simultaneously or as a next step in the optimization procedure.

Acknowledgements
Authors acknowledge the successful cooperation with Goliath Wind OÜ.

6. REFERENCES

[1] Manwell, J. F., McGowan, J. G., Rogers, A. L. *Wind Energy Explained: Theory, Design and Application.* Wiley, Chichester, 2009.

[2] Stander, J. N., Venter, G., Kamper, M. J. Review of Direct-Drive Radial Flux Wind Turbine Generator Mechanical Design. *Wind Eng.,* 2012, **15**, 459-472.

[3] de Vries, E. Wind Turbine Drive Systems: A Commercial Overview. In *Electrical Drives for Direct Drive Renewable Energy Systems* (Mueller, M. and Polinder, H. eds). Woodhead, Oxford, 2013, 139-157.

[4] Bang, D., Polinder, H., Shrestha, G., Ferreira, J. A. Possible Solutions to Overcome Drawbacks of Direct-Drive Generator for Large Wind Turbines. *Proc. EWEC 2009*, 2009, 1-10.

[5] McDonald, A. S. Structural Analysis of Low Speed, High Torque, Electrical Generators for Direct Drive Renewable Energy Converters. *Ph. D. Dissertation,* Edinburgh University, 2009.

[6] Polinder, H., de Haan, S. W. H., Dubois, M. R., Slootweg, J. G. Basic Operation Principles and Electrical Conversion Systems of Wind Turbines. *EPE Jour.,* 2005, **15**, 43-50.

[7] McDonald, A. S., Mueller, M. A., Polinder, H. Structural Mass in Direct-Drive Permanent Magnet Electrical Generators. *IET Renew. Pow. Gen.,* 2008, **2**, 3-15.

[8] Polinder, H. Principles of Electrical Design of Permanent Magnet Generators for Direct Drive Renewable Energy Systems. In *Electrical Drives for Direct Drive Renewable Energy Systems* (Mueller, M. and Polinder, H. eds). Woodhead, Oxford, 2013, 30-50.

[9] Kallaste, A. Low Speed Permanent Magnet Slotless Generator Development and Implementation for Windmills. *Ph. D. Dissertation,* TUT, Tallinn, 2013.

[10] Pyrhonen, J., Jokinen, T., Hrabovcova, V. *Design of Rotating Electrical Machines*. Wiley, Chichester, 2008.

[11] Li, H., Chen, Z. Design Optimization and Comparison of Large Direct-Drive Permanent Magnet Wind Generator Systems. *Proc. Int. Conf. on Elec. Mach. and Sys.*, 2007, 685-690.

[12] Gordon, P. Aspects of, and New Approaches to, the Design of Direct Drive Generators for Wind Turbines. *Ph. D. Dissertation,* University of Durham, 2004.

[13] Majak. J., Pohlak, M., Decomposition Method for Solving Optimal Material Orientation Problems. *Comp. Struct.*, 2010, **92**, 1839-1845.

[14] Majak. J., Pohlak, M., Eerme, M., Velsker, T. Design of Car Frontal Protection System Using Neural Networks and Genetic Algorithm. *Mechanika*, 2012, **18**, 453-460.

[15] Montgomery, D. C. *Design and Analysis of Experiments*. Wiley, New York, 1997.

[16] Pabut, O., Lend, H., Tiirats, T. Load Sensitivity Analysis of a Large Diameter Permanent Magnet Generator for Wind Turbines. *Proc. of the 9th Int. Conf. DAAAM Baltic Indust. Eng. Tallinn*, 2014, **1**, 59-64.

[17] Cervellera, P., Zhou, M., Schramm, U. Optimization Driven Design of Shell Structures Under Stiffness, Strength and Stability Requirements. *Proc. of 6th WCSMO,* 2005.

7. ADDITIONAL DATA ABOUT AUTHORS

MSc. Ott Pabut
TUT, Department of Machinery
Ehitajate tee 5, 19086 Tallinn, Estonia
Phone: +372 51 644 57,
E-mail: ott.pabut@ttu.ee
http://www.ttu.ee

10th International DAAAM Baltic Conference
"INDUSTRIAL ENGINEERING"
12-13 May 2015, Tallinn, Estonia

IMPROVEMENT OF MECHANICAL PRESS PRODUCTIVITY AND ACCURACY BY COMPENSATION OF FRAME OPENING

Raz, K.; Cechura, M.; Chval, Z.; Zahalka, M.

Abstract: *The paper presents a design of adaptive controlled compensation of deformation in frame of "C-shaped" forging press. For compensation are used hydraulic or mechanical principles. The aim of this frame deformation compensation is to reduce opening depending on the actual press force. In actual time is preloading force changed due to sensor of deformation. This leads to exact contact between both tools and also better tools guidance. The results are more accurate products, productivity and quality improvement and tools life increasing.*
Key words: force, deformation, compensation, open frame.

1. INTRODUCTION

During works, carried out by Centre for research of forming machines on university of West Bohemia in Pilsen are designed new forming and forging machines and also innovations of old machines are performed. Main problem with forming machines with open frame ("C" shaped frame) is deformation of this frame and also misalignment of forging tools. This has influence on accuracy and quality of products. Elimination of this frame opening during forming operations, and an adaptive change, depending on the size of the loading force is the subject of this paper.
For this problem in today known solution, but it is not adaptive and it is not easy changeable. The main aim is to use frame with minimum changes in structure and use maximum amount of standard parts.

2. HYDRAULIC PRESS WITH OPEN FRAME

The goal of innovation is to avoid widening of the frame. It means that the deformation of the frame y1 and y2 are equal to zero [1,2].

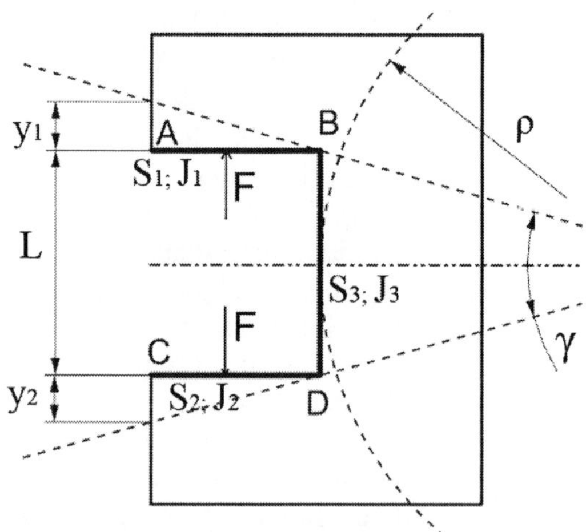

Fig. 1. Model of open frame

γ – angle of rotation
y_1,y_2 – deformation of the frame
F – loading force
A-B-C-D –computational model for analytical solution
ρ – radius of curvature
Sn, Jn – sections and quadratic moments of individual sections of the frame.

By using these values and also using simplification of the frame can be calculated the deformation of the frame. More modern approach is to use virtual modeling and analysis of stress and strain using FEM methods. It will tell us real values with considering ale shape changes.

2.1 Determination of preloading force

The preloading force Fp was determined from the torque balance between this force and the working force F relative to the neutral axis from the FEM calculation. More details can be seen from Fig. 2, where forces are displayed on the frame and the vertical line is neutral axis (the place where the deformation is equal to zero).

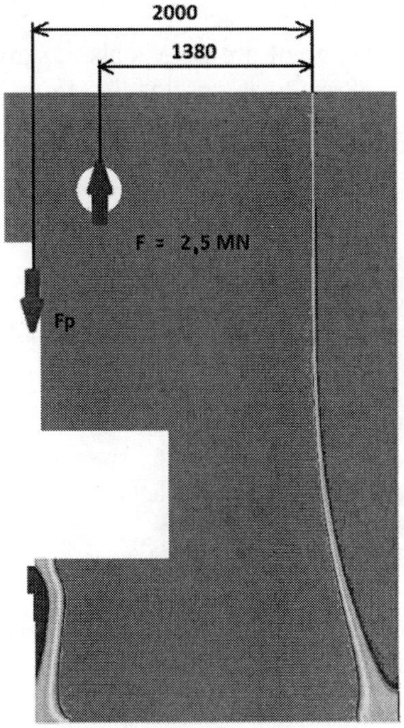

Fig. 2. Torque balance of forces

To obtain preloading force Fp is used wedge mechanism. To determine the preloading force is necessary to make a detailed analysis of the forces and provide the adjustment force Fmx according to the following scheme.

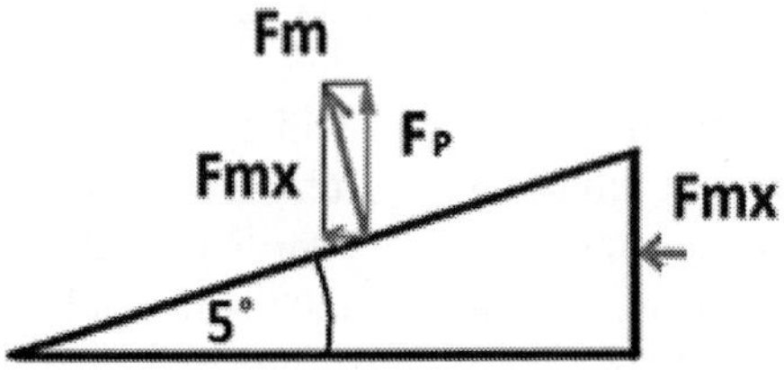

Fig. 3. Decomposition of forces on the wedge, angle 5 ° without friction effects

3. PRELOADING MECHANISM

There are many ways how to get preloading. Adaptive preloading force is performed by moving of the wedge surfaces between the vertical rod and frame. Changing of position (also preloading force) can be done in many ways, the easiest way is to use a screw connection, but it does not allow rapid adaptive changing of position. Better solution is hydraulic cylinder, adaptively controlled by deformation sensor. Force of hydraulic cylinder must correspond with force Fmx. Because vertical rods are not as tough as the frame, stress in material of vertical rods is above 100 MPa. To eliminate stress in the contact area between the rod and the frame is useful to increase the contact surface in order to achieve optimum contact pressure and stress.
According these facts were prepared following solutions of adaptive controlled frame deformation [3,4].

3.1 Variant 1- force on wedge mechanism caused by screw

The principle is to realize preloading force by pushing of 5° wedge. Controlling is carried out manually by screw. This screw is located on the side of the vertical rod.

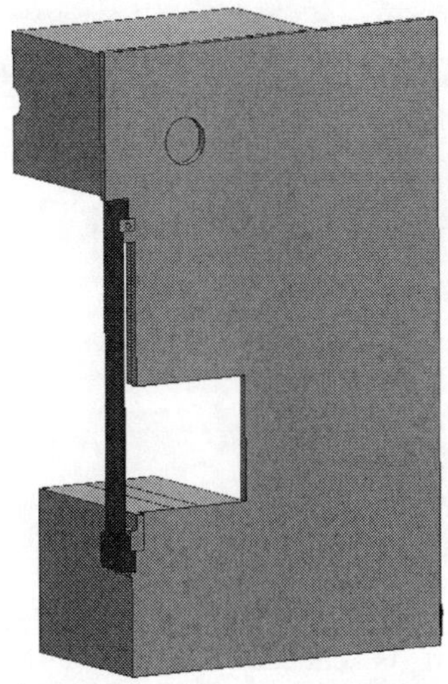
Fig. 4a. Vertical rod and press frame

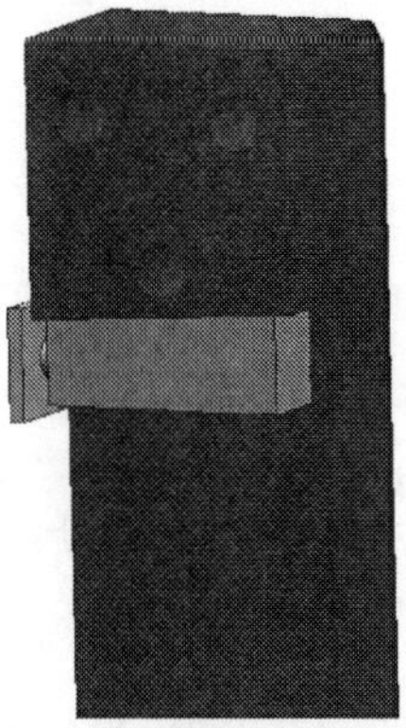

Fig. 4b. Screw- wedge mechanism

3.2 Variant 2- force on wedge mechanism caused by hydraulic mechanism

Because is necessary to make change of preloading force automatized, is better to use solution with hydraulic mechanism and also with controlling system. Principle is same as previous. The wedge is pushing it to the space between vertical rod and frame. As more is pushed, higher preloading force is. Force, which is moving with wedge, is caused by hydraulic mechanism. Cylinder of hydraulic mechanism is by its lower face attached to the frame. Vertical rod is attached by 3 bolts to the frame (on side of hydraulic mechanism), but holes are bigger in order to allow vertical movement.

Fig. 5. Variant 2- force on wedge mechanism is caused by screw (cutted view)

3.3 Variant 3- force on wedge mechanism caused by modified hydraulic mechanism

Preloading is caused by acting of a force on vertical rod. Again is used principle of wedge. Changeable force is caused by hydraulic cylinder. Difference between this variant and previous one is position of hydraulic cylinder. Here is attached to the front part of vertical rod.
Cylinder is in contact with vertical rod (dark). Piston is attached by screw to the frame. Change of pressure is also changing position of piston and preloading force [5].

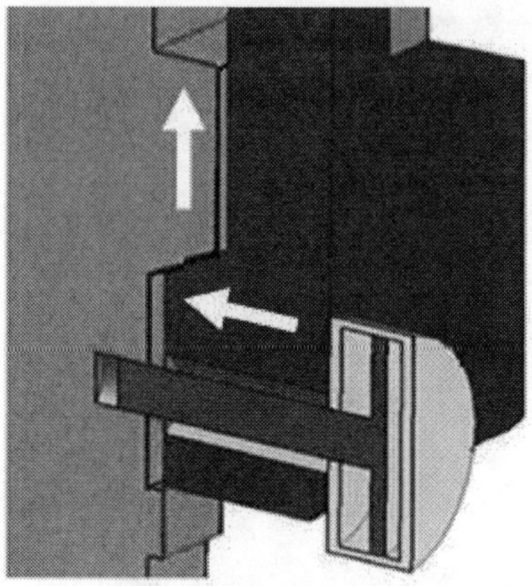

Fig. 6. Variant 3- force on wedge is caused by modified hydraulic; arrows are vectors of deformation (movement) during preloading

4. FEM ANALYSIS AND SIMULATIONS

For evaluating and comparing of preloaded and non-preloaded frame was used virtual FEM analysis. The result is minimization of frame deformation. Maximal stress in vertical rod is up to 130 MPa. Because contact pressure is higher, it was neccesary to make contact surface between vertical rod and frame wider. As an evaluation criterion was used displacement in vertical direction. Compared was preloaded and non-preloaded frame (Fig. 7a, 7b). Measuring point for deformation was on top contact surface of vertical rod [6].

Non-preloaded frame has maximal deformation up to 2mm.

Preloaded frame eliminated deformation and value is equal to zero.

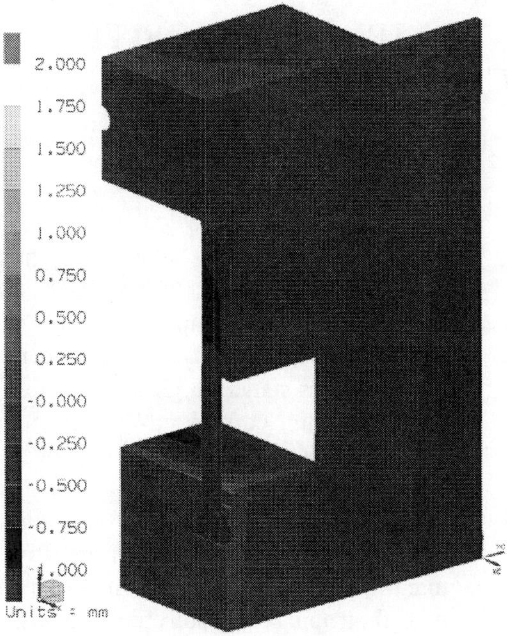

Fig. 7a. Displacement of preloaded frame

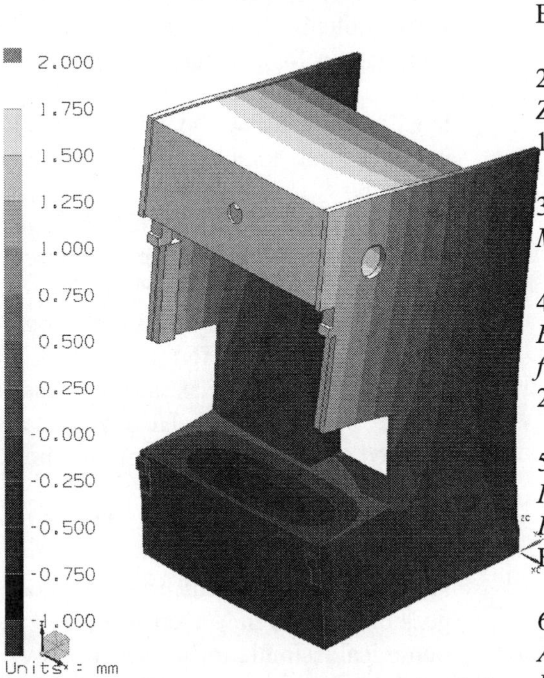

Fig. 7b. Displacement of non-preloaded frame

5. CONCLUSION

The design and modification of forming press frame helped to eliminated deformation. The range of usage is because of this solution much better. Lifetime of tools used for forming is higher because guidance is in proper position. This will save money and also time for tool changing. This topic should now continue into the real prototype.

6. ADDITIONAL DATA ABOUT AUTHORS

Ing. Karel Raz, Ph.D., researcher,
Regional Technological Institute,
University of West Bohemia, Univerzitni 8, Plzen, 306 14, Czech Republic.
Email: kraz@rti.zcu.cz.
Tel: +420 377 638 751.

7. REFERENCES

1. Cechura, M., Stanek, J. *Tvareci stroje-Hydraulické lisy.*Univestity of West Bohemia, Plzen, 1999.

2. Rudolf, B., Kopecky, M. *Tvareci stroje-Zaklady stavby a vyuziti.*SNTL, Prague, 1985.

3. Banabic, D. *Advanced Methods in Material Forming.*Springer, 2007.

4. Piscan, I., Janssens, T., Predincea, N. *Experimental Validation of FEM for frictional Contacts.*Annals of DAAAM for 2011, Vienna, 2011, 741-742.

5. Cechura, M., Smolik, J. *Developement and Innovations of Existing Design Solutions of Forming Machines.*Univestity of West Bohemia, Plzen, 2012.

6. He, W., Zeng, P., Lin, F. *Finite Element Analysis for 40 MN Die Forging Press.*Advanced Materials Research, 2012, 690-693.

10th International DAAAM Baltic Conference
"INDUSTRIAL ENGINEERING -
12-13 May 2015, Tallinn, Estonia

COMPARISON OF NUMERICAL SIMULATION AND EXPERIMENT OF A FLEXIBLE COMPOSITE CONNECTING ROD

Sedláček, F.; Lašová, V.; Kottner R. & Bernardin P.

Abstract: *The work deals with a comparison of numerical simulations and experimental tests of a flexible composite connecting rod in order to create an appropriate computational model for predicting the strength and stiffness. For the numerical simulation were used finite element analysis in the software Siemens NX 10 and SIMULIA Abaqus 6.14. Experimental specimens with different geometry were exposed to quasi-static loading. Zwick/Roell Z050 testing machine was used for tensile tests.*

Key words: flexible, composite connecting rod, FEM analysis

1. INTRODUCTION

Composite materials have an ever greater range of applications in industrial practice. Today, commonly used construction materials in aviation and aerospace engineering, are increasingly used in the automotive and railway industries. One possible application is in flexible elements, for which the main factor is strain energy, which could be simply described by the equation;

$$U = \frac{\sigma^2}{\rho E}$$

From which clearly follows that the material with a lower modulus of elasticity E and density ρ, will have a relatively higher strain energy U. And the composite materials fulfil this property excellently (composite materials have high strength with low weight compared with conventional steel, up to 80%). Other positive properties of composite materials are water resistance, high fatigue strength, resistance to corrosion, higher natural frequency, low friction, etc. However, the use of composite materials has many negative aspects, which include in particular complicated design and difficult manufacturing, which are mainly caused by orthotropic (or anisotropic) properties of the material.

The objective of this work is to design and verify potential applicability of composite materials in for example a flexible joint. It has the task of transferring tensile/compressive loads as well as bending and torsion. It is not such a problem to propose fundamental (functional) areas for this element, but the part that is used to connect to subsequent components and this paper focuses on this area. It specifically focuses on the integrated types of joints of glass fibre. Integrated joints are already addressed in several papers [4, 5, 6]. However, they are targeted at the use of carbon fibre joints.

2. EXPERIMENTAL TESTS

A simplified element was created for possible design verification using numerical simulations, specifically the wrapping loops.

E-glass fibres (Aeroglass 2400 TEX) and epoxy resin (LH298 + hardener H512) as were chosen the material. The

experimental samples were wound on a special form in two versions (16pcs with seven threads a 16pcs with four threads of the fibreglass).
Experimental tensile tests were carried out by quasi-static load (0.5 mm/sec) on the Zwick/Roell Z050 machine. The samples were attached according to verified methods [4, 5] where one side eyes were tightly attached using special jaws and the other side free fastening was used. The results can be seen on the charts.

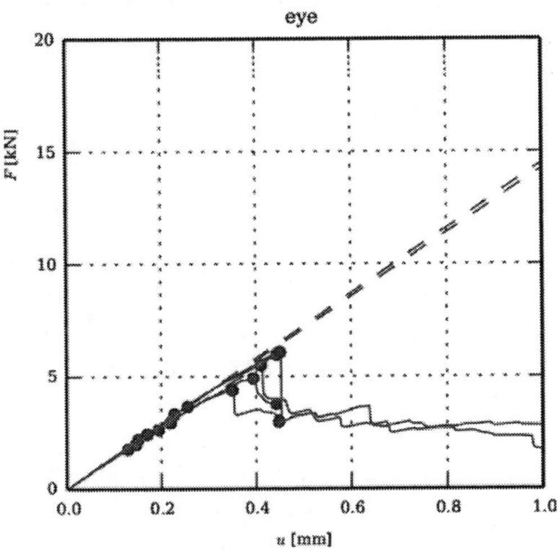

Fig.1 results of the experiment wrapping loops with seven threads of fibres

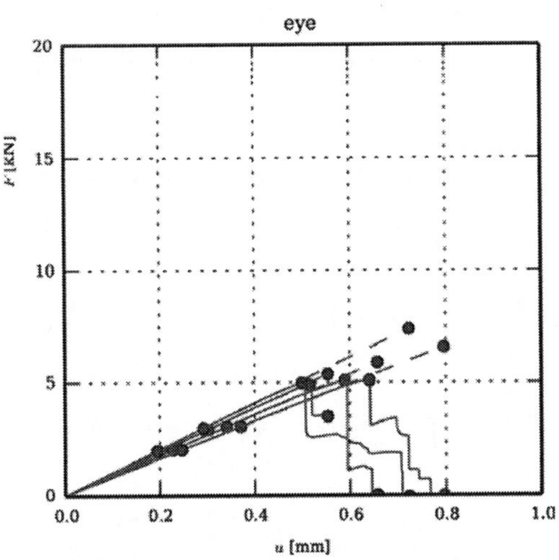

Fig.2 results of the experiment wrapping loops with four threads of fibres

// Fig.3 measured wrapping loops

E-Glass fibre + epoxy resin (H512+LH298)	
V_f [-]	0.73
V_m [-]	0.27
E_1 [MPa]	52640
E_2 [MPa]	8576.16
E_3 [MPa]	8576.16
G_{12} [MPa]	1986.75
G_{23} [MPa]	1986.75
G_{13} [MPa]	3264.20
ν_{12} [-]	0.295
ν_{23} [-]	0.31366
ν_{13} [-]	0.295
X_T [MPa]	2000
X_C [MPa]	1033
Y_T [MPa]	315
Y_C [MPa]	51
Z_T [MPa]	315
Z_C [MPa]	51
S or S_{12} [MPa]	48
S_{13} [MPa]	42
S_{23} [MPa]	48

Tab.1 Mechanical properties of tested composites

3. NUMERICAL SIMULATION

Numerical simulation was performed on the three-dimensional model by using finite element software Siemens NX 10 (with solver NX Nastran 10) and SIMULIA Abaqus 6.14.

3.1. FE model of wrapping loop

The meshes were created by brick elements CHEXA8 type (C3D8 in Abaqus) with 8 nodes. Subsequently, the elements were assigned to the appropriate orientation of material (see Fig. 4). The model was calculated as a symmetrical solution by transverse plane (YZ) with appropriate removal of degrees of freedom. The load was applied in the form of displacement of a pin according to the values obtained from experimental measurements.

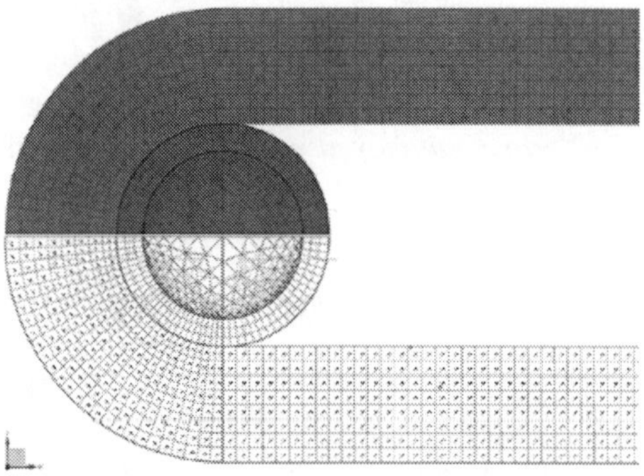

Fig.4 FE model, including a material orientation of the elements

3.2. Results of the numerical simulation

From the results of the individual components of normal stresses (L, T, T') and shear stress (LT, TT', LT') follows that the most critical is the index 3 (T′) and 6 (LT′), see Fig. 5 and 6.

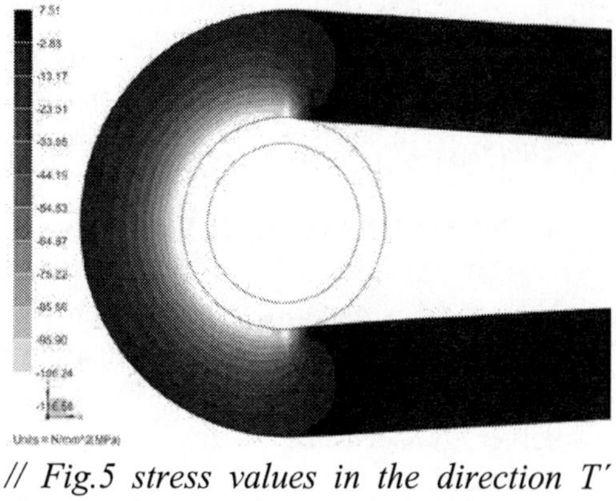

// Fig.5 stress values in the direction T′ (loop with seven threads of the fibres)

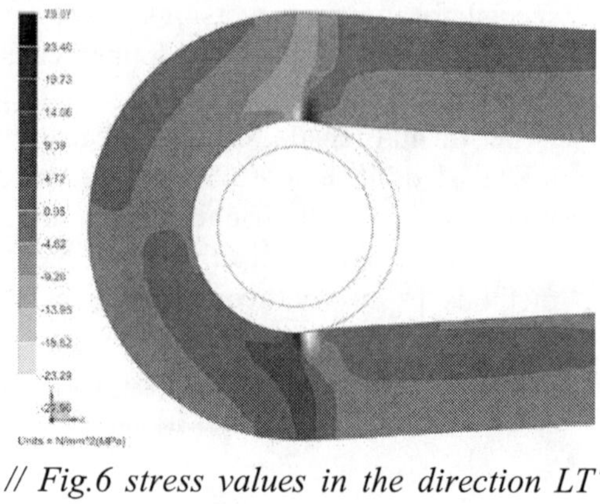

// Fig.6 stress values in the direction LT′ (loop with seven threads of the fibres)

Subsequently, the failure indexes from individual components of stresses were evaluated. Strength criterion 'Maximum stress' for the first approach was chosen. According to this theory, failure occurs if any component of stress has reached the ultimate strength of the material [7]. Failure indexes for individual directions of this strength criterion are listed in Table 2.
Where σ_L (σ_T, $\sigma_{T'}$) is normal stress in longitudinal (transverse) direction; X^T (Y^T, Z^T) is in Longitudinal tensile strength (transverse) direction; X^C (Y^C, Z^C) is compressive strength in longitudinal (transverse) direction; τ_{LT} ($\tau_{TT'}$, $\tau_{LT'}$) is shear stresses in LT (TT' or LT') plane and S_{LT} ($S_{TT'}$, $S_{LT'}$) is shear strength in LT (TT' or LT') plane.

Failure indexes	Maximum stress criterion
F_L	σ_L/X^T if $\sigma_L > 0$ σ_L/X^C if $\sigma_L < 0$
F_T	σ_T/Y^T if $\sigma_T > 0$ σ_T/Y^C if $\sigma_T < 0$
$F_{T'}$	$\sigma_{T'}/Z^T$ if $\sigma_{T'} > 0$ $\sigma_{T'}/Z^C$ if $\sigma_{T'} < 0$
F_{LT}	$\lvert\tau_{LT} / S_{LT}\rvert$
$F_{TT'}$	$\lvert\tau_{TT'}/S_{TT'}\rvert$
$F_{LT'}$	$\lvert\tau_{LT'}/S_{LT'}\rvert$

Tab.2 Failure indexes of Maximum stress criterion

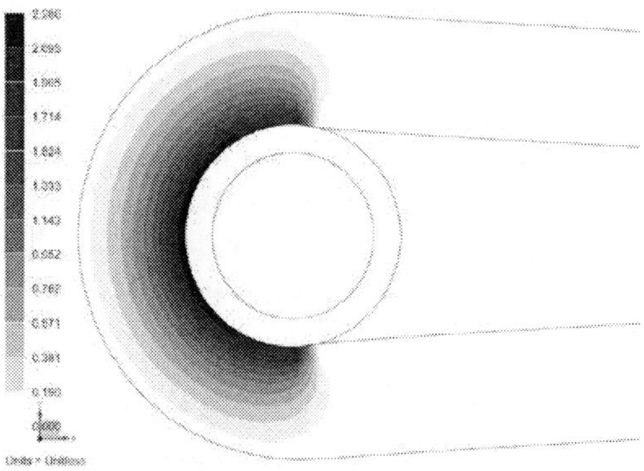

Fig.7 Failure index in the direction T′ (Maximum stress criterion)

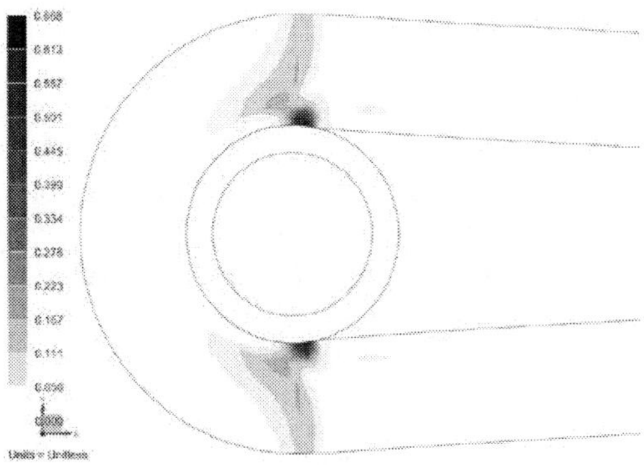

Fig.8 Failure index in the direction LT′ (Maximum stress criterion)

By using this strength criterion, the results of FI indexes do not correspond to values obtained by experimental tests. Therefore strength criterion LaRC04 (Laminates and Reinforced Composites/Langley Research Centre – invented in 2004 [8]) was used. LaRC04 is an interactive strength criterion for long fibre unidirectional composites. It is derived for the three-dimensional stress state and has incorporated therein correction for the nonlinear behaviour of the composite in shear area. For the modes LaRC04 #02 and LaRC04 #03 this strength criterion was modified, according to [5, 6]. The modification assumes that the strength of the matrix at loading pressure in the direction transverse to the fibre is also dependent on the tension in the direction of the fibers σ_L. However the stress σ_L causes hardening of the matrix. Further were used the adjusting parameters, parameter P_M for the LaRC04 #2 mode (matrix failure, if $\sigma_{T'} < 0, \sigma_L \geq 0$):

$$FI_M = \left(\frac{\tau^T}{S^T - \eta^T \sigma_n + \sigma_L P_M}\right)^2 + \left(\frac{\tau^L}{S^L - \eta^L \sigma_n + \sigma_L P_M}\right)^2 \leq 1$$

And parameter P_F for LaRC04 #03 mode (fibre failure, if $\sigma_L > 0$):

$$FI_F = \frac{\sigma_L}{\dfrac{X^T + X^T P_F}{Y^C + X^T P_M}\sigma_{T'} P_F + X^T} \leq 1$$

where τ_T and τ_L are stresses in the plain of the failure, S_T is the transverse shear strength and S_L is the longitudinal shear strength and η_T and η_L are coefficients of the friction. The failure criterion and the values of the adjusting parameters P_F and P_M were investigated using the comparison of the experiments and corresponding numerical simulations including results of previous works [3, 4, 5].

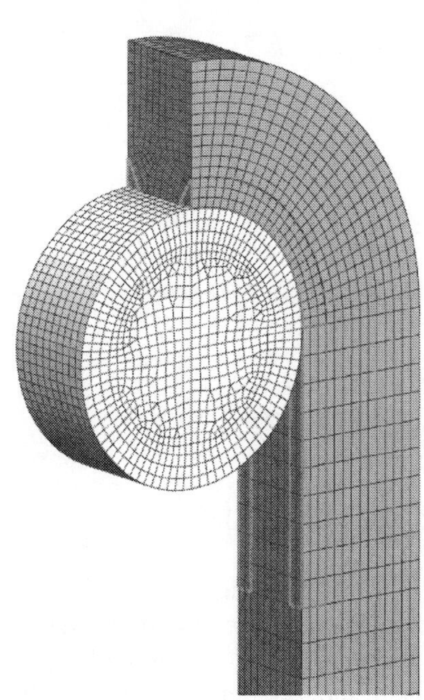

Fig.9 Mesh of the FE model with divide of cross-section

For numerical simulation of the matrix it is considered that it is already broken (if the failure of the matrix occurred before the failure of the fibres according LaRC04 # 2). For each geometry the cross-section was divided at an angle α in the place where according LaRC04 #2; $FI_M = 1 = FI_{Mmax}$. Between the separate parts and base part the contact of the type "touching" (without considering the friction) was defined (see Fig.9).

In the chart (Fig. 10) is possible to see comparing experimental data with numerical analysis using a modified strength criterion LaRC04 (for the most critical modes LaRC04 #02 and LaRC04 #03) and strength criterion Maximum stress (for the most critical index – in direction T′).

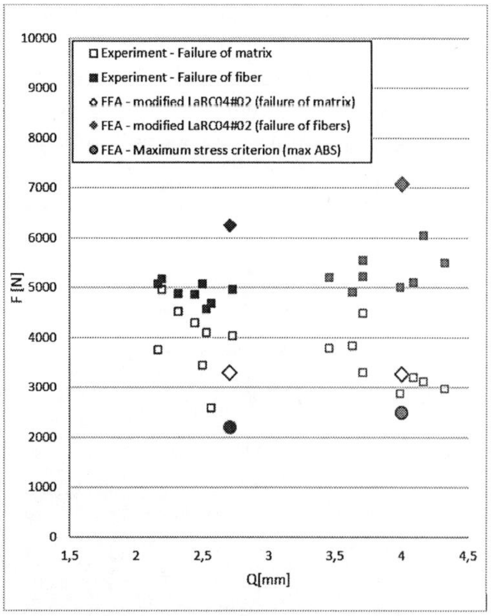

Fig.10 comparison of experimental data with numerical analysis

4. CONCLUSION

By using numerical simulation with the comparison of the resulting data from the experimental tests significant mismatch using the strength criterion 'Maximum stress' was revealed. Subsequently a modified strength criterion LaRC04 was used. By using this strength criterion the maximum difference of 27% was created to conformity compared with the experiments.

These data are the basis for further research in the area of integrated joints of glass fibre and we are still working on further experimental tests that provide more accurate results for a modified strength criterion LaRC04.

5. ACKNOWLEDGEMENTS

This paper is based upon work sponsored by project RTI - Regional Technological Institute reg. no. CZ.1.05/2.1.00/03.0093 and TAČR TA project TE01020075.

6. REFERENCES

[1] Krishan K. Chawla, *Composite materials; Science and Engineering*, 3rd Edition. Springer, New York, 2012.

[2] Laš, V.: *Mechanika kompozitních materiálů.* Západočeská univerzita v Plzni, Plzeň, 2. vydání, 2008.

[3] Kottner, R., Krystek, J., Zemčík, R., Lomberský, J., Hynek, R. *Strength Analysis of Carbon Fiber-reinforced Plastic Coupling for Tensile and Compressive Loading Transmission.* Collection of Technical Papers - Structures, Structural Dynamics and Materials Conference, 2011, roč. 2011-1982, č. April 2011, s. 1-12. ISSN: 0273-4508

[4] Kottner, R., Bek, L., Krystek, J., Kroupa, T., Lašová V. *Failure analysis of pin joint of carbon/epoxy composite plate.* Advanced Materials Research, 2012, roč. 634-638, č. 1, s. 2796-2799. ISSN: 1662-8985

[5] Krystek, J., Kottner, R., *Load capacity prediction of carbon or glass fibre reinforced plastic part of wrapped pin joint.* Materiali in Tehnologije, 2015, roč. 49, č. 6. ISSN: 1580-2949

[6] Kottner R., *Spojování kompozitních a kovových strojních částí z hlediska tuhosti a pevnosti.* Západočeská univerzita v Plzni, Disertační práce, Plzeň, 2007

[7] Vasiliev Valery V., Morozov Evgeny V.: *Advanced Mechanics of Composite Materials and Structural elements*, 3rd Edition. ISBN: 978-0-08-098231-1, Elsevier, Oxford, 2013.
[8] Pinho, S. T., Dávila, C. G., Camanho, P. P., Iannucci, L., Robinson, P., *Failure Models and Criteria for FRP Under In-Plane or Three-Dimensional Stress States Including Shear Non-Linearity*. Research report, NASA/TM-2005-213530, NASA Langley Research Center, 2005, 69 p.

7. ADDITIONAL DATA ABOUT AUTHORS

Ing. František Sedláček*;
Department of Machine Design, Faculty of Mechanical Engineering, RTI - Regional Technological Institute, University of West Bohemia; Univerzitní 8, 306 14, Czech Republic, fsedlace@kks.zcu.cz

doc. Ing. Václava Lašová, Ph.D.;
Department of Machine Design, Faculty of Mechanical Engineering, University of West Bohemia; Univerzitní 8, 306 14, Czech Republic, lasova@kks.zcu.cz

Ing. Radek Kottner, Ph.D.;
NTIS - New Technologies for the Information Society; University of West Bohemia, Univerzitní 8, 306 14, Czech Republic, kottner@kme.zcu.cz

Ing. Petr Bernardin;
Department of Machine Design, Faculty of Mechanical Engineering, University of West Bohemia; Univerzitní 8, 306 14, Czech Republic, berny@kks.zcu.cz

** Corresponding Author*

10th International DAAAM Baltic Conference
"INDUSTRIAL ENGINEERING -
12-13 May 2015, Tallinn, Estonia

LASER CLADDING TECHNOLOGY IMPLEMENTATION FOR IN-SITU REFURBISHMENT OF SHIP DIESEL ENGINE CRANKSHAFT JOURNALS

Stepans Sklariks and Toms Torims

Abstract: *This article provides an overview of state-of-the-art laser cladding technology which can be implemented in the in-situ refurbishment of ship diesel engine crankshaft journals. The envisioned laser cladding system arrangement consists of a carbon dioxide (CO_2) laser, a Neodymium-doped Yttrium Aluminum Garnet (Nd:YAG) laser or a high-power diode laser, a vertical-axis laser cladding powder feeder, a coaxial laser cladding nozzle as well as a process guidance and control system.*
Key words: Laser cladding; lasers; powder feeders; laser cladding nozzle.

1. INTRODUCTION

Laser cladding technology can be implemented for in-situ marine crankshaft renovation by fitting a laser cladding nozzle positioning and guidance device directly onto the crankshaft journal fillets. This novel in-situ concept of applying laser cladding for marine crankshaft repair provides clear economic benefits and many technological advantages.[1;2]
The purpose of this paper is to establish the state-of-the-art technological means of implementing the method to repair damage[1] caused by the gradual degradation of in-service performance of the diesel engine crankshaft journal. This article sets out the most relevant information on the laser cladding equipment and the actual laser cladding process, to be able to define the system specifications used in the in-situ laser cladding refurbishment process.

Laser cladding technology ensures that a thin layer of a desired metal is deposited on a moving substrate using a laser heat source. The material deposition on the substrate can be performed by several methods: powder injection, pre-placed powder on the substrate or wire feeding. Some of these methods and their variants are shown in Fig. 1. The most popular of the aforementioned methods is laser cladding by powder injection. During this process, the laser beam melts the powder particles and a thin layer of the moving substrate to deposit a layer of the desired material on the surface. Different kinds of materials can be deposited on a substrate with laser cladding by powder injection to generate a layer whose thickness can range from 0.05 to 6 mm[2] and width can be as narrow as 0.03 mm.[3]

2. OVERVIEW OF THE LASER CLADDING APPLICATION FOR COMPONENT REPAIRS

Conventional part repair technology relies on destructive, high temperature welding processes. Moreover, error-prone machining processes can have a negative impact on the timely execution of a repair. Laser cladding is a sufficiently safe and accurate technology to repair high-value parts, particularly for critical contact surfaces. The laser cladding repair not only offers complete restoration of the initial quality of a part, but also enhances it, resulting in a longer and more reliable service life. Thus it is possible to restore a high-value product which would otherwise have to be replaced. [4]

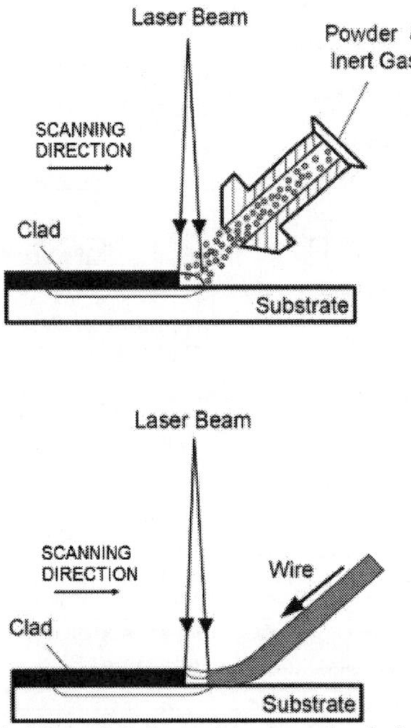

Fig. 1. Material deposition methods in laser cladding [5]

Nonetheless, laser cladding also has its drawbacks. The quality of the laser-cladded layer can vary in different areas due to the complex nature of disturbances in this process. Such occurrences can even be observed between processing cycles performed under the same operating conditions. Even if an optimal set of parameters is found experimentally and is subsequently used in an open-loop process, a surface layer formed may not be of good quality due to the very random or periodic nature of disturbances in the system.

The high investment cost, the lack of laser source efficiency as well as hard-to-control cladding processes are the main drawbacks that hinder a widespread use of this technology in industry.[6]

Laser cladding is used to repair and refurbish high-value components such as tools, turbine blades and military components. Using this process, a part that was over-machined due to an error in the design or manufacturing process can be saved and successfully employed for further operations.

It is possible to use a process such as welding to repair damaged parts, however given that these methods are usually of a destructive nature due to the highly distributed temperature over the area being repaired, the mechanical properties are impaired owing to cracks and pores, leading to a reduction in part life. In contrast to welding, laser cladding can provide a permanent structural repair and refurbishment on alloys that are generally considered unweldable due to the small heat-affected zone, rapid solidification, increased cleanness, lower dilution as well as increased control over the depth of the heat-affected zone.

One of the main areas where laser cladding has gained in popularity is turbine blade repair and refurbishment. Due to their high value and operational importance, these components have become a target for a repair technology application that can maintain their original mechanical and metallurgical features.[6]

3. LASER CLADDING PROCESS PARAMETERS

A large number of operational and physical parameters determine the quality of the obtained clad. A summary of these parameters is given in Fig. 2.

Laser cladding parameters may be defined as dimensionless variables, in terms of the energy balance between the absorbed power Aq and the power used to heat and melt the clad material, in order to successfully establish the operational process window as well as achieve a constant, normalized clad thickness:

Dimensionless beam power:

$$q^* = \frac{Aq}{r_B \lambda (\mathrm{T_m} - t_0)} \tag{1}$$

Dimensionless clad thickness:

$$l^* = \frac{l}{r_B} \tag{2}$$

Dimensionless traverse rate:

$$v^* = \frac{v r_B}{a} \tag{3}$$

Dimensionless volumetric latent heat of melting:

$$L_m^* = \frac{L_m}{\rho c(T_m - T_0)} \quad (4)$$

Dimensionless surface temperature:

$$T_p^* = \left(\frac{1}{\pi}\right)^{\frac{3}{2}} q^* \tan^{-1}\left(\frac{b}{v^*}\right)^{\frac{1}{2}} \quad (5)$$

Thermal diffusivity:

$$a = \frac{\lambda}{\rho c} \quad (6)$$

Where:

A – material absorptivity;
q – beam power $[Js^{-1}]$;
$2r_B$ – beam diameter (spot size) [mm];
l – clad thickness [mm];
v – beam traverse rate $[ms^{-1}]$;
ρ – density of the cladding material $[kg\ m^{-3}]$;
c – specific heat capacity $[J\ kg^{-1}K^{-1}]$;
T_m – melting temperature of the clad [K];
T_0 – initial temperature of the clad $[K]$;
L_m – volumetric latent heat of melting $[Jm^{-3}]$;
λ – thermal conductivity of the clad material $[Js^{-1}m^{-1}k^{-1}]$.[7]

Given the complexity of interactions between the laser beam, the alloy deposition mechanism and the molten region, as well as the fact that no previous knowledge is available for these specific application purposes, further research is needed in order to obtain a successful numerical model of coherent input, process and output parameters. Nevertheless empirical adjustments as well as the consideration of many inflectional factors must be taken into account for the obtained model such that the process result can be predicted, which will offer a direction for further research.[7]

4. MAIN COMPONENTS OF A LASER CLADDING SYSTEM

Of all the numerous laser systems available in the market, those most commonly used for laser cladding are CO_2 lasers, lamp-pumped and diode-pumped Nd:YAG lasers and high-power diode lasers (HPDL).[8]

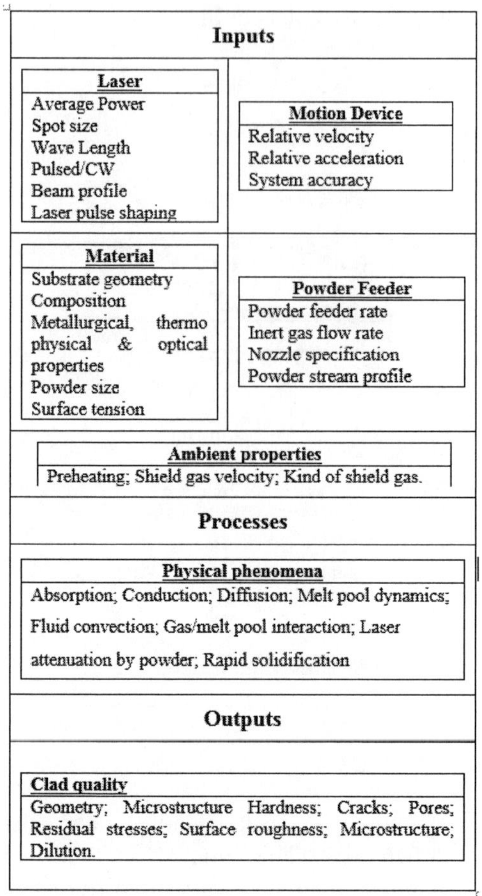

Fig. 2. Laser cladding parameters [7]

Metals are more reflective at 10 μm than at 1 μm, thus Nd:YAG and HPDL lasers which produce light at wavelengths of 1.024 μm and ~0.85 μm respectively are more efficient for metal processing compared with a CO_2 laser (10.6 μm). Aluminium is highly reflective when using a CO_2 beam, whereas the Nd:YAG or HPDL laser is perfectly absorbed. On the other hand, most carbon and stainless steels absorb CO_2 and Nd:YAG beams in a very similar manner. CO_2 laser beams are focused on smaller spots and are more symmetrical, which improves clad width. HPDL lasers provide a wide beam distribution and have a low beam quality, such that HPDL lasers in today's market cannot be used for materials with a high melting temperature.

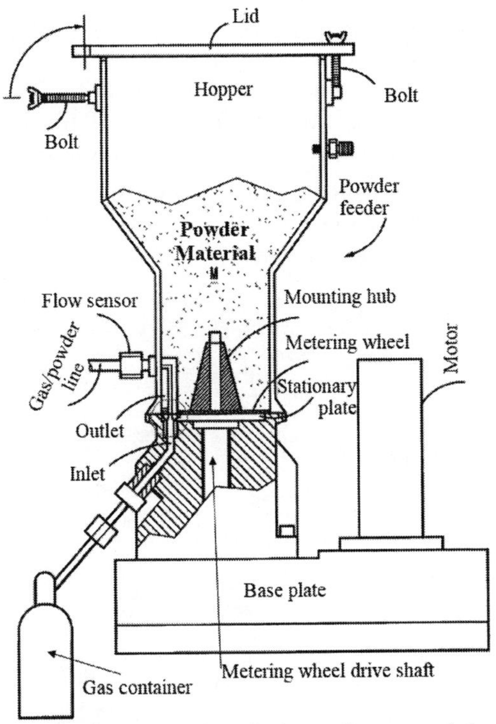

Fig. 3. Cross-sectional view of a potential powder feeder system [9]

Another important issue when selecting a laser for in-situ refurbishment in a confined space is the beam delivery and manoeuvrability of the laser nozzle, which in the case of a CO_2 laser would be very limited, due to the impossibility of transporting a CO_2 laser beam along optical fibre owing to its wavelength of 10.6 μm, which is inappropriate for this technology. Nd:YAG and HPDL lasers on the other hand can be passed along a fibre optic cable and as a result, can be connected to the end effector of a robot with any degree of freedom. However with HPDL, it is necessary to use a standard lens to achieve an appropriate working distance from the focus point. But there is also a very high risk that the protective glass and the lens will quickly become dirty or even damaged by the powder particles. Thus the processing of complex shaped surfaces (such as crankshafts) becomes complicated.[6]

The main components of any type of powder feeder are the powder material hopper with an orifice, as well as a load cell based electronic weighing mechanism, a powder carrier and a supplementary back-pressure gas system for the stability of the powder stream. In order to increase the controllability of gravity-based powder feeders, different devices such as a metering wheel can be integrated into the powder feeder.[6] Fig. 3 illustrates a possible design for a vertical-axis, rotating wheel powder feeder system.

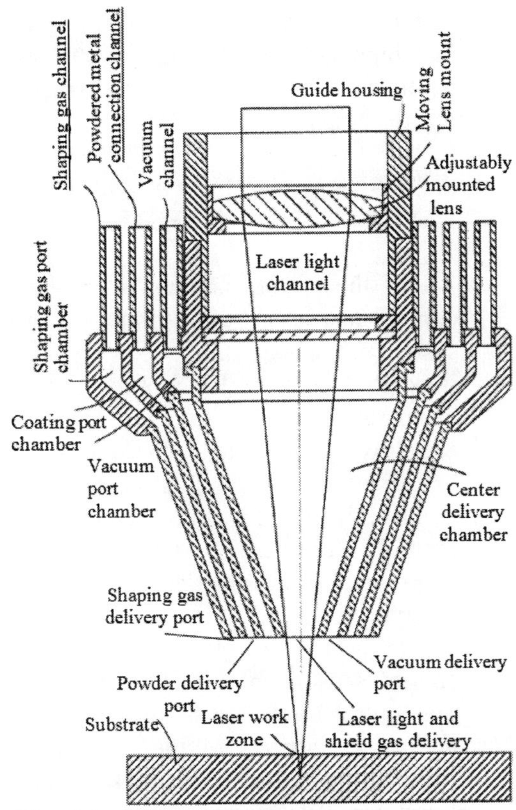

Fig. 4. Possible embodiment of a coaxial laser cladding nozzle [10]

The two basic nozzle configurations for the laser cladding application are coaxial and lateral nozzles.

The coaxial nozzle has gained the most widespread popularity when choosing a nozzle type for a specific laser cladding process, due to various advantages:

• Obstruction of the laser beam is not so critical compared with pre-placed or paste methods, resulting in good bonding formation without the need for binders;

• Deposition is not orientation-dependent, which makes the process omnidirectional and results in stable track formation;
• The combination of material in powder form is relatively easy, which makes it suitable for in-situ alloy deposition.[5]

A possible design for such a nozzle is given in Fig. 4.

In spite of the advantages listed, the integrability of the given component design must be investigated further, to successfully establish all of the limiting factors that pertain to the confined space of the shipboard engine compartment, for the application of in-situ marine crankshaft refurbishment.

5. GENERAL ARRANGEMENT OF A LASER CLADDING SYSTEM

All of the aforementioned laser cladding components can be integrated within a system, the schematic of which is illustrated in Fig. 5. The motion of the laser beam relative to the substrate is provided either by a CNC table, a CNC turning stage or a robotic arm. In such an embodiment the overall system is controlled by a master control computer, which provides coordination information to as well as receiving data from all of the components shown in Fig. 5. Many other secondary control sensors may be integrated within the control system to provide information on various aspects of the laser cladding system's operation to the master control computer.[10]

This system can be implemented for monitoring the marine crankshaft in-situ renovation device. Such a device would allow repairs of a damaged marine crankshaft whilst the ship is at sea far away from on-shore repair facilities, as well as eliminating the need for very complex, time-consuming and a very high-cost crankshaft dismantling from the ship.

The device comprises two guide-ways and two opposite guide-ways to position it on the crankshaft fillets and two frame parts, each of which are fixed to the respective guide-way.

The device also comprises two upper rods, positioned in the upper part of the frame part, and two lower rods in the lower part of the frame part, by means of which both frame parts are rigidly connected to each other (cf. Fig. 6).

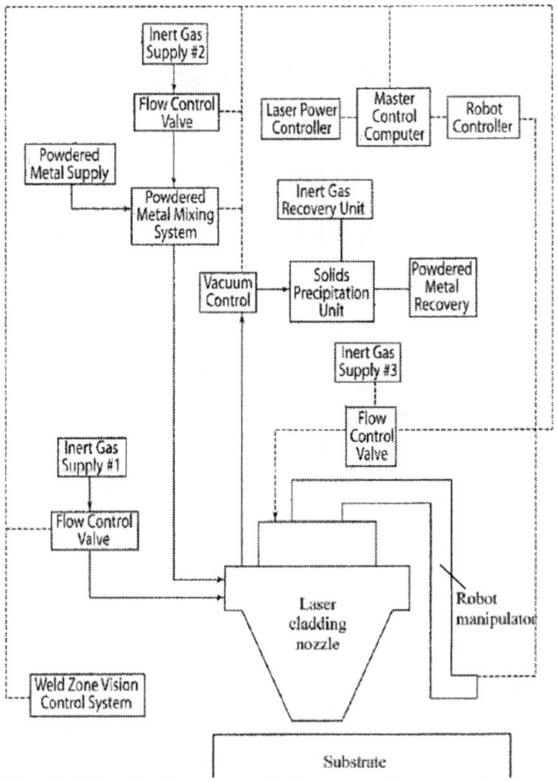

Fig. 5. Block diagram of the overall control system and related components [10]

The device further comprises two carriages which are installed on the upper rods and lower rods so that both carriages can slide along these rods. A laser nozzle is installed operatively between both carriages. The device includes two servo motors, the first of which is installed in the first carriage and is operatively connected to the laser nozzle to control its pivoting angle (see Fig. 6).

A second servo motor is installed on the second carriage and is operatively connected to one of the two lower rods by means of a gearing transmission, to control the laser nozzle's longitudinal position.

To ensure the device's positioning and controlled rotation around the crankshaft journal, it comprises the two aforementioned guide-ways and two opposite guide-ways. Two supporting plates are permanently fixed on the opposite guide-ways. Furthermore, these supporting plates are connected to each other by two opposite rods so that both perpendicular guide-ways are in fixed connection to each other. When installed on the crankshaft journal, the guide-ways and opposite guide-ways are connected and fixed to each other by means of four adjustable arms. The adjustable arms are connected to the guide-ways and opposite guide-ways by eight guidance-screws (cf. Fig. 6).

While the crankshaft is being rotated around its main axis, the laser head top-down position is maintained by eight pneumatic cylinders. These cylinders are connected to the guide-ways and opposite guide-ways by the aforementioned eight guidance- screws. The pneumatic cylinders can rotate freely around these eight guidance-screws.[7]

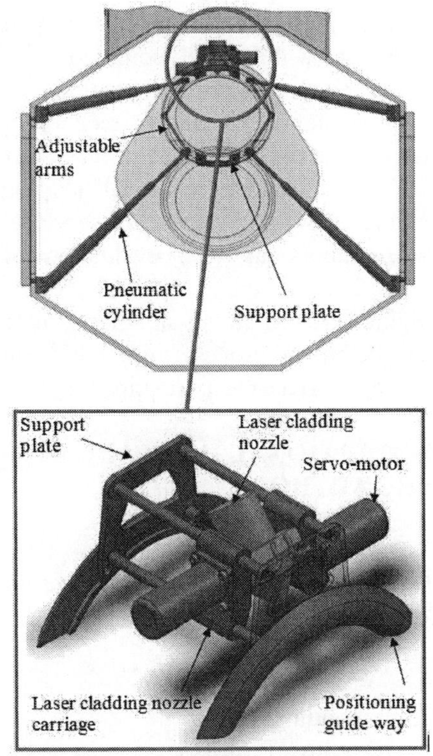

Fig. 6. 3D model of an in-situ marine crankshaft renovation laser cladding device [7]

6. CONCLUSIONS

Laser cladding offers numerous advantages as a repair process, providing complete restoration of a part's initial quality, as well as its overall enhancement, resulting in improved and longer service life. It thus demonstrates sufficient suitability for the refurbishment of a high-value component such as a marine crankshaft.

Further research must be undertaken in terms of establishing input, process and output parameter coherence as well as necessary empirical adjustments that will lead to development of a knowledge-based controller which is of crucial importance for the in-situ marine crankshaft refurbishment process, given its complexity and lack of available information on the subject.

CO2, Nd:YAG and HPDL are the lasers currently used for laser cladding, with Nd:YAG laser displaying the most advantages for in-situ crankshaft refurbishment processes, in terms of compact design and high manoeuvrability potential within the confined working space of a ship engine's housing.

Implementation of current technology for the in-situ refurbishment process requires further identification of all ship diesel engine crankshaft serviceability requirements, as well evaluation of the suitability of the proposed ship diesel engine crankshaft in-situ laser cladding refurbishment process with the identified primary serviceability requirements.

6. ADDITIONAL INFORMATION ON THE AUTHORS

Full title of the Manuscript: Laser Cladding Technology Implementation for the In-Situ Refurbishment of Ship Diesel Engine Crankshaft Journals.

Corresponding author:

1) Dr. sc.ing., Professor
2) Toms Torims
3) Kalku Str. 1, Riga, LV-1658, Latvia
4) E-Mail: Toms.Torims@rtu.lv
5) Tel.: +371 20200195

1) Master's degree candidate;
2) Stepans Sklariks
3) Full address: Atmodas Street 68-45, Jelgava, LV-3007, Latvia
4) E-Mail: stepsh@inbox.lv
5) Tel.: +371 27165198

8. REFERENCES

1. **Toms Torims, Guntis Pikurs, Andris Ratkus, Andris Logins, Janis Vilcans, Stepans Sklariks,** *Development of Technological Equipment to Laboratory Test In-Situ Laser Cladding for Marine Engine Crankshaft Renovation,* Vienna, Elsevier, 2014

2. **Toms Torims et al.,** *Development of the in-situ laser renovation technology for marine diesel engine crankshafts,* Vienna, DAAAM International, 2012, ISBN 978-3-901509-91-9

3. **Fraunhofer-Gesellschaft,** Powder nozzles for lateral feed, *Fraunhofer IWS Official website,* [Cited 04 19 2015] http://www.iws.fraunhofer.de/en/business_fields/surface_treatment/laser_cladding/system_technology/cyclone_powder_nozzle.html

4. **M.U. Islam, L. Xue, and G. McGregor,** *Process for manufacturing or repairing turbine engine or compressor components. Patent 6269540* USA, August 7, 2001

5. **Juansethi Ramsés Ibarra Medina,** *Development and Application of a CFD Model of Laser Metal deposition [Thesis],* Manchester School of Mechanical, Aerospace and Civil Engineering, 2012

6. **Ehsan Toyserkani, Amir Khajepour and Stephen Corbin,** *Laser Cladding,* Florida, CRC Press LLC, 2005, 0-8493-2172-7

7. **Toms Torims et al.,** *The Application of Laser Cladding to Marine Crankshaft Journal Repair and Renovation,* Copenhagen, ASME 12th Biennial Conference on Engineering Systems Design and Analysis, 2014

8. **F.-W. Bach, A. Laarmann and T. Wenz,** *Modern Surface Technology,* Weinheim, Wiley-VCH, 2006

9. **Francisco P. Jeantette et al.,** *Method and System for Producing Complex-Shape Objects, Patent 6046426* USA, April 4, 2000

10. **Ronald Peter Whitfield,** *Laser Cladding Device with an improved nozzle, Patent 8117985B2* USA, February 21, 2012

11. **Karl F. Renk,** *Basics of Laser Physics,* Heidelberg, Springer, 2012, ISBN 978-3-642-23564-1

12. **William M. Steen, and Jyotirmoy Mazumder,** *Laser Material Processing (Fourth Edition),* London, Springer, 2010

10th International DAAAM Baltic Conference
"INDUSTRIAL ENGINEERING"
12-13 May 2015, Tallinn, Estonia

BACKSTOP AND BULLET TRAP DESIGN

Tikal, F.; Špirk, S. & Šedina, J.

Abstract: *Taking into consideration the lack of standards for dynamic bullet and bullet fragments trap with difficult-defined energy, this issue is now solved by individual proposals of these devices. The priority requirement of the construction is certainly its security.*
The device commonly ignores the investment cost and also operating cost due to lack of information on sizes of dynamic operational load.
Optimised bullet trap reflects these criteria and also respects the issue of assembling of special difficult to weld used materials with respect to long-life joints.
Finally, the concept is designed for environmental friendly approach to waste management.
Key words: bullet trap, design, ballistic protection, explicit FEM analysis

1. INTRODUCTION

When analysing the current state of dynamic bullet trap constructions, an absence of standards and design recommendations for this issue was identified.
In practice, it is common to use only rough design knowledge without any verification by an analytical calculation, let alone numerical simulation and also the lack of knowledge of characteristics of the materials used. Safety and durability are solved only by oversizing constructions and thickness of materials is estimated. Furthermore, it does not take into consideration optimum joints. The material is often welded without regard to the lifetime of such a joint under dynamic impact stress.
They are also commonly applied in combination with non-metallic materials such as rubber, usually in the form of discarded tires, respectively conveyor belts etc. The above mentioned factors lead to high purchase price and expensive and environment-unfriendly bullet trap operation.
The motivation was therefore a systematic identification of a suitable construction along with both conventional and special materials and tests them for use in pre-selected defined load.
The procedure was as in the field of numerical simulations and practical testing. The concept of optimized dynamic bullet trap was verified by using an explicit finite elements solver. The explicit analysis allows simulating of extreme nonlinear behaviour in extreme situations, which lies on the limits or outside the area of solvable typical FEM programmes.
For testing of selected materials was at first performed a set of dynamic tests on specimens to determine material input data for numerical simulation. Subsequently was on the simulation model performed a set of analyzes of impacts and included the effects of possible penetration. Based on the simulation results the type of material was evaluated depending on the spatial orientation of the sample sheet metal and its damage the degree of protection. To validate the concept of bullet trap the set of experiments was conducted with using the test stand under real operating conditions.

2. CONDITIONS STATEMENT

When analyzing requirements of the bullet trap was at first created statistics according to the regional requirements. This statistics was the further basis. The most frequently used limit caliber bullets, resp. their energy for civil rifles, were identified.

Subsequently were chosen materials with regard to the future intention to use a sandwich design and was defined speed and energy of bullets for a numerical simulation of the falling distance, which according to the producer information complied with the conditions of the experimental testing.

Used boundary conditions

Cartriges:
SAKO .308 Win 141A Racehead with the bullet weighing 10.9 grams, bullet type Racehead HPBT. It is a full metal jacket bullet with the lead core. The bullet is excellent in very thin casing and in bottom shaped as "boat tail", which gives the bullet very good ballistic characteristics.

Tested materials:
Hardox 500 with a thickness of 8mm, S355 (CSN 11523)
a thickness of 10 mm and 12 mm

The dimensions of the samples of materials:
A4

Rake angles of samples on the stand:
90 ° and 45 ° to the horizontal orientation

Distance:
50 meters

3. NUMERICAL SIMULATION

3.1 Explicit FEM Theory

The beginning of the development of explicit solvers goes back to the sixties. At that time began the development particularly at universities. The HEMP programme, which had freely available code, was the basis of gradually developed these days used software. Explicit time integration is suitable for a simulation of processes with large deformations and reshaping. There is a better chance of capturing nonlinear behavior of material and fracture. The explicit solvers are generally better suited for tasks with complex contact situations. Thanks to these characteristics are the explicit solvers determined to solve conflict tasks, so called crashes, bullet holes etc.

Explicit code basically comes from Newton's second law of motion. This is the equation of motion incorporated in the matrix form (1). This equation is defined at the given time. In order to keep the balance of the dynamic forces, the relationships described below must be met [1].

$$\{a_t\} = [M]^{-1}\left(\{F_t^{ext}\} - \{F_t^{int}\}\right) \tag{1}$$

Here is $\{a_t\}$ the acceleration vector (at time t), $[M]$ is the weight matrix, F_t^{ext} a vector of external forces acting on the body and F_t^{int} is the vector of internal forces.

After definition of the internal forces and adding some basic elements, it is possible to create an equation for the numerical solution in the following format (2). The element $\{F^{houg}\}$ was added to prevent hourglassing and $\{F^{cnt}\}$ as vector of contact forces. Furthermore $\{\sigma_n\}$ is internal tensions matrix, $[B]$ is matrix of reshaping elements.

$$\{F_t^{int}\} = \sum\left(\int_\Omega [B]^T\{\sigma_n\}d\Omega + \{F^{houg}\}\right) + \{F^{cnt}\} \tag{2}$$

The solvers based on the explicit code are conditionally stable. It means that they are stable only under certain conditions. This is mainly about the time step size. It is related to propagation of voltage waves in the material (see following relationship (2)). Here is c the wave's propagation speed in the material, l is the characteristic size of the element, E is the module of the material elasticity and ρ the density of the material.

$$t \leq t_{krit} = \frac{l}{c} = l\sqrt{\frac{\rho}{E}} \quad (3)$$

The big advantage of the explicit method is the use of elements with a single integration point. The advantages of this method are, however, redeemed by reduced stability of the calculation. If the element is deformed symmetrically, there is no corresponding change in internal energy. In the calculation result are therefore the typical imbalance between kinetic and internal energy of the system. This numerical error is called hourglassing. It is obvious that during the dynamic calculations must be always controlled the total energy. As critical it is considered the increase in hourglassing energy over 5% of the total energy of the system. In extreme cases of hourglassing increase may even collapse calculation. To limit the hourglassing occurrence are used different methods.

3.2 Explicit FEM solver used

Pam-Crash is a FEM solver that is part of the software package VPS (Virtual Performance Solution) produced by ESI Group. [2] The software is used for crash simulation and safety assessment. This software is most commonly used in the automotive industry.

The software has been developed since 1978 and is connected with the early car crash simulations. The software is based on the finite element method (FEM) and allows modeling of complex geometry with a broad range of different types of finite elements. The programme offers a wide range of linear and nonlinear materials including visco-plastic, foam and multi-layered composites, including models of failure [3].

Thanks to the usage of the explicit formulation FEM, is the software suitable for simulation of nonlinear tasks with a large number of contacts (mainly based on the penalty algorithm).

This software was selected based on references from the field of defense, because it allows solving tasks such as the performance of munitions with respect to explosion, cratering and the simulation of kinetic energy penetrators.

3.3 Description of numerical simulations

Initially was completed a simulation model, subsequently were set boundary conditions and implemented material data from dynamic tests on specimens. The following description of the simulation model is valid for one situation of boundary conditions. Other models were created as variants. These variants were modifications of this debugged pilot case.

The bullet consisting of a lead core and a brass surface casing crashed into a metal sheet of material S355 with the thickness of 10 mm. Initial bullet speed was 779 m/s [4]. To create the model, we used linear quadratic elements with eight nodes. The average size of the element is about 0.5 mm. Material model number 19 "elastic-plastic-with-damage-failure" was chosen as a representative.

All three materials were defined based on this model. Flexible behavior was defined on the basis of stress-strain curves. Various stress-strain curves are defined for various values of strainrate. Thanks to that, the strengthening of the material during rapid deformation was taken into account.

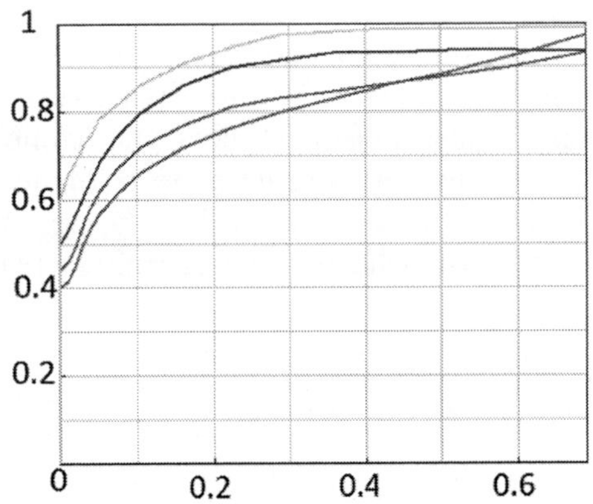

Fig. 1. Material S355 True stress [GPa] - True strain [1], while strainrate 0; 1; 140;

2100; s-1

As a criterion, it was considered maximum plastic strain element for elimination. Critical time step is defined by characteristic element size, young modulus and stiffness Δt=8,9e-6 ms. Time of simulated process duration is 0.14 ms.

3.4 Results of numerical simulation
The results of simulation give a good knowledge of the resistance of the plate of steel material. It is very good visible a hole forming in the entrance of a bullet. For a board with the thickness of 10 mm at a right angle to the bullet trajectory there is the complete penetration of the projectile. The bullet speed is due to the loss of energy during penetration reduced by 70%.

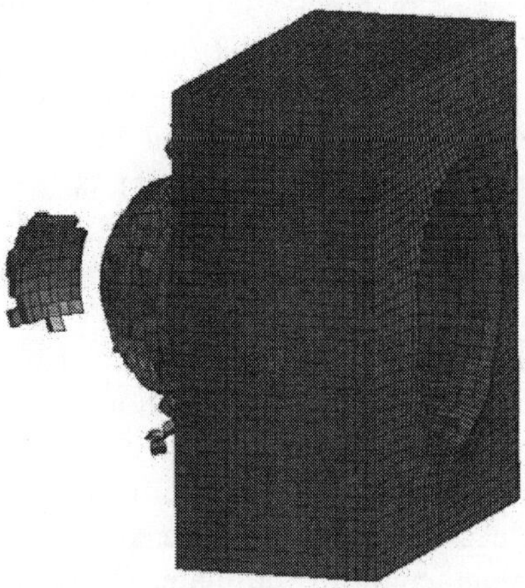

Fig. 2. Results of the simulation – the board inclined by 0°

When simulating bullet impact with a plate inclined by an angle of 45 °, with respect to the bullet trajectory, does not come to penetration. This result corresponds to the result of the experiment. It is apparent that the board inclination significantly increases its resistance.

Fig. 3. The results of simulation – board inclined by 45 °

4. EXPERIMENTS

For the validation of numerical simulations was performed set of experiments using the testing stand in real operating conditions. The shooting was conducted at a distance of 50 m with a sniper rifle of following parameters:
Remington 700 Police (short action) [5]
Caliber .308 Win., Barrel Length 26" Standard Twist 12".
For the shooting experiments were used specially made rectangular metal sheets of paper size A4 positioned immovably in the special stand, which allows to select the angle of inclination of the material samples.
On the plate were shot two times for single variant of combination of metal sheet material and the angle of inclination. Each score was documented. See Figure Nr. 4, 5 and 6
During the shooting experiment were achieved results that confirmed the assumptions ballistic-resistant of metal sheet calculated by using numerical simulations.

5. CONCLUSION

Numerical simulations provided the assumptions that were subsequently verified by the shooting experiments. See Table Nr. 1

These findings will be used for a selection of materials and design of optimized dynamic bullet trap.

Material	Thickness	Angle	Results
Hardox 500	8 mm	90°	2x micro deformation
S355	10 mm	90°	2x penetration
S355	12 mm	90°	2x cup deformation
Hardox 500	8 mm	45°	2x abrasion only
S355	10 mm	45°	2x cup deformation
S355	12 mm	45°	2x cup deformation

Table 1. Results of experiments

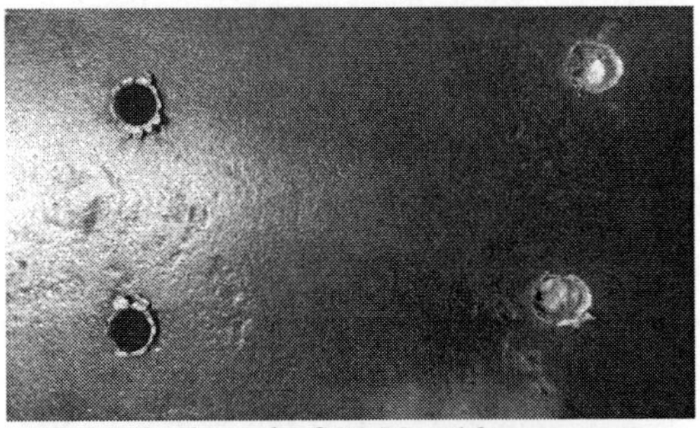

Fig. 4. The board of S355 – 10 mm

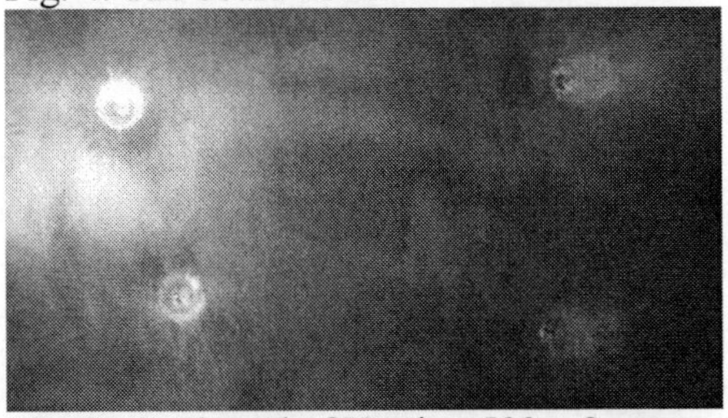

Fig. 5. The board of Hardox 500 – 8 mm

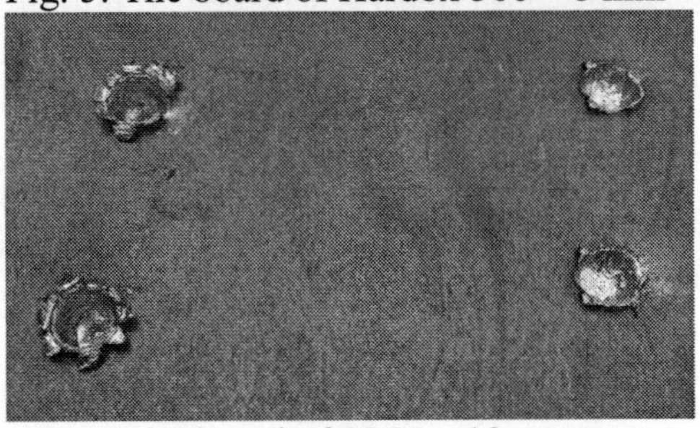

Fig. 6. The board of S355 – 12 mm

For future research is planned to extend the analysis to other caliber and inclination angles of material samples.
The tests of other materials will be included. The materials will be selected with regard to price. It also assumes the development and use of special rolled sandwich materials.
Further to the shooting conduct will be necessary to include an electronic gate to measure the speed of bullets (chronograph) and install to the testing stand a speed thermal imaging camera for more detailed evaluation of impacts dynamic process with simultaneous temperature monitoring.
The results achieved will be prepared for the evaluation software, which enables effective proposals of bullet trap mainly for civilian and military indoor shooting ranges and ballistic departments.

6. ACKNOWLEDGEMENT

This paper is based upon work sponsored by project "Regional Technological Institute" reg. no. CZ.1.05/2.1.00/03.0093.

7. REFERENCES

[1]Belytschko T, Lin JL, Tsay CS. *Explicit algorithms for the nonlinear dynamics of shells. Computer Methods in Applied Mechanics and Engineering* 1984; 42:225–251.

[2] Impact and high velocity impact [WWW] http://virtualperformance.esi-group.com/applications-impact-and-high-velocity-impact-analysis (29.04.2015)

[3] E. Mestreau, R. Lohner: *Airbag Simulation Using Fluid/Structure Coupling.* 34th Aerospace Sciences Meeting & Exhibit, Reno, NV, January 1996.

[4] Sako Ballistics [WWW] http://luoti.sako.fi/Ballistics/index.jsp (29.04.2015)

[5] Remington 700 Police [WWW]

10th International DAAAM Baltic Conference
"INDUSTRIAL ENGINEERING"
12-13 May 2015, Tallinn, Estonia

DESIGN OPTIMIZATION OF GLASS CANOPY PANEL SUBJECTED TO SNOW LOAD

Õunapuu, E.; Velsker, T.; Eerme, M.

Abstract:

The scope of this study is design optimization of glass canopy panel with prespecified strength properties and geometrical dimensions. The panel is subjected to snow load. Maximum Von Mises stresses and maximum deflection of the panel are considered as objective functions.

The stress-strain relationship of the glass canopy panel with point fixings is determined by use of FE analysis. A mathematical model for evaluation of objective functions is developed. Optimal set of design variables is determined based on preliminary analysis.

1. INTRODUCTION

Fast development in architecture has transformed glass from window material into load carrying element material. Glass canopy panels, floors and full glass facades are common in modern city buildings. Application of the point supported glass and FEM (finite element method) analysis have been the main reasons of expeditious development in the field.

The considered glass canopy panel represents thin glass layers with relatively large surface area. The glass layers have a certain amount of holes for point fixings. The main problems for designing such panels are large deflections of the panel and high stresses around fixing holes. The solution procedure proposed in the current study is based on combining the hybrid genetic algorithms (HGA), meta-modelling techniques and FE analysis.

Among a large number of meta-modelling techniques available like regression methods [1, 2], kriging models [3, 4], splines & radial based functions [5, 6], etc., the feedforward artificial neural network techniques [7, 8] are selected due to their high accuracy and simplicity. The optimization procedure introduced can be regarded as extension of the optimization algorithms developed by workgroup for design of composite and sheet metal structures [9-12]. These multicriteria optimization algorithms are based on pioneering study in this area by Deb [13], Coello-Coello [14]. Deb et al. introduced nondominated sorting based NSGA2 algorithm with reduced computational complexity in [15]. Latter algorithm is widely used approach in design of optimization methods up to now. One new trend in design optimization of advanced engineering materials and structures is design of stiffness variables [16-20] (caused by advances in manufacturing technology).

The aim of the current study is to determine optimal configuration of glass canopy panel subjected to snow loading. Current study can be considered as extension of paper [21] containing preliminary design based on Taguchi design of experiment.

2. PROBLEM FORMULATION

The current study is focused on design optimization of glass canopy panel subjected to snow loading (Figure 1).

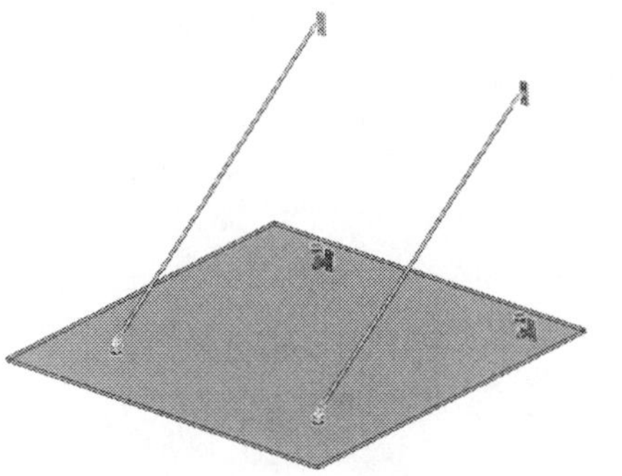

Fig. 1. Glass canopy with point supports

Design of such a structure is a complex task problem due to presence of mixed integer variables, rather specific stress-strain behaviour of glass and multiple constraints. The main characteristics considered in design of glass canopy panel are maximum stresses, maximum deflection and maximum strain energy density. These criteria are conditioned by the thickness of glass panel and the location and diameter of fixing holes.
The dimensions of the panel are given by the manufacturer of glass canopies which are 1700 mm in width and 2000 mm in length. The objective is search for an optimal set of design variables X1, X2, X4 and X5.

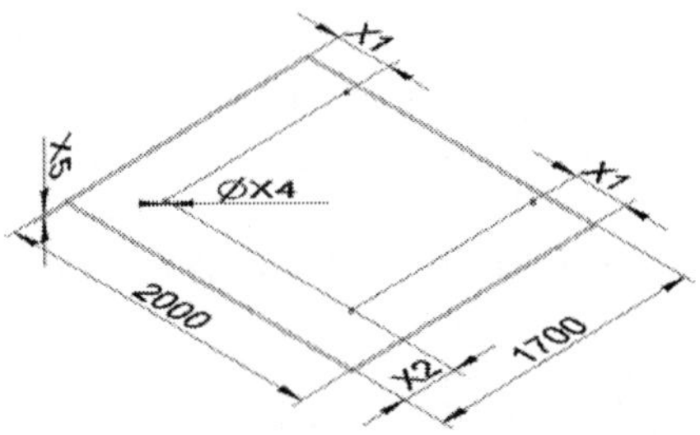

The design variable X3 is fixed in the current study based on results of preliminary design [21] and technological constraints. Panel is made of tempered soda lime silicate glass. It is assumed that the panel is made from monolithic solid glass, although common practice is to use two thinner glass panels laminated together. The variables X1 and X2 stand for the coordinates of the fixing holes, X4 is the diameter of the hole and X5 is the thickness of the panel. Panel is subjected to gravity and pressure caused by snow load, no wind load is considered (Figures 2-3).

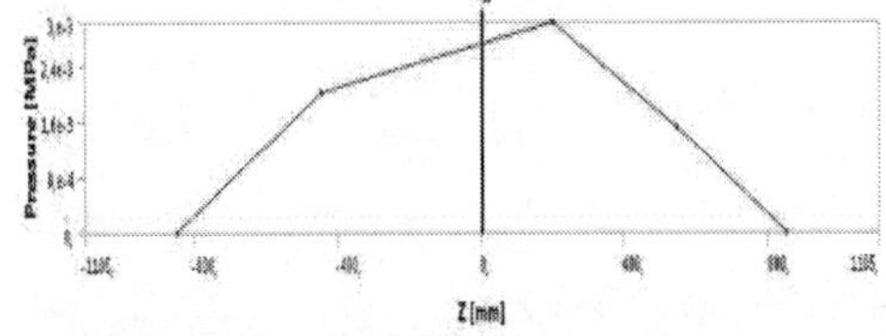

Fig. 2. Distribution of snow load

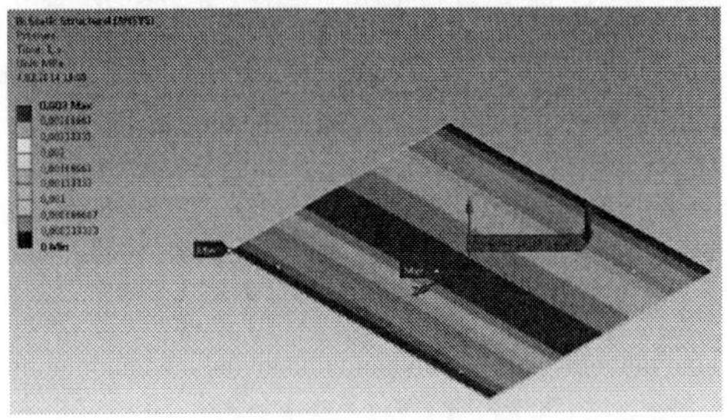

Fig.3. ANSYS Variable load pressure model

3. FINITE ELEMENT ANALYSIS

To analyse glass panel with point fixings, a three dimensional FEM model is applied. The deformation of the glass canopy panel subjected to snow load may exceed the thickness of the panel. Thereby the behaviour of the panel cannot be described accurately by linear theory. A nonlinear plate theory is applied. The stress-strain state of the glass panel is analysed with FEA software (ANSYS) where simulation model with solid elements has been developed. For avoiding long calculation times, the general mesh element size of the model is delimited with 20 mm. Due to fact the maximum stresses occur around the fixing holes, the size of mesh elements around the holes is decreased to 3 mm. Such mesh is employed around holes in 40 mm diameter sphere.
The values of the design variables are chosen based on manufacturing and structural limits. Four levels have been considered for the design variables X1, X2, X4 and three levels for X5, respectively.

The value of the variable X3 is fixed as mentioned above. Parametrical model according to the values of design variables was created in ANSYS Workbench.

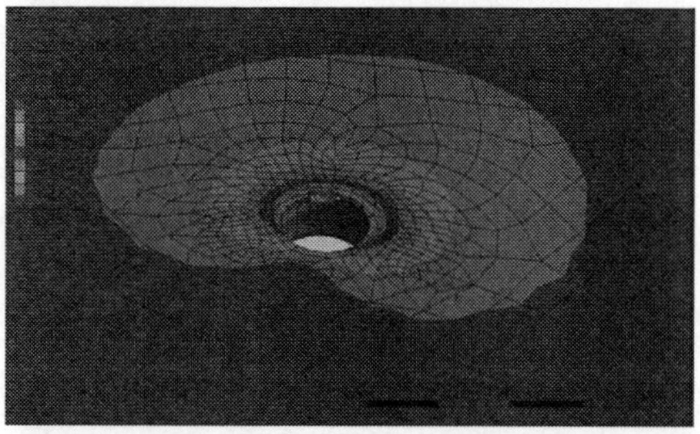

Fig. 2 Stresses around fixing holes

The maximum stresses occur on the edges of fixing holes.

4. DESIGN OPTIMIZATION

In the following, the output data obtained from FE analysis are treated as response values. However, the evaluation of the objective functions by means FE simulations is time consuming. A common technique for reducing computational cost in optimization algorithms is using surrogate models for approximation of the objective functions.

In the current study the artificial neural networks (ANN) based surrogate models are employed to guide the search towards a global optimum. The Marquardt-Levenberg algorithm has been applied for training of a two-layer backpropagation artificial neural network. This algorithm combines the advantages of the steepest descent (wider convergence domain) and Newton (higher convergence speed) methods. The configuration of the ANN has been tuned for particular problem. The radbas neurons and purelin neurons are used in first and second network layers, respectively.

The considered optimization problem is a mixed integer problem. The design variables X1 and X2, describing positions of the holes are real valued, but the radius of the hole and thickness have integer values. An analysis of the optimality criteria has been performed and different formulations of the optimization problem are examined.

The minimization of the maximum value of the Von Mises stress and maximum deflection are considered as two criteria

$$f_1(\bar{x}) = \min(\sigma_{\max}^{VonMises}(\bar{x})), \quad (1)$$

subjected to

$$x_i \le x_i^*, \qquad -x_i \le x_{i*}, , \qquad i = 1,...,n, \quad (2)$$

$$\varepsilon_{\max}(\bar{x}) \le \varepsilon_{\max}^*(\bar{x}), \quad (3)$$

$$C(\bar{x}) \le C^*(\bar{x}) \quad (4)$$

and

$$f_2(\bar{x}) = \min(\varepsilon_{\max}(\bar{x})) \quad (5)$$

subjected to (2), (4) and

$$\sigma_{\max}^{VonMises}(\bar{x}) \le \sigma_{\max}^{VonMises*}(\bar{x}). \quad (6)$$

In (1)-(6) x_i stand for normalized design variables, $\bar{x}$ is a vector of design variables, $\sigma_{\max}^{VonMises}(\bar{x})$ and $\varepsilon_{\max}(\bar{x})$ stand for normalized maximum values of Von Mises stress and deflection, respectively. The lower and upper limit values of the variables/functions are denoted by asterisk. These two objectives are nonconflicting and their application results a quite similar solutions shown in table (rows 2-3).

Table 2. Objective functions and optimal solutions

Objective	X1	X2	X4	X5
f_1	394.56	450.00	35	20
f_2	399.51	450.00	34	20
f_3	396.52	450.00	35	20

The objective $f_3(\bar{x})$ shown in last row of the Table 2 is defined as weighted sum of the objectives $f_1(\bar{x})$ and $f_2(\bar{x})$ as

$$f_3(\bar{x}) = w_1 f_1(\bar{x}) + w_2 f_2(\bar{x}). \quad (7)$$

Here the weights w_1 and w_2 determine the significance of the corresponding criteria

$$w_1 + w_2 = 1. \quad (8)$$

In Table 2 the values of the design variables are given in original dimensions in order to simplify understanding their real meaning.

The hybrid genetic algorithm (HGA) developed by workgroup for analysis of composite structures [7] is adapted for solving above considered mixed integer optimization problem. The genetic algorithm is applied for global search. Local search is performed by combining steepest descent and Branch and Bound methods.

5. DISCUSSION

The FEA results as well as optimization results are given in section 2 and 3, respectively. Three optimality criteria are examined. As it can be expected the first two criteria minimization of the maximum Von Mises stress and maximum deformations are not conflicting. Thus, the use of physical programming techniques for combining these criteria is justified (for conflicting criteria can be applied Pareto concept). Employing the weighted summation criterion (7) allows to consider two objectives but perform optimization with one combined objective. In comparison with objectives (1) and (5) here is less constraints necessary (on design variables only). The weights w_1 and w_2 can be assigned according to significance of the objectives (higher weight to more critical objective).

It can be seen from Table 2 that the values of the design variables corresponding to combined objective remains between the values of the single objectives (X1) or coincide with these values. Also, the optimal solutions corresponding to different criteria are close.

The values of the thickness (X5) are equal to upper limit since the cost is not considered as objective function and constraints on value of the cost (4) applied appears satisfied in the case of all optimal designs considered (in general increasing thickness improve mechanical performance but increase cost). The cost can be considered as objective in future study. The position parameter X5 also achieved upper limit value in the case of optimal designs determined. Here can be considered reconfiguring of the design domain for this variable.

6. CONCLUSION

A nonlinear FE analysis of the glass canopy panel subjected to snow load is performed. Based on FEA results the ANN based mathematical model is composed. The full factorial design of experiment technique has been applied for selection of input dataset. The HGA based multilevel optimization procedure is developed/ adapted for particular problem. Optimal configuration of the glass canopy panel subjected to given load case is determined. The results obtained by applying three different optimality criteria are close (see table 2). Introduction of additional optimality criterion – cost of the structure seems reasonable. The simulation models and optimization procedures developed can be applied for solving wider class of design problems.

7. REFERENCES

1. A. C. Rencher, W. F. Christensen. Methods of Multivariate Analysis. Third Edition. Wiley Series in Probability and Statistics. John Wiley & Sons, Inc. 2012.
2. T. P. Ryan. Modern Regression Methods. Wiley Interscience. 2011.
3. I. Pan, M. Babaei, A. Korre, S. Durucan, A multi-period injection strategy based optimisation approach using kriging meta-

models for CO2 storage technologies. Energy Procedia. 2014, 63 3492 – 3499.
4. K. Elsayed, C. Lacor, Robust parameter design optimization using Kriging, RBF and RBFNN with gradient-based and evolutionary optimization techniques. Applied Mathematics and Computation 2014, 236, 325–344.
5. H.Cohen.Numerical Approximation Methods. Springer Science + Business Media LLC. 2011.
6. J. Kers, J. Majak. Modelling a new composite from a recycled GFRP. In: Mechanics of Composite Materials, 2008, 44(6), p 623-632.
7. H.Herranen, O.Pabut, M.Eerme, J.Majak, M.Pohlak, J.Kers, M.Saarna, G.Allikas, A.Aruniit. Design and Testing of Sandwich Structures with Different Core Materials. Materials Science-Medziagotyra, 2012, 18 , 1, 45-50.
8. J.Majak, M.Pohlak, M.Eerme, T.Velsker. Design of car frontal protection system using neural networks and genetic algorithm. 2012, Mechanika, 4, 453-460.
9. J.Majak, M. Pohlak. Optimal material orientation of linear and non-linear elastic 3D anisotropic materials. 2010. Meccanica, 45, 5, 671-680.
10. J.Majak, M.Pohlak. Decomposition method for solving optimal material orientation problems. Composite structures. 2010, 92, 8, 1839-1845.
11. J.Majak, S.Hannus, Orientational design of anisotropic materials using the Hill and Tsai-Wu strength criteria. Mechanics of Composite Materials, 2003, 39, 6, 509-520.
12. J.Lellep, J. Majak, Nonlinear constitutive behaviour of orthotropic materials. Mechanics of Composite Materials. 2000. 36, 4, 261-266.
13. K.Deb. Multi-Objective Optimization Using Evolutionary Algorithms. Wiley, M. 2009.
14. CA. Coello Coello, GT.Pulido, A Micro-Genetic Algorithm for Multiobjective Optimization. Evolutionary Multi-Criterion Optimization. Lecture Notes in Computer Science, 2001, 1993, 126-140.
15. K. Deb, A.Pratap , S.Agarwal, T. Meyarivan, A Fast Elitist Multi-Objective Genetic Algorithm: NSGA-II. IEEE Transactions on Evolutionary Computation. 2002, 6(2), 182-197.
16. Y.Guo, M.Ruess, Z.Gürdal, A contact extended isogeometric layerwise approach for the buckling analysis of delaminated composites. Composite Structures. 2014, 16, 55-66.
17. J.Sliseris, K.Rocens, Optimal design of composite plates with discrete variable stiffness. Composite Structures, 2013, 98, 15-23.
18. J.Sliseris, G.Frolovs, K. Rocens, V.Goremikins, Optimal Design of GFRP-Plywood Variable Stiffness Plate. Procedia Engineering, 2013, 57, 1060-1069.
19. J.Sliseris, H.Andrä, M.Kabel, B.Dix, B.Plinke, O.Wirjadi, G. Frolovs, Numerical prediction of the stiffness and strength of medium density fiberboards. Mechanics of Materials, 2014, 79, 73-84.
20. A. Khani, M.M. Abdalla, Z. Gürdal, Optimum tailoring of fibre-steered longitudinally stiffened cylinders, Composite Structures, 2015, 122, 343-351.
21. T.Velsker, H.Lend, M.Kirs. Design of glass canopy panel. Proceedings of the 8th International Conference DAAAM Baltic Industrial engineering,2012, 759 – 764.

II MECHATRONICS & SYSTEM ENGINEERING

10th International DAAAM Baltic Conference
"INDUSTRIAL ENGINEERING"
12-13 May 2015, Tallinn, Estonia

INTEGRATION OF HOUSEHOLD APPLIANCES TO THE EXISTING INTERNET INFRASTRUCTURE

Airola, A.; Jousimaa, O.; Niemi, K.; Vuokko, S.; Kiviluoma, P. & Kuosmanen, P.

Abstract: *The Internet of Things (IoT) is an interconnection between uniquely identifiable computer systems and existing Internet infrastructure. The field is quickly growing while large companies are starting to take interest in it. To study possibilities and challenges of Internet connected devices, an ordinary coffee maker was modified to receive commands and send status over the Internet as a first experiment. A smartphone application was built as a controlling user interface. Adding Internet connectivity was forthright to accomplish, but delay of message delivery did real time control unfeasible. This study approached the problem by considering how to add value to basic devices with relatively low-priced components. The ultimate outcome would be a future household where appliances can communicate with each other and the user.*

Key words: Internet of Things, WiFi, Home Automation

1 INTRODUCTION

The Internet connected computer systems can currently be small embedded microcontrollers, which can be installed virtually to any possible object. These devices can vary in size from a large industrial paper machine to tiny heart monitoring implant or biochip. Their function can be either passive or active depending on whether they have only sensors to measure current function of the machine or also actuators so that the device can be remotely controlled. [1] It is common between many of these applications to utilize an IP address (IPv4 or IPv6) as a handle [2]. Thus each machine can be monitored or controlled individually.

In recent years, IoT has been making a significant breakthrough in every industrial field. In addition, home automation utilizing IoT is increasingly adapting to our household appliances. It is a concept of utilizing cloud services to improve user experience of everyday devices. Because the field is quite novel, there are a lot of elementary research to be done. [3]

Smart household devices have existed for many years [4] but still none of them has become commercially successful. Many of them are too futuristic or highly priced to fit every household. Rapid development of technology and lack of multi-vendor-platform have limited lifecycle of intelligent household devices [3] [5] [6].

In this research paper, new ways to incorporate Internet to existing, everyday household appliances will be investigated. As a part of the project, an expandable platform and a few demonstration applications were built. The main ideology was integration: same user interface can be used to control multiple different household devices [7]. Second key-concept was to utilize existing infrastructure such as WiFi connection and smartphones.

The first goal was to build a coffee maker, which is remotely controllable via smartphone application. The main idea was that the coffee maker is able to dose the correct amount of water and coffee ground based on the user input. The coffee maker is controlled by a WiFi connected microcontroller, which communicates with a cloud server. The server is a proxy between the coffee maker and the smartphone.

The secondary goal was to add more appliances to be controlled with the same smartphone application. LED lighting was decided to be incorporated to the platform. This way a smart wake-up feature incorporating wake-up alarm, coffee maker and lighting could be implemented: When the alarm shuts off, lights and the coffee maker turn on. Data security of IoT is excluded from this research.

The basic working principle of the whole system is shown in Figure 1.

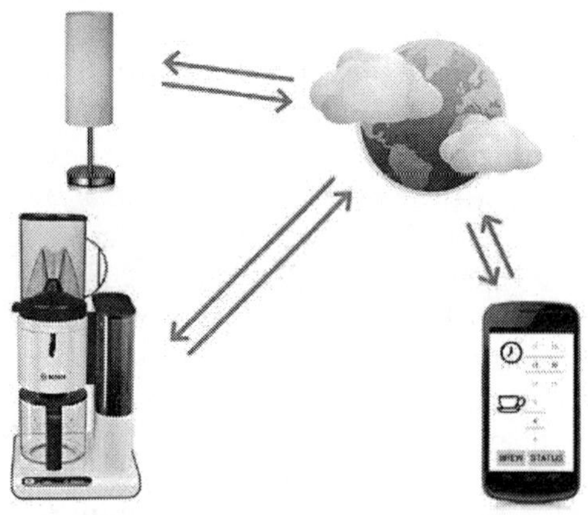

Fig. 1. Basic working principle of the system.

2 METHODS

2.1 Coffee maker

Safety is one of the highest priorities. To ensure safety, the supply voltage circuitry was decided to be kept as simple as possible. This led to use on-off relays as controlling the AC power to the coffee maker. Water supply is strictly limited to separate water tank to avoid water damages.

Arduino Mega microcontroller platform was chosen for its easy connectivity and prototyping features. It also offers great expandability to integrate all desired features but lacks performance to have modern stand-alone user interface. If the device would be mass produced, a custom solution would be justified.

ESP8266 WiFi module [8] was used to enable internet connectivity. The third key-component was relay board which was used for on-off switching. Devices connected to the relay board were a dosing pump and a coffee maker. Coffee-grind-dispenser brought a significant factor of uncertainty. That was caused by its mechanical properties. Required operating torque was discovered to be variable and unknown which led us to select a stepper motor with high torque. Coffee-dispenser was driven with stepper motor attached directly to the rotor, instead of its original crank mechanism. To control the stepper motor, a stepper shield was used.

The inlet water hose from the external water tank was attached to the same water line which moves water from coffee maker's own water tank to the coffee filter during brewing.

The target was that all the components would be hidden into the base to achieve more finished appearance. The coffee-grind-dispenser was the only visible component due to its location on top of the coffee maker. To fit the dispenser and the stepper motor sturdy to the coffee maker's frame a separate support was manufactured by 3D printing.

The basic schematic of the coffee maker unit is shown in Figure 2.

2.2 LED lighting

In addition to coffee maker, a separate control box was developed for LED lighting. Similarly, Arduino Nano and ESP8266 were chosen for control and communication.

The control box was designed to support PWM output. Thus the brightness of lights could be adjusted in the user interface. In theory, any other low-voltage and low-power device could also be connected to this expansion box.

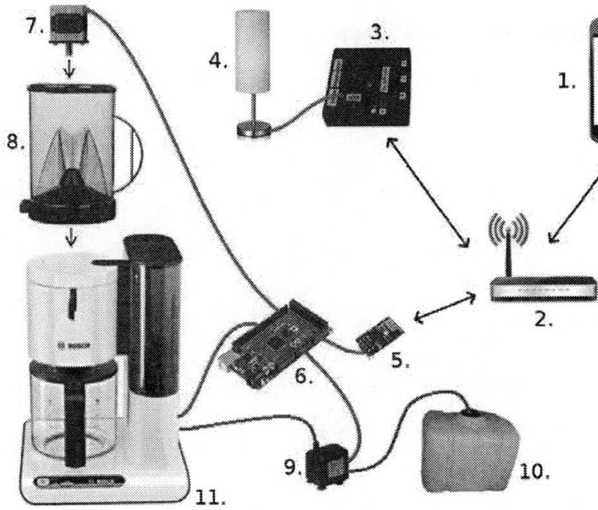

Nr.	Component
1	Smart phone
2	Cloud
3	Control box for LEDs (inc. Arduino Nano and WiFi module)
4	LED lights
5	WiFi module
6	Arduino Mega
7	Stepper motor
8	Coffee grounds dispenser
9	Water pump
10	Water tank
11	Coffee maker

Fig. 2. Basic schematic of the system.

2.3 Programming microcontrollers

The Arduinos were programmed using C++ based Arduino programming language [9]. For each subsystem, a separate function was created to ease the task of coding the control sequences in the main loop.

First the codes were implemented to control coffee maker and LED lights through the local serial monitor via PC. Later, the WiFi module was incorporated to the code to read the commands directly from the cloud-based server.

2.4 Android application

Android application was designed with Android Studio [10] to give an easy and intuitive way for the user to control connected devices. It allowed separate control of the coffee machine and the lamp or simultaneous control of both of the devices.

The separate control allowed the user to choose the amount of coffee cups and the time when it starts brewing. Lamp control enabled the user to choose brightness of the lamp on scale from zero to hundred, zero meaning it was switched off. The application also had a wake-up mode which allowed the user to choose a time to wake up. The mode set a wake-up alarm, brewed selected amount of coffee and turned the lights on at that time.

2.5 Cloud server

A public Internet facing server acted as a mailbox between the devices. This solved the problem on how different devices could find each other regardless of possible firewalls, NATs or other issues. The server itself didn't have any control logic, it only took data in and presented it in suitable format to user.

Having a publicly available connection allowed to have devices controlled from anywhere, it led to a unpredictable lag between sending and receiving commands. Since security is outside the scope of this paper, even basic authentication or encryption in commands wasn’t implemented.

3 RESULTS

3.1 Wireless communication

The wireless communication between the microcontrollers and the server worked well most of the time. The microcontrollers were pre-programmed to connect to the server every minute and check whether there are any new commands. However, in crowded networks the WiFi modules had some difficulties in connectivity and it took

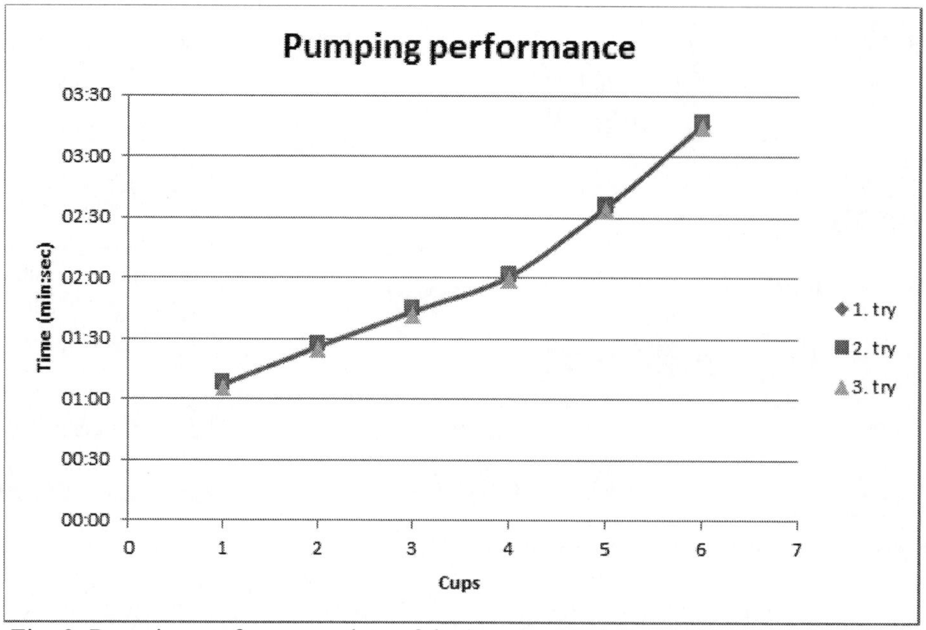

Fig. 3. Pumping performance data of dosing pump.

Wireless connectivity between server and device had also significant latency and jitter in response times. In practice this means that any real-time control over TCP/IP is not feasible, and local microcontroller had to control every time-sensitive task such as pumping and dosing.

3.2 Coffee dispensing

The automated water dispensing system proved to be reasonable accurate. The pumping data can be seen from Figure 3. As it is presented in the graph, the dispersion of the flow rate is quite low so the closed-loop control wasn't required in the system.

On the other hand, the coffee ground dispenser had some issues with moisture condensing inside the dispenser and contaminating the unused coffee grounds. Also the spinning rotor in the dispenser got occasionally jammed because of small coffee ground particles getting wedged between surfaces. However, by grinding wider clearance to the rotor, the problem was solved.

The software of the coffee maker was rather simple, linear program. The program polled for new instructions from server roughly at one minute interval, and acted on any commands received. Figure 4 shows the program flowchart of the coffee maker.

3.3 Light dimming

The light dimming using PWM turned out to be straight forward to implement. The device controller had three distinct modes: WiFi controlled, Manual controlled and off. The modes selected the source of data: on WiFi mode, data was polled from the server. On Manual mode, user could press the buttons on controller to adjust the duty cycle of PWM outputs. Off mode was for the convenience, it set all the outputs to 0 %.

Device also had a watchdog timer to guard against software lockups. In each cycle of the program, a watchdog timer was reset. If the watchdog timer was not reset in eight seconds, the software would be reset and started from initializing the hardware peripherals.

Figure 5 shows the program flowchart of LED lights.

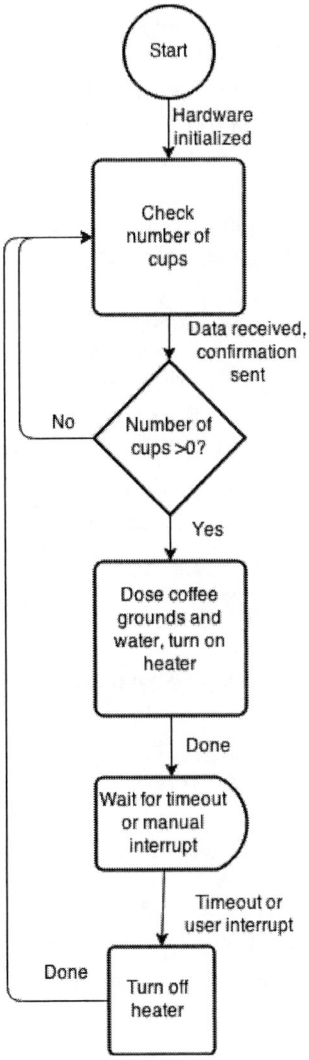

Fig. 4. Software structure of coffee maker.

While using the coffee maker and the LED lights in the same network, there seemed to be some interference between the devices and the other device occasionally refused to work. By further software tuning, this issue could be achievable to overcome.

3.4 User interface

The Android application (Figure 6) turned out to be simple and easy to use. The application had an individual button for each function (coffee, lights, wake-up) and easy sliders to set the right time and amount of cups.

With the status page, the user could also easily check the current command send to the coffee maker and LED control box.

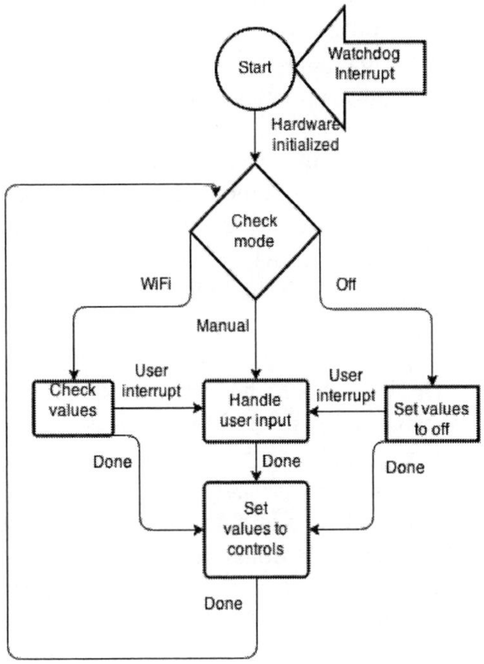

Fig. 5. Software structure of LED lights.

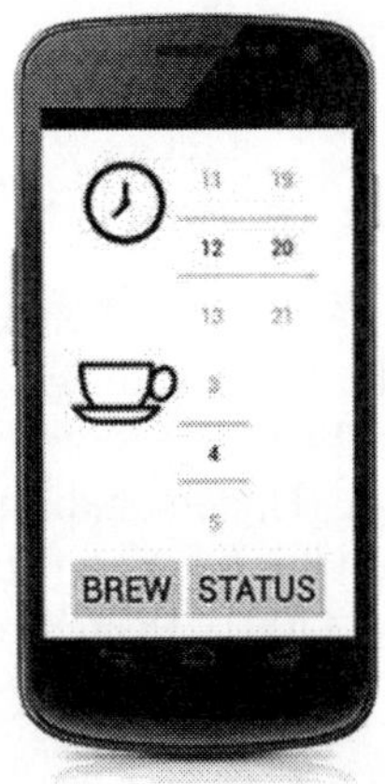

Fig. 6. User interface.

4 CONCLUSIONS

Internet connection could be used to relay high-level commands, such as "Brew 6 cups of coffee at 14.30". However, unpredictable lag in communication prevented real time control of the system, so local device had to control time-critical tasks such as dosing water.

Intelligence in device is important to verify the safety limits of commands, for example to avoid dispensing water amount exceeding water tank capacity and causing water damage to floor structures.

If remote operator does not have a line of sight to device being controlled, fault conditions can lead to catastrophic failures. Therefore Internet connected devices should have even more robust safety electronics, such as sensors, than most current household devices.

User interface design is extremely important to maintain easy user experience Devices should have a way of presenting status for example to remind user for filter changing.

Overall, the project showed that IoT has potential to adapt to everyday household devices. Additional component cost is not the greatest challenge decelerating generalization of IoT but lack of device-unifying software and hardware platform. By further developing and standardisation, safe and useful concepts could be created.

5 REFERENCES

[1] Katasonov, A. & Kaykova, O. & Khriyenko, O. & Nikitin, S & Terziyan, V. Smart Semantic Middleware for the Internet of Things, 2008. [Referenced 29.4.2015]

[2] 3Com Corporation. Understanding IP Addressing: Everything You Ever Wanted To Know, 2001. [Referenced 22.2.2015]

[3] Adiono, T. Challenges and Opportunities in Designing Internet of Things, 2014. [Referenced 23.3.2015]

[4] Withanage, C. & Ashok, R. & Yuen, C. & Otto, K. A Comparison of the Popular Home Automation Technologies, 2014. [Referenced 23.3.2015]

[5] Wang, W. & Lee, K & Murray, D. Building a Generic Architecture for the Internet of Things, [Referenced 21.10.2014]

[6] Nasrin, S. Novel Protocol Enables DIY Home Automation, 2014. [Referenced 23.3.2015]

[7] Hodges, S. & Taylor, S. & Villar, N. & Scott, J. Prototyping Connected Devices for the Internet of Things, 2004. [Referenced 20.10.2014]

[8] Espressif Systems. ESPRESSIF SMART CONNECTIVITY PLATFORM: ESP8266, 2013. [Referenced 22.2.2015]

[9] Arduino Reference page. https://www.arduino.cc/en/Reference/HomePage [Referenced 22.4.2015]

[10] Android Studio home page. https://developer.android.com/sdk/index.html [Referenced 22.4.2015]

CORRESPONDING ADDRESS

Panu Kiviluoma, D.Sc. (Tech.),
Senior University Lecturer
Aalto University School of Engineering
Department of Engineering Design and Production
P.O.Box 14100, 00076 Aalto, Finland
Phone: +358504338661
E-mail: panu.kiviluoma@aalto.fi
http://edp.aalto.fi/en/

ADDITIONAL DATA ABOUT AUTHORS

Airola, Antti, B.Sc. (Tech.)
E-mail: antti.airola@aalto.fi

Jousimaa, Otso
E-mail: otso.jousimaa@aalto.fi

Niemi, Kalle
E-mail: kalle.niemi@aalto.fi

Vuokko, Sami, B.Sc. (Tech.)
E-mail: sami.vuokko@aalto.fi

Kuosmanen, Petri, D.Sc. (Tech.), Professor
Phone: +358 500 448 481
E-mail: petri.kuosmanen@aalto.fi

10th International DAAAM Baltic Conference
"INDUSTRIAL ENGINEERING"
12-13 May 2015, Tallinn, Estonia

TWO DEGREE-OF-FREEDOM UPPER LIMB EXOSKELTON TRAINER FOR ELDERLY PEOPLE

Bhatt, N.; Räsänen, N.; Lehtinen, J.; Tervamäki, T.; Salerto, S.; Kiviluoma, P. & Kuosmanen, P.

Abstract: *The rehabilitation of people with mobility difficulties is an issue that deserves attention. Conventional methods such as face-to-face physiotherapy offer a solution but demand huge resources in form of man-hours. In addition, face-to-face therapy is time consuming for the patient. An exoskeleton trainer, which can be used from the comfort of the patient's home, might provide safe method for training.*

This paper describes a concept of two degree-of-freedom (DOF) exoskeleton trainer. The 2-DOFs Exoskeleton trainer, described in this paper, is designed for upper-limb rehabilitation purposes for elderly people. The device provides the operator with controlled and safe training of upper limb muscles. The purpose of the trainer device is to create resistive force that acts against the operator's movements.

Key words: muscle rehabilitation, upper-limb exoskeleton, isokinetic velocity, exoskeleton trainer.

1. INTRODUCTION

The population of people aged 60 or above, is predicted to increase by approximately three times of the current population by 2050 [1]. Increase in population requires innovation in the healthcare sector in order to keep up with the problems experienced by the ageing population. The effect of ageing on muscles includes loss of muscle mass, muscle fibres, and decrease in the water content of tendons [2]. These effects result in reduced strength, longer response time, loss of motion and flexibility of joints in elderly people. These effects are further accelerated by a sedentary lifestyle and can be slowed down significantly by regular exercise involving basic movements of different body parts [3,4]. However, certain amount of caution is required in the case of elderly people to ensure that the muscles and joints are not overstressed during exercising.

There is a significant amount of research about exoskeletons [5], but it mainly focuses on military and muscle assistance applications. There is little information about the use of exoskeletons in rehabilitation, or about how these technologies could be applied into rehabilitation. The study referred [6], which focus on exoskeletons that assist muscles had similar overall mechanical design, very similar parts and mechanical concepts as the exoskeleton presented in this paper. The structural implementations from these applications were used as a basis in the designs of the current paper. Only few research papers [6][7] discussed exoskeleton robots for rehabilitation and had similar focus and concept as our project.

This paper describes a two-degree-of-freedom (DOF) upper-limb exoskeleton trainer for elderly people rehabilitation purposes. The device described in this paper was developed for controlled training of the arm and shoulder muscles. The article addresses the mechanical design, electrical design and control design of the 2-DOF exoskeleton trainer.

2. SYSTEM REQUIREMENTS

The requirements for the 2-DOFs exoskeleton trainer are set according to rehabilitation purposes, safety and elderly people's limitations. The requirements of the 2-DOFs exoskeleton trainer are:

i) Safety. The device should be safe to operate and cause no harm to the user in any situation. The range of motion of the device should be limited according to restrictions set by human physiology.
ii) Two degrees of freedom. The device should have rotating joints at the elbow and at the shoulder to enable exercising and rehabilitation for certain upper-limb muscles: deltoid muscle, biceps and triceps.
iii) Ergonomic design. The device should be fast and easy to wear, it should fit on any person and it should be mobile. In addition, the device should be lightweight.
iv) Motion resisting operational principle. The device should exercise upper-limb muscles by resisting the movements and not provide additional force.
v) Position sensing technology.
vi) Load sensing technology.

The exoskeleton trainer described in this paper is designed for rehabilitation of upper limb muscles. The controlled movements implemented for the upper-limb are: lateral raise for deltoid muscle, bicep contraction for arm and triceps extension for arm.

It was decided that the device should enable 2 to 3 exercising movements and that it should be mobile and that the arm would be a good focus point, because the arms are used extensively in everyday life. The concept was further simplified to include only the shoulder and elbow joints. Design work began on building the idea around the two joints, the controller and actuators were chosen and, lastly, also the sensors. The aim was to keep the device compact and simple but also useful, accurate and generally functional.

The maximum torque values for the shoulder joint and elbow joints are approximately 75 Nm and 65 Nm respectively in healthy individuals aged between 40 to 50 years. However these maximum values are not being targeted since they would be rarely required for muscle training. Such high powered actuators would also increase the overall complexity significantly in terms of design and power requirements which are unnecessary for this prototype. The goal is to only demonstrate basic muscle training operation.

3. SYSTEM DESIGN

3.1 Mechanical design

The mechanical structure of the 2-DOF exoskeleton trainer (Figure 1) consists of:

i) Aluminium frame with PTFE slide bearings
ii) 2 geared DC motors with optical encoders
iii) Microcontroller
iv) Wearable vest.

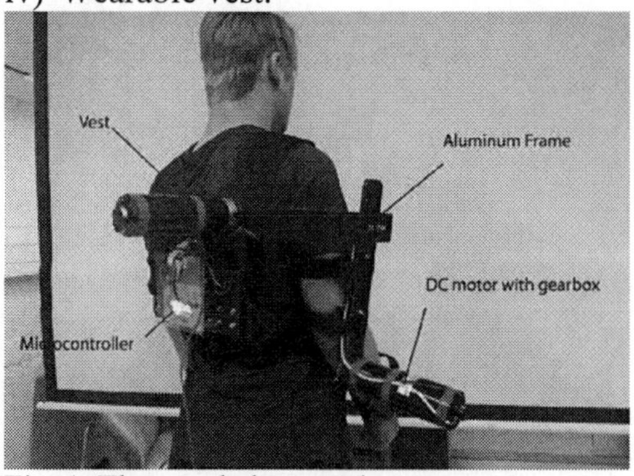

Fig. 1. The exoskeleton trainer.

The shoulder axis of the device for the lateral raise movement is located behind the operator. The elbow axis is located on the side of the operator. The geared DC-motors are attached into the joints. The joints of the exoskeleton are designed to align with the human's shoulder and elbow joint's turning axes (Figure 2). Therefore, lengths of the lever arms do not alter during the operation of the device. However, alignment of the axes of human and exoskeleton set some requirements. The misalignment of the axes in exoskeleton can result into uncontrolled forces and cause extra strain on the muscles. [5] For ensuring the alignment of the axes of the operator and exoskeleton, the described exoskeleton has adjustability in horizontal position of shoulder and elbow axis. In addition, the length of the upper-arm lever arm is adjustable according to user's dimensions.

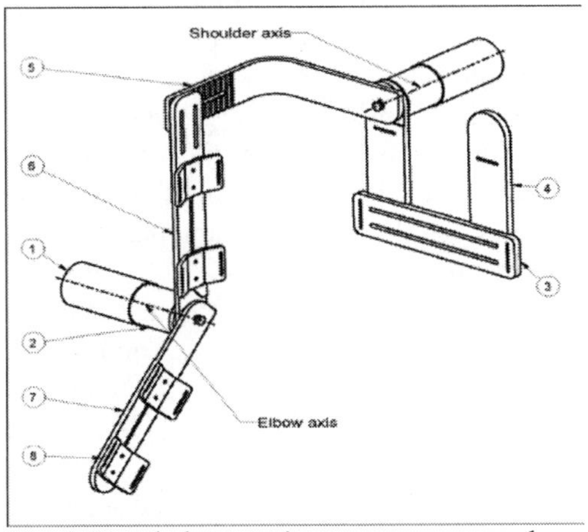

Fig. 2. Exoskeleton trainer components and rotational axes.

ITEM NO	COMPONENT
1	DC motor
2	Gearbox
3	Back plate
4	Support plate
5	Shoulder frame
6	Upper arm frame
7	Lower arm frame
8	Arm attachment pad

Table 1. The main components of exoskeleton trainer.

3.2 Control system design

The system was developed with minimal components and as simple as possible in order to test the proof of concept since this is the first prototype. Electric motors were chosen to be used as the actuators for generating resistance required for muscular exercise. Based on a study conducted by Thomas Harbo et. al. [9], the electric motors with a max torque rating of 8 Nm and stall torque of 54 Nm selected.

The motors were controlled using a microcontroller with a compatible motor driver. The microcontroller was programmed to drive the motors to generate torque in the direction opposite to the motion while maintaining the speed constant, thus obtaining isokinetic movement.

A PID control algorithm was used for controlling the motors. The input given to the control system is motor speed determined using rotary optical encoder with a resolution of 300 pulses per second. The output is varying torque at constant speed. The activation of motors was based on a threshold velocity set and thus the motors stop as soon as the user stops the movement. This ensures safety to a certain extent.

Finally motor current is measured and logged during the exercise and can be used to determine torque throughout the range of motion. This can be used to evaluate muscle performance and improvement over time. Motor current characteristics as given in the data sheet are 0.4 A at no load to 2.2 A at rated speed of 2500 rpm. This information can be used to calculate torque using rated torque value of 8Nm since the current increases linearly with torque. The control system flow chart is shown in Fig. 3

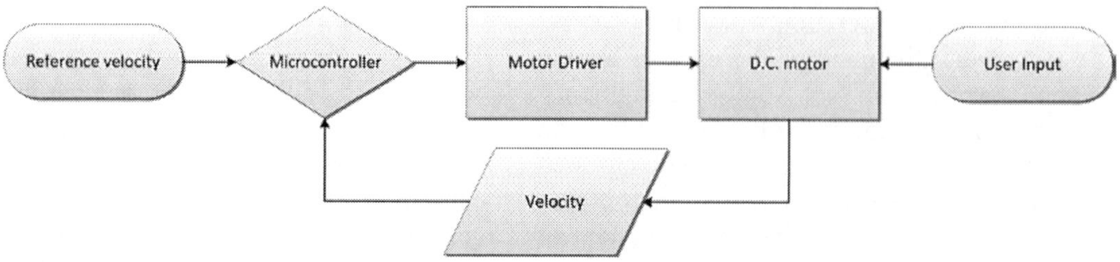

Fig. 3. Control system

The device assists the operator in exercising three different muscles: lateral deltoid, bicep and triceps. The device is programmed to produce isokinetic contraction, so that tension in the muscle remains constant as the muscle's length changes, i.e. the velocity remains constant throughout the entire range of motion. The device assists elderly people to rehabilitate their muscles routinely. The shoulder movements are shoulder abduction and adduction and elbow movements are elbow flexion and extension.

4. RESULTS

The initial plan was to tune to PID parameter by testing of the device but in practice, this proved to be too time consuming. The parameters were tuned as much as possible in the available time but the control system is remained oscillatory in nature. However, it responds to the change in muscle power applied by varying torque. The final output demonstrated the proof of concept of a wearable muscle training exoskeleton.

The mechanical design of the exoskeleton trainer proved the concept working. For the future research, improvements in the areas of user-friendliness, adjustability and overall weight are needed. The greatest challenge in the final prototype was aligning the motion axes of the device and the operator. The device had capability for adjustments in the length of upper-arm part, horizontal position of shoulder joint and horizontal position of upper-arm frame i.e. in elbow joint position. The described adjustability capacity proved to be insufficient. Lack of adjustability resulted in limited range of motion of joints during operation. The overall weight of the device was rather high considering elderly people as the target user group. However, the main source for increased mass of the device was the geared DC-motors, which were over dimensioned. In addition, the integration of the device and the vest requires more attention. Attachments of the device caused flexibilities in the system, which led to misalignment of the motion axes and uncomfortableness in user experience.

Aluminium was the material chosen for the exoskeleton frame. The advantages of aluminium as construction material for described device were lightness, easy manufacturability, relatively low costs compared to more advanced metals and composite materials and satisfactory mechanical properties. The disadvantages of material selection were softness of aluminium and constructions stiffness. Softness of the aluminium resulted in abrasion in parts subjected to forces. Abrasion was significant especially in joints, which implemented keyway design. Abrasion in shoulder part and upper arm part generated gaps between key and keyway. Furthermore, position errors due to increased gaps impeded control design.

The joints of the exoskeleton trainer utilized cotter joint design with PTFE-plastic slide bearings. The principle of applying slide bearings proofed to be simple, cost-effective and functional. The coefficient of friction between PTFE and aluminium parts was low enough for enabling proper functionality of the device. Challenges occurred in connections between steel key and key grooves of aluminium parts, which resulted in abrasion.

The presented device also met the set safety requirements. The device had mechanical limitations for the range of motions, which prevented overextensions of operator's joints. The control system was designed to act as resistance, which intrinsically included the safety aspect i.e. the control system did not allow the device to produce torque, which would harm the operator. In addition, a manual kill switch was included into the device, which would switch off the power from the device in case of malfunction.

5. CONCLUSION AND DISCUSSION

This paper described the concept of a 2-DOF exoskeleton trainer for rehabilitation purposes for elderly people. Actuator selection, mechanical design and control design of the device were described. Mechanical design of the device proved the concept functional with certain limitations. Further research is needed for improving user-friendliness and ergonomics and decreasing overall weight. Especially material selection, adjustability and integration of the device to the user's body requires more research. For the commercial products the overall design has to be optimized for comfortable and fast dressing-up. The challenges are the weight and different body sizes, which require attention for aligning motion axes. For the future, more merged design is needed to address these challenges.

6. REFERENCES

1. World Population Ageing 1950-2050. United Nations New York, 2001. http://www.un.org/esa/population/publications/worldageing19502050/pdf/80chapterii.pdf. 10.2.2015.
2. Dorrens, J. M. J. R., and M. J. Rennie. "Effects of ageing and human whole body and muscle protein turnover." *Scandinavian journal of medicine & science in sports* 13.1 (2003): 26-33.
3. Marks, Ray. "The effect of ageing and strength training on skeletal muscle." *Australian Journal of Physiotherapy* 38.1 (1992): 9-19.
4. Goodpaster, Bret H., et al. "The loss of skeletal muscle strength, mass, and quality in older adults: the health, aging and body composition study." *The Journals of Gerontology Series A: Biological Sciences and Medical Sciences* 61.10 (2006): 1059 1064.
5. Yan, Tingfang, et al. "Review of assistive strategies in powered lower-limb orthoses and exoskeletons." *Robotics and Autonomous Systems* 64 (2015): 120-136.
6. Chong, Yu Zheng, Kwok Hong Chay, and Yee Hern Lim. "Design and development of wearable upper extremities powered exoskeleton-A cost-effective rehabilitation system." *Biomedical Engineering and Sciences (IECBES), 2014 IEEE Conference on.* IEEE, 2014.
7. Rahman, M. H., et al. "Development and control of a wearable robot for rehabilitation of elbow and shoulder joint movements." *IECON 2010-36th Annual Conference on IEEE Industrial Electronics Society*. IEEE, 2010.
8. Effects of Aging. American Academy of Orthopaedic Surgeons. http://orthoinfo.aaos.org/topic.cfm?topic=A0019. 10.2.2015.
9. Harbo, Thomas, John Brincks, and Henning Andersen. "Maximal isokinetic and isometric muscle strength of major muscle groups related to age, body mass, height, and sex in 178 healthy subjects." European journal of applied physiology 112.1 (2012): 267-275.
10. Andersen, Henning. "Motor dysfunction in diabetes." Diabetes/metabolism research and reviews 28.S1 (2012): 89-92. 10.2.2015.

11. Gopura, R. A. R. C., Kiguchi, K., & Bandara, D. S. V. (2011, August). A brief review on upper extremity robotic exoskeleton systems. In Industrial and Information Systems (ICIIS), 2011 6th IEEE International Conference on (pp. 346-351). IEEE.
12. Stienen, A. H., Hekman, E. E., Van der Helm, F. C., Prange, G. B., Jannink, M. J., Aalsma, A. M., & Van der Kooij, H. (2007, June). Dampace: dynamic force-coordination trainer for the upper extremities. In Rehabilitation Robotics, 2007. ICORR 2007. IEEE 10th International Conference on (pp. 820-826). IEEE.
13. Macaluso, Andrea, and Giuseppe De Vito. "Muscle strength, power and adaptations to resistance training in older people." *European journal of applied physiology* 91.4 (2004): 450-472.

7. CORRESPONDING ADDRESS

Panu Kiviluoma, D.Sc. (Tech.), Senior University Lecturer
Aalto University School of Engineering
Department of Engineering Design and Production
P.O.Box 14100, 00076 Aalto, Finland
Phone: +358504338661
E-mail: panu.kiviluoma@aalto.fi
http://edp.aalto.fi/en/

8. ADDITIONAL DATA ABOUT AUTHORS

Räsänen, Niko, B.Sc.
E-mail: niko.rasanen@aalto.fi

Lehtinen, Jere, B.Sc.
E-mail: jere.j.lehtinen@aalto.fi

Bhatt, Nimish, B.Eng.
E-mail: nimish.bhatt@aalto.fi

Tervamäki, Tuomas, B.Sc.
E-mail: tuomas.tervamaki@aalto.fi

Salerto, Samu, B.Sc.
E-mail: samu.salerto@aalto.fi

Address: Same as above

10th International DAAAM Baltic Conference
"INDUSTRIAL ENGINEERING"
12-13 May 2015, Tallinn, Estonia

HIGH ACCURACY FILTER TRANSMISSION MEASUREMENT FOR DETERMINATION OF THE DETECTION EFFICIENCY CALIBRATION OF Si-SPAD DETECTORS

Dhoska, K.; Hofer, H.; López, M.; Rodiek, B.; Kübarsepp, T. & Kück, S.

Abstract: *A high accuracy filter transmission measurement for the determination of the detection efficiency of a Si-SPAD detector has been carried out using an integrating sphere with attached detector. The measurement method and the improvement of total measurement uncertainty of the Si-SPAD detection efficiency calibration are described in this paper.*

Key words: Si-SPAD detector, integrating sphere, filter transmission measurement, detection efficiency.

1. INTRODUCTION

Over the last decade, silicon single-photon avalanche diodes (Si-SPADs) have become increasingly important in different application fields such as quantum operations, vision systems, astrophysics telecommunications, biology and medicine. Such a variety of usage fields of Si-SPADs is based on their high detection efficiency at few photon levels in a wide spectral range; from the visible to the near infrared. The detection efficiency is typically measured by sending few photons onto the detector at a known repetition rate and recording the number of detection events [[1]]. Typically a strong attenuated laser or incandescent lamp is used as a light source. Thus, in order to achieve reliable measurements for the detection efficiency calibration, the Physikalisch-Technische Bundesanstalt (PTB), the national metrology institute of Germany, established recently a compact setup for Si-SPAD calibration that uses traceable transfer standards and a measurement procedure based on filter transmission technique, see Fig. 1 and ref. [[2, 3]]. However, the previous calibration results of the Si-SPAD detection efficiency have shown that one of the major uncertainty contributions in the measurement uncertainty budget comes from the neutral density filter transmission measurement [[3]]. For this reason, a high accuracy filter transmission measurement is required to improve the measurement uncertainty of the Si-SPAD detection efficiency. The novelty of this research work is that instead of using a single silicon detector (Si-Diode), an integrating sphere is employed for the filter transmission measurement. The integrating sphere is a device used for collecting and spatially integrated radiant flux [[4]]. The systematic errors that may be introduced due to specular reflections between the Si-Diode and the objective used during the filter transmission measurement are practically eliminated by using the integrating sphere.

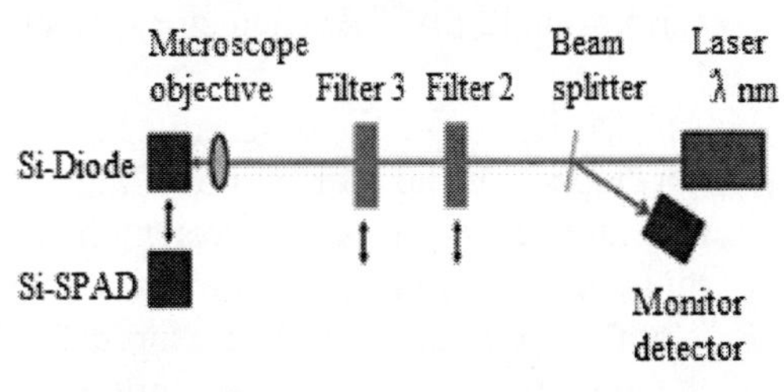

Fig. 1. Schematic view of the calibration setup for Si-SPADs [[2, 3]].

2. MEASUREMENT METHOD AND SETUP

For the determination of the Si-SPAD detection efficiency applying this technique, the transmission of the filters is required to calculate the optical power impinging on the Si-SPAD detector. However, since a very low filter transmission is needed, which is not possible to be measured directly with an analogue detector, a two step measurement procedure for the filter transmission determination is required. This is carried out as follows: in a first step, the filter transmission is measured individually for each filter (T_{F2}, T_{F3}) by using high accuracy translation position stages and the integrating sphere with attached detector as a light sensor. In a second step, the filter transmission of the two filters is measured as a filter package ($T_{Combined}$); i.e. both filters are simultaneously placed in the beam path. From these three measurements a deviation between individual filters and total filter combination can be calculated, that is,

$$Dev = 1 - \frac{T_{F2} \cdot T_{F3}}{T_{Combined}}. \quad (1)$$

This deviation can be taken as an overall uncertainty contribution of the filter transmission measurement for the determination of the detection efficiency of Si-SPAD detectors, as described in [3].

The measurement setup for the filter transmission measurement and the determination of the quantum detection of Si-SPAD detector is shown in Fig. 2. A tunable laser source with a wavelength range from 766 nm to 781nm is used. The laser beam is focused through a microscope objective APO M-PLAN 20x with an 0.42 numerical aperture and a working distance of 20 mm. We have used neutral density filter NG9 D 2.6 for Filter 2 and neutral density filter NG9 D 3.0 for testing the measurement procedure.

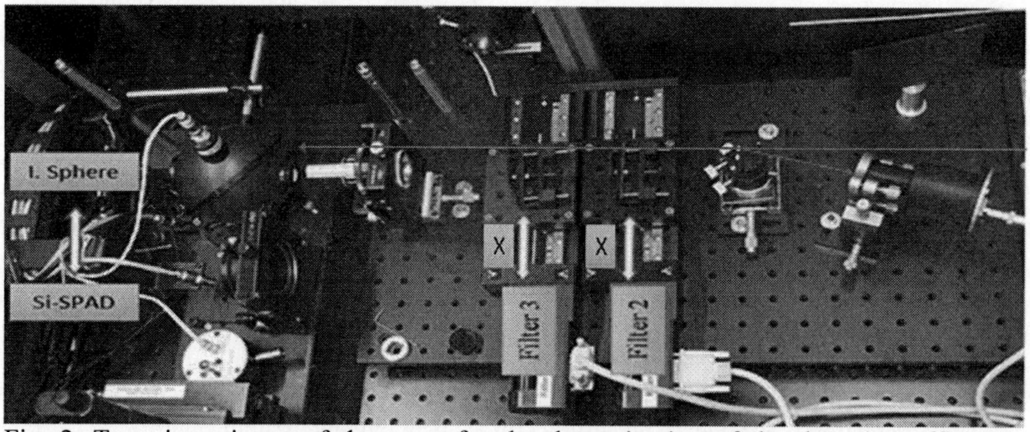

Fig. 2. Top view picture of the setup for the determination of the detection efficiency calibration of Si-SPAD detector by using integrating sphere with attached detector.

An Agilent VEE program has been developed for the realization of automated measurements. At each wavelength, in the range from 766 nm to 781nm with steps of 2 nm, were realized 100 measurements. In the measurement procedure, Filter 2 and Filter 3 are moved in x-direction of the translation stage. The position repetition of the translation stages is ≤ 5 μm, which avoid errors due to the spatial non-homogeneity of the filter transmission. The translation range of the stages is 80 mm. The measurement program is composed of 4 modules. In module 1 the optical power of the laser is measured with the integrating sphere without any filter. The mean value of the measurement obtained with module 1 is used as a reference value for calculation of the filter transmission in relation to the results obtained in the next measurement modules. Afterwards, the measurement module 2 only Filter 2 is moved in the beam path (x = 37 mm). Module 3 continues with the measurement of the individual filter transmission by moving Filter 2 to position x = 0 mm and Filter 3 to position x = 37 mm. Finally, module 4 completes the measurement procedure by measuring the total transmission of the filter combination, i.e. Filter 2 and Filter 3 are positioned in the beam path (position x = 37 mm) simultaneously.

3. MEASUREMENT RESULTS

The summary of the filter transmission measurement results with their deviations for different wavelengths are shown in Table 1 and dispersions of the deviations are depicted in Fig. 3. The deviation between the transmission measurement of the single and combined filters is ≤ 0.05 % for the whole measured wavelength range.

Nr	λ (nm)	Filter 2	Filter 3	Combined Filters	Deviation (%)
1	766	0.0186882	0.0086968	0.0001626	0.020
2	768	0.0188345	0.0087792	0.0001654	0.012
3	770	0.0189695	0.0088636	0.0001681	-0.023
4	772	0.0190674	0.0089108	0.0001699	0.016
5	774	0.0191954	0.0089864	0.0001724	-0.044
6	776	0.0193084	0.0090481	0.0001748	0.030
7	778	0.0194161	0.0091136	0.0001769	-0.012
8	780	0.0195284	0.0091665	0.0001791	0.035
9	781	0.0195746	0.0091978	0.0001800	-0.011

Table 1: Summary of the filter transmission measurement results and deviations calculated by equation (1).

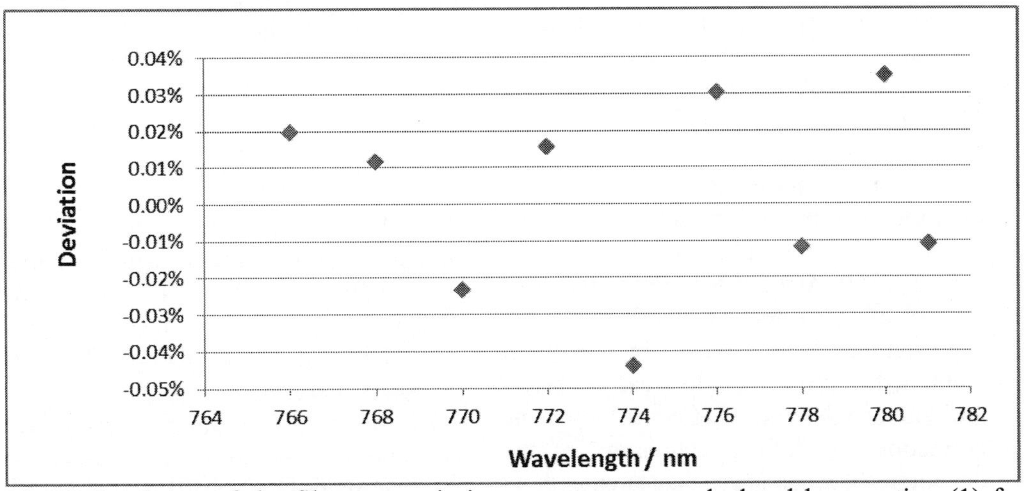

Fig. 3. Deviation of the filter transmission measurements calculated by equation (1) for different wavelengths.

4. DETECTION EFFICIENCY UNCERTAINTY

The mathematical model for determination of the detection efficiency of the Si-SPAD accompanied with all possible contribution factors for evaluation of the measurement uncertainty is given by [3],

$$\eta = \frac{hc}{\lambda}\frac{A_2 A_3}{A_1}\frac{Q_1 Q_4}{Q_2 Q_3} s_{Si} F_{filt}, \qquad (2)$$

where η is the detection efficiency of the SPAD; i.e. the measurand value, h is the Planck constant, c is the speed of light, λ is the wavelength, A_1, A_2, A_3 are the signal amplification factors, Q_1, Q_2, Q_3 are the ratios of the signal of the Si-diode attached to the integrating sphere and the monitor detector signal, Q_4 is the ratio of the counter and the monitor detector signal, s_{Si} is the spectral responsivity of the integrating sphere with the attached Si-diode, and F_{filt} is the factor taking into account the use of two filters.

The estimation of uncertainty of the detection efficiency measurement is carried out following the guide to the expression of uncertainty in measurement (GUM) [5]. Based on the propagation law of uncertainty and uncertainty of input quantities, we can evaluate the combined uncertainty as [3]:

$$u_c^2(\eta) = c_1^2 \cdot u_h^2 + c_2^2 \cdot u_c^2 + c_3^2 \cdot u_\lambda^2 + \\ + c_4^2 \cdot u_{A_1}^2 + c_5^2 \cdot u_{A_2}^2 + c_6^2 \cdot u_{A_3}^2 + c_7^2 \cdot u_{Q_1}^2 + \\ + c_8^2 \cdot u_{Q_2}^2 + c_9^2 \cdot u_{Q_3}^2 + c_{10}^2 \cdot u_{Q_4}^2 + \\ + c_{11}^2 \cdot u_{s_{Int.Sph}}^3 + c_{12}^2 \cdot u_{F_{Filt}}^2 \quad , (3)$$

where the $u(i)$ are the standard uncertainties of the input quantities and c_i are the sensitivity coefficients which are calculated from the partial derivatives of all input quantities.

In continuity on the improvement of the total measurement uncertainty we will not focus to the estimation of all possible contributor factors but only at the factor of the use of two filters as one of the major contribution. The deviation between combined and individual filters measurements will influence in the correction factor. The expression of the deviation between two measurements is estimated from equation (1) and correction factor by equation (4):

$$F_{Filt} = \frac{T_{Combined}}{T_{Individual}} = \frac{T_{Combined}}{T_{F2} \cdot T_{F3}}, \qquad (4)$$

where T_{F2}, T_{F3} are the filters transmission of each individual filter and $T_{combined}$ is the filter transmission of combined filter.

These results have shown that the largest deviation ~ 0.05 % comes from the transmission measurements carried out at λ=774 nm. This deviation is used, for simplicity, for the estimation of the uncertainty of the correction factor F_{Filt} for each wavelength. Thus, the standard uncertainty of the correction factor that uses two filters is estimated:

$$u_{F_{Filt}} = \frac{T_{Individual} - T_{combined}}{T_{combined} \cdot \sqrt{3}} = 3.3 \cdot 10^{-4}. \quad (5)$$

The uncertainty contribution of the factor which uses two filters is obtained:

$$u_c(F_{Filt}) = \sqrt{c_{12}^2 \cdot u_{F_{Filt}}^2} = 2.2 \cdot 10^{-4}, \qquad (6)$$

where sensitivity coefficient is c_{12} = 0.64. The final result has shown that uncertainty contribution is 8.1 %. Referring to the previous estimated results we have improved one of the major uncertainty contributions by significantly reducing measurement uncertainty from 54.8 % [3] to 8.1 %. The new contribution value of the factor using two filters is included into the uncertainty budget and is used to estimate the combined uncertainty $u_c(\eta)$. Finally, the obtained detection efficiency of the Si-SPAD detector is:

$$\eta_{SPAD} = 0.6359 \pm 0.0014,$$
$$\eta_{SPAD} = 0.6359 \pm 0.22\ \%.$$

5. CONCLUSIONS

In this paper, the accurate filter transmission measurement method and the improvement of its measurement uncertainty have been developed for the calibration of the detection efficiency of Si-SPAD detector.

The use of an integrating sphere instead of a single Si-Diode for the filter transmission measurement has significantly reduced the measurement uncertainty contribution of the filter factor from 54.8 % to 8.1 %. The relative standard uncertainty of the detection efficiency of the Si-SPAD was improved from < 0.5 % to < 0.25 %.

Future work will be focused on reducing the uncertainty contribution of the absolute spectral responsivity on the integrating sphere as the last larger contribution factor for the determination of detection efficiency of Si-SPAD detectors.

6. REFERENCES

1. Ch. J. Chunnilall, I. P. Degiovanni, S. Kück, I. Müller and A. G. Sinclair, Metrology of single-photon sources and detectors: a review, *Optical Engineering*, 2014, **53**, 01-17.
2. S. Kück, H. Hofer, S. Peters, M. López, Detection Efficiency Calibration of Silicon Single-Photon Avalanche Diodes Traceable to a National Standard, In: *Proceedings of NEWRAD: 12th International Conference on New Developments and Applications in Optical Radiometry,* (Park, S. and Ikonen, E., eds.) Espoo, Finland, 2014, 93-94.
3. M. López, H. Hofer, S. Kück, Detection efficiency calibration of single-photon silicon avalanche photodiodes traceable using double attenuator technique, *Journal of Modern Optics*, 2015, **62**, S21-S27.
4. A technical guide to integrating sphere theory and applications. http://www.labsphere.com/uploads/technical-guides/a-guide-to-integrating-sphere-theory-and-applications.pdf.
5. *ISO/IEC Guide 98:1995.* Guide to the expression of uncertainty in measurement (GUM). International Organization for Standardization: Geneva.

7. ADDITIONAL DATA ABOUT AUTHORS

Klodian Dhoska, doctoral student,
Department of Mechatronics, Tallinn University of Technology, Ehitajate tee 5, 19086, Estonia.
Email: klodian.dhoska@ttu.ee

Helmuth Hofer, engineer,
Division of Optics, Physikalisch-Technische Bundesanstalt (PTB), Bundesallee 100, 38116 Braunschweig, Germany.
Email: helmuth.hofer@ptb.de

Beatrice Rodiek, doctoral student,
Division of Optics, Physikalisch-Technische Bundesanstalt (PTB), Bundesallee 100, 38116 Braunschweig, Germany.
Email: beatrice.rodiek@ptb.de

Marco López, doctor,
Division of Optics, Physikalisch-Technische Bundesanstalt (PTB), Bundesallee 100, 38116 Braunschweig, Germany.
Email: marco.lopez@ptb.de

Toomas Kübarsepp, professor,
Department of Mechatronics, Tallinn University of Technology, Ehitajate tee 5, 19086, Estonia.
Email: toomas.kubarsepp@ttu.ee

Stefan Kück, professor,
Division of Optics, Physikalisch-Technische Bundesanstalt (PTB), Bundesallee 100, 38116 Braunschweig, Germany.
Email: stefan.kueck@ptb.de

8. ACKNOWLEDGEMENT

This research work has been supported by Physikalisch-Technische Bundesanstalt in Braunschweig, the national metrology institute of Germany and by the project "Single-Photon Sources for Quantum Technology" (SIQUTE) of the European Metrology Research Programme (EMRP). The EMRP is jointly funded by the EMRP participating countries within EURAMET and the European Union.

10th International DAAAM Baltic Conference
INDUSTRIAL ENGINEERING
12-13 May 2015, Tallinn, Estonia

CUBESAT MISSION FOR MULTISPECTRAL EARTH OBSERVATION FROM LOW EARTH ORBIT

Karmin, M.; Ehrminger, R. ; Shatov, J. ; Mzhavia T. ; Vaher, O. ; Dihuliya D. ; Hiiemaa, M. & Tamre, M.

Abstract: *An international team of Master students from the Tallinn University of Technology in cooperation of MEKTORY Space Center designed a three unite sized CubeSat for low Earth observation mission. The data acquired during the mission can be used for educational, scientific, space technology development and knowledge transfer purposes. The contributing team from the department of Mechatronics designed the subsystems for the optical payload, attitude determination and control system. The preliminary design of the optical system consists of a multispectral line scan sensor, shutter and a Ritchey-Chrétien reflector. Compact deployable mechanism design for the secondary mirror is considered. The requirements for the attitude determination and control system are determined according to the chosen line-scan sensor and the optical system.*

Key words: CubeSat, Multispectral-line scan- sensor, Ritchey-Chrétien reflector, deployable optics

1. INTRODUCTION

The main goal of the low earth orbit (LEO) mission is to take multispectral images with relatively high ground resolution of 25 m/pixel. The limited volume, weight and power requirements as well as the relatively low spectral power intensity are the main drivers for this design concept. The data gathered during the mission can be used for general land observation tasks, e.g. agricultural applications, thermal mapping and other scientific purposes.

In order to determine the available backscattered spectral power intensity for the CubeSat at the altitude an atmospheric model was developed. Based on the model data a line scan sensor with the suitable sensitivity was chosen. After the sensor selection, the requirements for the optical system and the altitude determination and control system have been identified.

The current paper presents the final stage of the first conceptual design study. One of the results of this study is the identification of requirements for the subcomponents for each module. Additionally the operational range of latitude and the detectable terrain types can be differentiated and specified. Further steps of the project are optimisation and verification of the spectral models and the integration with other subsystems. Furthermore, prototyping and testing of components and modules is considered.

2. GENERAL CUBESAT CONCEPTS AND MISSION REQUIREMENTS

The mechanical outer frame structure is based on the CubeSat standard with three units. The predefined dimensions of 100x100x340.5 mm limit the maximal available volume for all components. The illustration of the conceptual solution of the CubeSat is presented in figure 1.

Fig. 1. Illustration of the overall design

The general top-level requirements are the theoretical ground resolution of 25 m/pixel for RGB and NIR bands from the 600 km low earth orbit and the image size of 2 k pixels. The design process started with the spectral light intensity estimation according to atmospheric conditions and backscattering characteristics of different landscapes including seasonal light conditions. In the second phase the optical system subsystem has been developed within a volume of 100x100x120 mm³ and the overall weight of around 4,25 kg. The power budget of the optical system is around eight Watts (peek).

Parallel to the development of the optical system the specifications for the second sub module "altitude determination and control" have been determined. According to the timeline and technical requirements the iACDC-100 from Berlin Space Technologies has been considered as the best option. In a long-term perspective, a self-developed system is pursued.

The onboard computer has been developed by the communication team. It enables high reliability image data pre-processing and communication with space graded components.

2.1 SPECTRAL POWER INTENSITY ESTIMATION

For determining the power spectrum of light available at given conditions, the model of the light path was created. Each stage of the light's path acts like a filter removing various amount of power in different area of wavelength power spectrum from prior one. As the light enters the atmosphere at some zenith angle it travels through corresponding air mass (AM), which will deform the power spectrum intensity. The loss of incoming light is subtracted from extraterrestrial irradiation. As light reaches surface energy will be absorbed and converted into heat as well as reflected. Power spectrum intensity which reached Earth will be multiplied with corresponding albedo characteristics for different scenarios. The remaining light will travel once more through the atmosphere (perpendicular) and outgoing losses are subtracted from remaining power spectrum. The light intensity, which reaches the satellite and enters the optical system changes its characteristics one more time before the sensor can detect an object.

The light intensity model is based on different landscape scenarios. The model returns the value of the power spectrum of the light reached to sensor dependant on the zenith angle and one specific terrain type which is based on albedos characteristics. The results provide an estimation about possible operational ranges and restrictions for a chosen sensor.

The spectral reference used is ASTM E 490 [1] extraterrestrial spectrum by the American Society for Testing and Materials.

To determine the spectral losses through atmospheric distortion, previously mentioned reference extraterrestrial spectra is compared to ASTM G173-03 (ground spectrum) [1]. This light path corresponds to AM coefficient 1.5.

The situation where both incoming and outgoing light paths are with zenith angle 0 deg is possible only at position where satellite is aligned to the same line with the sun and the target. Since the plan is to take images from several positions on orbit, it is necessary to calculate various AM coefficients of incoming light rays and the latitude of the target. The light intensity depends also on the zenith angle. To complete the model the Pickering equation is implemented [2].

To determine the increased loss of power spectrum of incoming light through atmosphere, due the increased cross-section of the atmosphere, the AM coefficient of zenith angles from 0 to 90 deg were compared to losses of AM1.5 and each spectral characteristics was modified accordingly.
The remaining power reached Earth will be either absorbed and converted to heat or reflected from different earth's surface materials. To determine reflectance of different scenarios the spectral albedos from U.S. National Renewable Energy Laboratory's SMARTS model [3] are used. A number of sample albedos were chosen and used to estimate the backscattered energy intensity from Earth.
One result of the spectral power estimation is a degradation of visible and infrared light in connection with higher zenith angle. The main reason for this effect is the exponential growth of atmospheric distortion. The following graphs 2 to 4 present the results of the spectral power estimation. On the x-axis the power intensity in $nJ/cm^2/sec$ and on the y-axis the spectrum from 300 to 1400 nm. The z-axis represents the zenith angle form 0 to 90 deg.

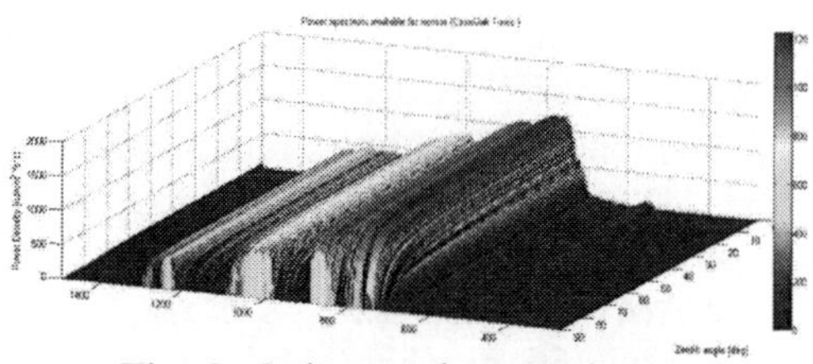

Fig. 2. Oak tree characteristics

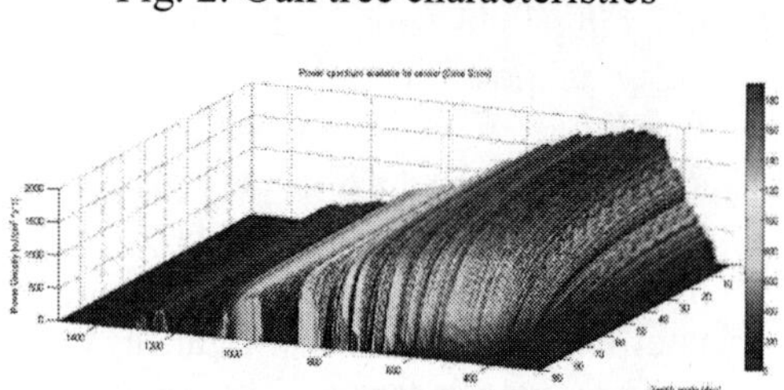

Fig. 3. Snow reflectance patterns

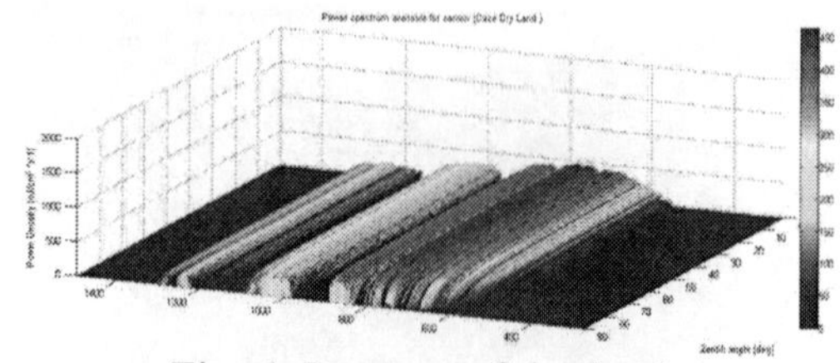

Fig. 4. Patterns of dry land

2.2 OPTICAL SYSTEM AND INTERFACE

The optical system consists of four main components. The line scan sensor, shutter, the deployable system and the telescope.
The integrated configuration has three main stages. In stage a) the CubeSat can be transported and launched. After the launch and the tumbling phase the springs with the sunshade boards can be released. This process is presented in stage b). At the 3rd stage the legs are fully deployed and a slight rest load in the leaf springs provides the stability for in orbit maneuvering. The sunshade boards have a reflective characteristic at the outer surfaces. The inner sides absorb light in order to avoid reflectance in the optical path. The three stages of the deployable optical system are presented in Figure 5.

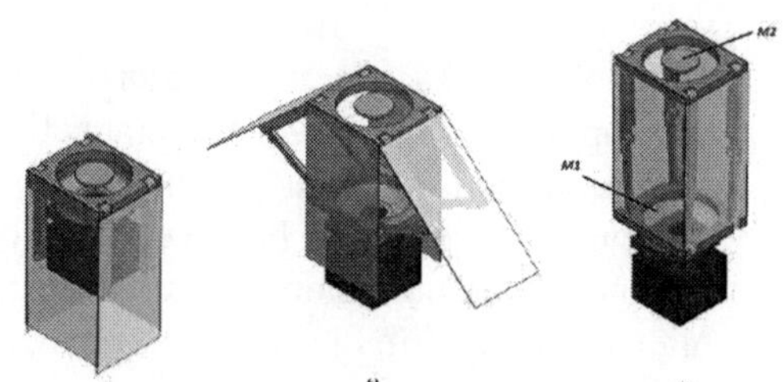

Fig. 5. Deployable optical system

2.2.1 RITCHEY–CHRÉTIEN REFLECTOR

The main reasons for choosing the Ritchey–Chrétien telescope are minimal weight and non apochromatic correction. With ULE as base material, the thermal distortion is kept to the minimum. In order to obtain the required theoretical ground resolution the focal length of 370 mm has been considered. To fit the sensor and the shutter a back focal length of 30 mm is estimated.

2.2.2 DEPLOYABLE MECHANISM

In order to take images with the required ground resolution within the available volume a deployable mechanism of the secondary mirror is considered. The current design incorporates a deployable secondary mirror and mounting frame which is passively actuated by four preloaded springs inside the interfolding legs. The process of deployment is presented in Fig. 5.

2.2.3 SUN PROTECTION SHUTTER

To ensure that the line scan sensor will not receive too much heat an optical semitransparent filter is placed between the optical system and the sensor. The sun protection shutter is mainly needed during the tumbling phase so it will only open with the orientation to Earth.

2.2.4 LINE SCAN SENSOR

The sensor chosen for the mission is a line scan sensor. Line scan offers much more cost-effective implementations of very high spatial resolution image capture and the dynamic range that is relatively higher than area array technology. The scanner is capable of having smear-free images of fast moving objects without expensive strobing. Line scanning eliminates the frame overlaps required to build seamless images. Frame overlap represent redundant data that uses up precious processing bandwidth. Applications like satellite, require high-speed and high-resolution solutions.

The camera sensor works on power supply of 8 W at 12 V with 6 pin Hirose connector. Spectral range of it is 400 to 950 nm and dynamic range of 60 dB. The sensor provides a maximum line rate of 70 kHz (2.8 kHz is required) [[4]].

The software interface level for the line scan sensor is chosen with consideration of versatility and flexibility of design. The solution has to be compatible with industry standards in order to use it with variety of sensors. Requirements consisted of: processing the image and appending the results to the image data stream, controlling external hardware, and operating during the real-time part of the application.

The preferred GenICam interface, standard is hosted by the European Machine Vision Association (EMVA). The goal of GenICam is to provide a generic programming interface for all kinds of cameras and devices. No matter what interface technology (GigE Vision, USB3 Vision, CoaXPress, Camera Link HS, Camera Link, 1394 DCAM, etc.) is used or what features are implemented, the application programming interface (API) should always be the same [[5]].

2.3 ALTITUDE DETERMINATION AND CONTROL SYSTEM

After de-orbiting the satellite will be in microgravity environment and without the application of force; it will be in a free tumble. Objects in low-earth orbit also experience disturbances by atmospheric drag, solar radiation pressure and other forces. In order for a satellite to be able to accomplish its mission, it must be stabilized and pointed to the desired direction. Attitude determination and control subsystem (ADCS) have to measure the position of satellite in Earth coordinate system and control it by applying forces to get the required pointing direction. There are many ways to determine the position of satellite. It must

be pointed in the desired direction with at least 1 degree accuracy. The 1 degree would mean a linear shift of ~10 km near the nadir (and more if the camera is pointed out of the nadir).
Considering the complexity of the ADC subsystem itself, high mission requirements and its importance for our mission, it is considered to use a commercial off-the-shelf module. Attitude determination and control module iACDC-100 from Berlin Space Technologies [6] is chosen as the best option. It has a star tracker for attitude determination and three reaction wheels plus three magnetorquers for attitude control. It gives pointing accuracy less than one degree, which meets the mission requirements. This module also comes with the algorithms for target-, nadir- and sun pointing, spin mode and De-Tumble mode.

2.4 SYSTEM PERFORMANCE ESTIMATION

Theoretical ground resolution is around 25 m/pixel. The static field of view is about 50 km x 98 m. If the main mirror diameter is 100 mm, the area of the primary mirror would be 78.54 cm². The pixel size of the sensor is 14.08 µm x 14.08 µm and the corresponding effective sensor area is 0.1624 cm². To illuminate the whole sensor at the sensor's focal length the illuminated diameter is reduced to a circle with approximate diameter of 31 mm. That leaves approx. 1 mm reserve to each side to compensate possible misalignment during the transport and launch phase of the satellite. The sensor uses approximately 0.02 % of the main mirror's effective area, hence 0.02 % of total possible energy is left after light passing the reflective surfaces. Based on the datasheet of the sensor, the spectral responses of the sensor at its bands are following: Red 700 nJ/cm², Green 800 nJ/cm², Blue 600 nJ/cm², NIR 300 nJ/cm² [4] For example to get a reading of DN = 1 at NIR band, it is necessary to get the energy intensity at the sensor level of 1/300 nJ/cm². With the 10 bit color format each channel has 1024 shades. For example at NIR band it takes 3.41 nJ/cm² to saturate the sensor to DN value 1024. Energy needed for full saturation (DN = 1024) for four bands are: Red 1.462 nJ/cm², Green 1.280 nJ/cm², Blue 1.706 nJ/cm², NIR 3.413 nJ/cm².
To stay in 25 m/pix ground resolution boundaries with satellite orbiting at altitude of 600 km it is necessary to determine if it is possible to saturate the image in a certain period of time that is equal or less than time needed for satellite to travel 25 m relative to ground. For this, it is important to find out the time needed to get maximum saturation for each channel and to see if it fits in to boundaries. If yes then there is enough light to operate in those conditions. At 600 km altitude the satellite is orbiting with the approximate speed of 27000 km/h. The radius of the orbit is about 6.98 * 106 m, circumference ≈ 4.3857 * 107 m and time taken for one orbit ≈ 5.747551126 * 103 seconds. That yields to the angular velocity of ≈ 0.062635 deg/s. At Earth's surface this makes ≈ 6974 m/s. 25 m fits into 6974 m around 280 times. To stay in the bounds of 25 m/pixel the ground resolution the maximum saturation time should be less or equal to 1/280 sec.

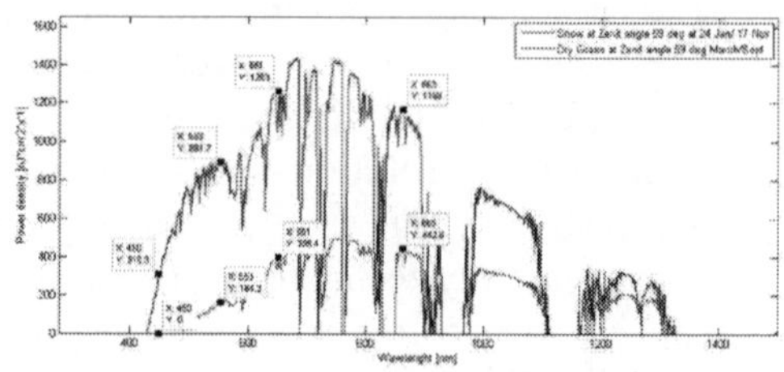

Fig 5. Spectral characteristics of dry grass

The image 5. shows the backscattered spectral characteristics of dry grass at the zenith angle of 59 degrees (corresponds approximate latitude of Estonia at spring and autumn) and characteristics of the snow (same latitude, dates 24 Jan/ 17 Nov). Based on the spectral model estimation at the case corresponding to albedo characteristics of dry grass, it seems that there is lack of light corresponding to the blue channel. The green channel receives approx. 170 nJ/cm²/s, red channel 395 nJ/cm²/s and NIR channel 420 nJ/cm²/s. With these values it is possible to determine the saturation times for channels where light is present.

3. CONCLUTION

The aim of this research is to push the possibility boundaries of easily accessible and research friendly format of CubeSat missions. Implementing an advanced optical subsystem, attitude determination and control solutions will provide stronger basis for future daring ideas. It is a great challenge to design and implement these systems according to the available room, weight, power limitation and required performance. However, with careful testing and verification the possibility to get valuable data for the future research is very high.

4. REFERENCES

[1] Reference Solar Spectral Irradiance: ASTM G-173. NREL Renewable Resource Data Center.
URL http://rredc.nrel.gov/solar/spectra/am1.5/ASTMG173/ASTMG173.html

[2] Keith A. Pickering. "The Southern Limits of the Ancient Star Catalog". DIO The International Journal of Scientific History. Vol 12. Sep 2002. pp 22.
URL http://www.dioi.org/vols/wc0.pdf

[3] Model of the Atmospheric Radiative Transfer of Sunshine, NREL Renewable Resource Data Center.
URL http://www.nrel.gov/rredc/smarts/

[4] Teledyne Dalsa,
URL http://www.teledynedalsa.com/public/mv/RoHS/03-097-20070-00_Piranha4_2k_4k_Material_Composition_Product_Declaration.pdf

[5] Fritz, D., Basler GenICam Standard Generic Interface for Cameras Version 2.0, 2009,
URL http://www.emva.org/cms/upload/Standards/GenICam_Downloads/GenICam_Standard_v2_0.pdf

[6] Berlin-Space-Technology,
URL http://www.berlin-space-tech.com/index.php_id=43.html

10th International DAAAM Baltic Conference
"INDUSTRIAL ENGINEERING"
12-13 May 2015, Tallinn, Estonia

A METHOD FOR ESTIMATING THE RELAXATION OF A PERSON IN AN EDUTAINMENT CONTEXT

Kokko, O.; Ahmad, B.; Aula, J.; Korpijärvi, J.; Kiviluoma, P.; Kuosmanen, P.; Sepponen, R. & Linnavuo, M.

Abstract: *This paper describes an estimation method which can be used to determine relative changes in a person's relaxation/stress state for edutainment purposes. As the edutainment context implies sensor measurements might be inaccurate and subject to different alignment and reading errors. The method accounts these problems by supporting variable number of sensors while maintaining the scale of the resultant value.*

The device to implement the proposed method includes a single electrode EEG headset, a resistance measurement sensor and a pulse oximeter. The method yields dimensionless resultant values which are comparable between multiple test subjects. The resultant value also reacts quickly to different disturbances in the person's stress and relaxation levels. The applications of this research can be utilized in different relaxing exercises and for example as an athlete's mental exercise tool.

Keywords: meditation, calmness, brain activity, EEG

1. INTRODUCTION

Educational entertainment, or edutainment, is any content that is designed to educate as well as to entertain. Games can be designed to have educational value and to teach people about certain subjects. The effects of different situations on a person's own mind and body are an interesting field of study and a game between two persons' minds can prove to be an educating and fun experience.

This article describes a method for estimating the relaxation of a person in an edutainment context. The idea is to formulate a relaxation index which represents the changes of a person's relaxation state and estimates it. This is done in a way that is accurate enough for comparison between two people in edutainment use, but not necessarily exact enough for medical use. For this purpose, the estimation based on three parameters: GSR (galvanic skin response), EEG (electroencephalography, brain activity) and heart rate was created.

The term relaxation, which is used in this paper, is a word describing a mental state where one is not feeling any agitation or excitement. It is possible to do measurements about the relaxation of a person because the state of mind affects many bodily functions. Disturbances in relaxation causes heart rate to rise above the resting heart rate and perspiration to increase. Excitement and anxiety also causes changes in brain activity.

All of these measurable factors are dependent on the individual who is the target of this examination. This means that without any foreknowledge it is very hard to measure how calm someone is. Consequently, measuring the absolute relaxation of someone is difficult without doing multiple measurements in different conditions. Calibration of the measurements is therefore required for every test subject.

The choice of parameters was based on previous research and suitability for edutainment applications. EEG can be filtered to different frequency ranges and previous studies have shown that when a person calms down it increases EEG activity in the frequency range of 7-13 Hz (alpha activity). [1] This can be determined reasonably well with a single electrode EEG headset. [2] In addition, being under mental stress person's skin will produce sweat which increases the skin conductance. [3] Simple resistance measurement can be utilized to determine these changes.

Studies have also shown the usefulness of skin conductance, heart rate, respiration, temperature and muscle electrical activity measurements for determining emotions and physiological events. [4]

2. METHODS

The three chosen parameters (GSR, heart rate and EEG) were chosen for their affordability, responsiveness to mental changes and simple usage. [5] All three measurements are also non-invasive to support edutainment applications. Lesser number of measurements could have also been utilized, but three different parameters provide redundancy, different response times and value stabilities in order to provide more stable and responsive results.

2.1 Galvanic skin response (GSR)

Galvanic skin response (GSR) is the measurement of electrical conductance of skin. In this study it is measured through two electrodes attached to fingertips, but there are also several other methods available. Usually, GSR is measured using a part of the skin with more sweat glands and another part with less as a reference. It is also possible to measure it on the same area through active electrodes. [6]

In this case, a simple voltage drop measurement is used to determine the conductance of the skin between the electrodes. First, the baseline of the voltage drop is calibrated and further values are then measured relatively to this baseline value. When a sudden change in nervous system will occur, like a deep breath or any painful experience, the nervous system will respond quickly by increasing the perspiration, thus, increasing the conductivity of the skin.

The decrease in conductivity is much slower than the initial increase due to sweating. Therefore only the changes in GSR are used in the calculation of the relaxation index. If the person calms down the derivative value of the GSR is slightly positive as the fingers dry and when the person sweats the derivative value is highly negative as the resistance decreases.

The value used for the relaxation index calculation is the derivative of skin conductance.

2.2 Heart rate

Heart rate can be measured by using pulse oximetry. Pulse oximetry is based on light absorption characteristic of the blood and is essentially a multiple wavelength plethysmograph. [7] Heart rate is a commonly acquired low-cost physiological measure that changes according to the subject's mental state [5]. Pulse oximeter is often connected either to fingertip or earlobe of the test subject.

Pulse oximeter consists of a LED and a photodiode or a phototransistor. In the sensor used in this study, both components are on the same side of the finger, so this pulse oximeter uses the changes of reflection caused by the blood. Measurement of the light passing through the finger is also possible.

The signal which is received at this stage is very weak and is passed through an amplification circuit and a low-pass filter to remove ambient noise. [8] The resting heart rate is individual so the absolute value cannot be used for the relaxation index. Therefore, the baseline of the heart rate is also calibrated before measuring the changes. When the subject is relaxing, the heart rate decreases and stays lower. Agitation, stress and excitement rises the heart rate. In contrary to GSR, continual differences in heart rate can be used as well, since these represent subject's current state.

The value used for the relaxation index calculation is the difference between current heart rate and the calibrated baseline.

2.3 EEG

Brain activity (EEG) is a voltage measurement from the scalp of a person. This requires at least one electrode to be in contact with the person's head and one contact that works as a reference point for the voltage measurement. The reference point can be another point on the scalp or a clip on the person's ear, as it is in this study. For positioning the electrodes, a system called "10-20 system" has been developed. This system defines names for the possible electrode points. [9]

EEG signal has numerous frequency components with designated names. For relaxation measurements, the most interesting frequency range is called Alpha band (8-15 Hz). Meditation has been shown to increase the intensity of Alpha activity in the frontal regions of the scalp. [10] Simple meditation measurement can be made for example from point Fp1 (10-20 system designation) which is used by many commercial EEG headsets. In this study, the measurement system consists of single electrode placed on the Fp1 point on the scalp and a reference point on the ear.

The voltage amplitude of brain waves measured through the scalp is around 10-100 μV and the frequency content varies from 0.5 to 100 Hz. Amplification circuit with high gain is required to convert the voltages into measurable range and filtering is needed to filter out noise below and above the frequency range of interest. The configuration consists of high-pass filter (cut-off around 1.5 Hz) and low-pass filter (cut-off around 45 Hz). The high-pass filter should have as steep cut-off as possible to remove lowest frequencies (near DC components) while preserving frequencies above 2 Hz. The low-pass filter is used to remove any residual 60 Hz interference. Fast Fourier Transform (FFT) is then used to transform brain waves into frequency domain to examine the intensity at different frequencies. The intensity around the Alpha band is then estimated and a value representing the relative content of activity is generated. This value is then passed on to the relaxation index estimation. [10]

2.4 The relaxation index

The relaxation index is a dimensionless scalar value which is calculated from the afore-mentioned parameters. Its purpose is to enable the comparison of two test subjects' relaxation states. As such the relaxation index is a way to display an estimation of a person's relaxation in a comparable way.

The three parameters used in this study for the calculation of the relaxation index are called M_1 from heart rate, M_2 from GSR and M_3 from EEG. The values of these variables are determined by the measurements described earlier.

The measured values need to be scaled accordingly in order to have different sensors effect the final index so that getting more relaxed increases the final index. This need is caused by differences in these three parameter values and their changing ranges. For M_1 (heart rate) we use K_1, for M_2 (GSR) we use K_2 and for M_3 (EEG), K_3 is used. The values K_1, K_2 and K_3 are constant factors and they are acquired from empirical testing of different measured parameter inputs. This way the final form for calculating the relaxation index is

$$R = K_1 M_1 + K_2 M_2 + K_3 M_3.$$

The relaxation index R is calculated out of every parameter that is used on that measurement time. Higher relaxation index corresponds to a more relaxed state.

In case only EEG and HR are measured, the formula used to calculate relaxation index will have the constant multiplier K_2 set to zero and the result will be multiplied by 3/2. As the changes in different parameters are set to cause approximately same kind of changes in the relaxation index, this 3/2 multiplier will make the resulting index comparable with an index that has been calculated from all three parameters. In the same manner, if only one parameter is used, the indexes constant multiplier is 3 and the other multipliers will be zeros. This way, the relaxation index becomes more reliable with more parameters, but is somewhat usable even with one measured parameter.

The resulting relaxation index value can then be compared to different subjects' values in order to compare the relaxedness of the subjects. The subject with higher relaxation index is estimated to be more relaxed.

3. RESULTS

The described sensors and calculations were used with real test subjects in order to determine the constant factors and to validate the correlation between relaxation index and the actual effects on the test subject. After the determination of the constant factors, the following results were obtained to validate the correlation.

Figure 1 contains data from a person who is sitting down and actively relaxing with closed eyes. This figure shows a relaxed baseline for heart rate HR (K_1M_1), GSR (K_2M_2) and EEG (K_3M_3) values and the resulting relaxation index (R). In this relaxed state the index rises over 100.

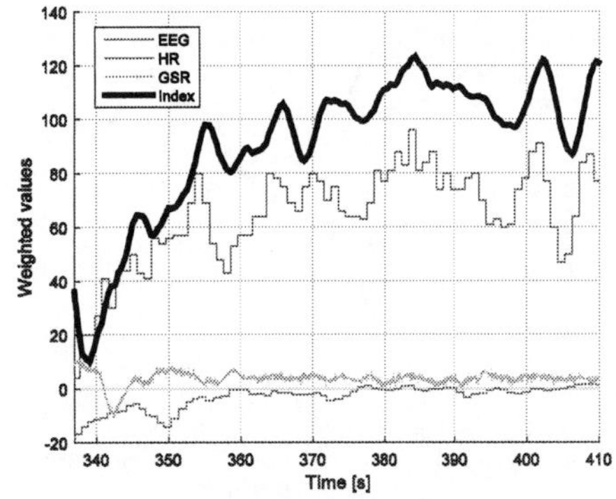

Fig. 1. Active relaxing with closed eyes.

Figure 2 shows the test subject reading a book. This concentration causes the index to stay around 80.

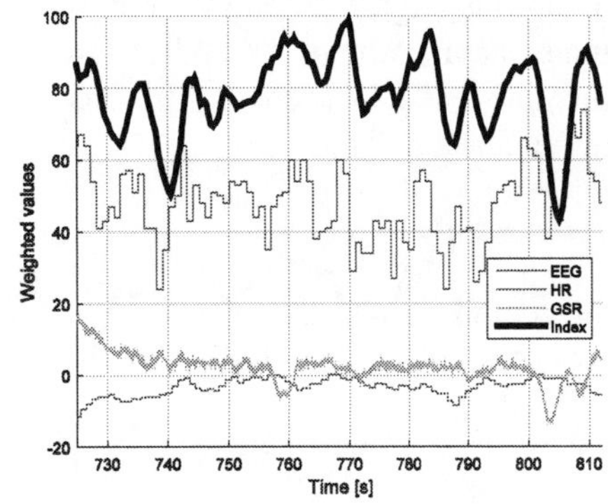

Fig. 2. Reading a book.

Figures 3 and 4 show the same parameters after different disturbances in subject's relaxation. In both figures the relaxation index drops momentarily after the disturbance.

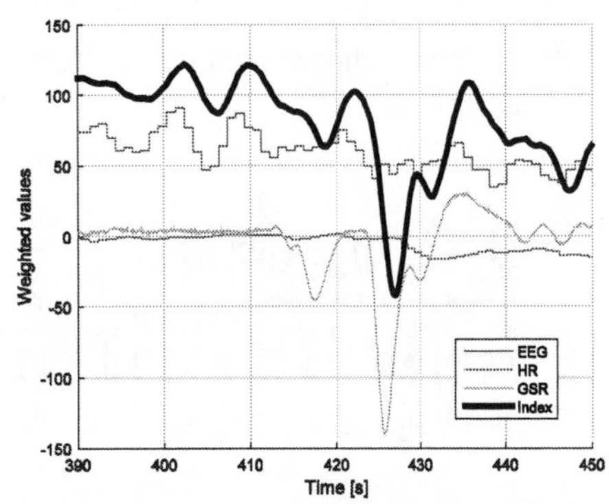

Fig. 3. After eating very spicy snack.

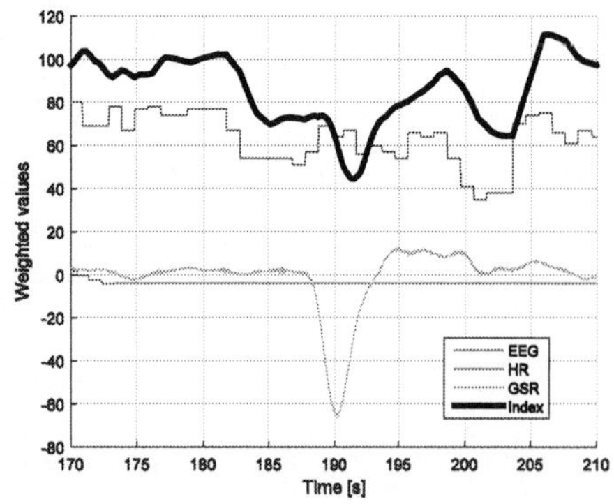

Fig. 4. Loud noise during relaxation.

To test the possibility to rise the index values by effort, the test subject held breath to activate the diving reflex. While not breathing, figure 5 shows the relaxation index staying above 100. After resuming of breathing at 885 seconds, the index drops momentarily.

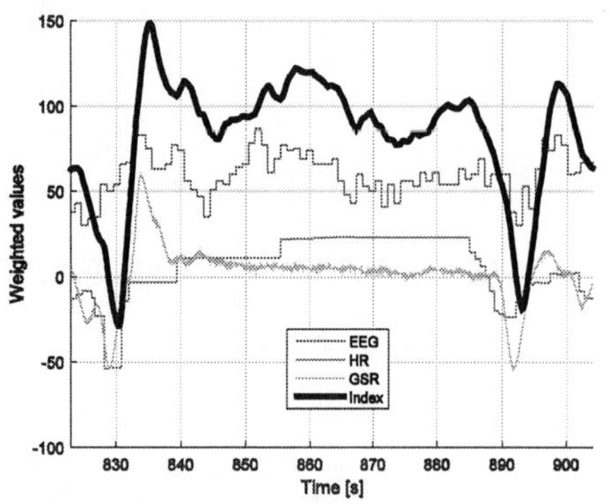

Fig. 5. Holding breath to activate the diving reflex.

Figure 6 shows a test subject watching a video and how that affects the measurements.

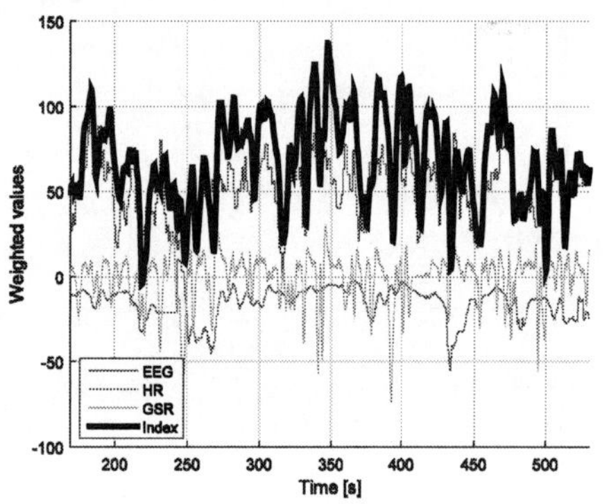

Fig. 6. Watching a video.

4. DISCUSSION

Figure 1 shows that a person is capable of affecting the relaxation index value by actively relaxing. The same can be seen in Figure 5 where the subject performed a task, which caused him to relax. While actively relaxing every parameter indicates that the person is becoming more relaxed and the index correlates with the change. These two figures also illustrate the index value to be around 100 while being in a clearly relaxed state.
Figures 3 and 4 show how different disturbances cause a momentary drop in the index value. This indicates that the index does react to changes in the person's state of mind. As the disturbances are short, the index returns to the original state quite quickly. Figures 3 and 4 also show that these short changes are mostly caused by GSR (K_2M_2). As mentioned earlier the GSR reacts quickly to changes in the mental state. These quick disturbances do not affect other parameters so clearly. This seems to also be the cause why the index returns back to the state before the disturbance.
The influence of active thinking and concentrating can be seen in figures 2 and 6. Both of the actions lower the relaxation index value clearly below 100. The video causes more variation as the GSR and HR (K_1M_1) vary. Reading a book keeps the values with considerably less variation. Both of these actions keep the steadier parameters EEG (K_3M_3) and HR in a stressing state. This active state keeps the index from rising.
These results show a clear link, between the relaxation index and a person's relaxation state.

5. REFERENCES

1. Arambula, P. & Peper, E. & Kawakami, M. & Gibney, K. H. The physiological correlates of Kundalini yoga meditation: A

study of a yoga master. Applied Psychophysiology and Biofeedback. 2001.
2. Crowley, K. & Sliney, A. & Pitt, I. & Murphy D. Evaluating a Brain-Computer Interface to Categorise Human Emotional Response. 10th IEEE International Conference on Advanced Learning Technologies. 2010.
3. Sun, F. & Kuo, C. & Cheng, H. & Buthpitiya, S. & Collins, P. & Griss, M. Activity-aware Mental Stress Detection Using Physiological Sensors. Mobicase. 2010. p. 211-230.
4. Szwoch, W. Using Physiological Signals for Emotion Recognition. In: Proc. 6th International Conference on Human Systems Interaction. 2013. p. 556–561.
5. Ikehara, C. S. & Crosby, M. E. Assessing Cognitive Load with Physiological Sensors. Proceedings of the 38th Hawaii International Conference on System Sciences. 2005.
6. Peuscher, J. Galvanic skin response. Version 3.0 November 2012. URL: http://www.tmsi.com/products/accessories?task=callelement&format=raw&item_id=43&element=fe0c95f3-af08-4719-bc51-36917715660d&method=download. (Referenced 20.12.2014)
7. Mihm, F. G. & Halperin, B. D. Noninvasive Detection of Profound Arterial Desaturations Using a Pulse Oximetry Device. Anesthesiology. 1985.
8. Bailey, B. & Ebel, B. & Galvin, S. & Grantham, J. & Little, S. & Morey, E. & Sankman, J. & Stemple. C. Development of a remote sensing pulse oximeter. May. 2010. URL: http://arizona.openrepository.com/arizona/bitstream/10150/146648/1/azu_etd_mr20100136_sip1_m.pdf (Referenced 20.12.2014)
9. Trans Cranial Technologies. 10/20 System Positioning Manual. Hong Kong. 2012. URL: http://www.trans-cranial.com/local/manuals/10_20_pos_man_v1_0_pdf.pdf (Referenced 19.12.2014).
10. Newandee, D. A. Measurement of the electroencephalogram (EEG) coherence, atmospheric noise, and Schumann resonances in group meditation. Biomedical Engineering Committee. 1996. p. 7, 15, 32. URL: http://archives.njit.edu/vol01/etd/1990s/1996/njit-etd1996-013/njit-etd1996-013.pdf (Referenced 19.12.2014).

6. CORRESPONDING ADDRESS

Panu Kiviluoma, D.Sc. (Tech.)
Aalto University School of Engineering
Department of Engineering Design and Production
P.O. Box 14100, 00076 Aalto, Finland
Phone: +358 9 470 23558
panu.kiviluoma@aalto.fi
http://edp.aalto.fi/en/

7. ADDITIONAL DATA ABOUT AUTHORS

Ahmad, Bilal, B.Sc. (Hons)
bilal.ahmad@aalto.fi

Aula, Janica
janica.aula@aalto.fi

Kokko, Olli
olli.kokko@aalto.fi

Korpijärvi, Jari
jari.korpijarvi@aalto.fi

Kuosmanen, Petri, D.Sc. (Tech), Professor
petri.kuosmanen@aalto.fi

Sepponen, Raimo, D.Sc. (Tech), Professor
raimo.sepponen@aalto.fi

Linnavuo, Matti
matti.linnavuo@aalto.fi

10th International DAAAM Baltic Conference
"INDUSTRIAL ENGINEERING"
12-13 May 2015, Tallinn, Estonia

DEVICE FOR WOOD FIBRE YARN BASED PRODUCT MANUFACTURING

Laitinen, H.; Vahtila, V.; Liski, T.; Komsi, P.; Klar, V.; Kiviluoma, P. & Kuosmanen, P.

Abstract: *The interest in new wood fibre based products in addition to paper has grown in the past decade. Successful implementation of continuous non-dissolving wood fibre yarn production raised interest to produce new textile-like structures directly from papermaking grade wood pulp. These new structures can be utilized in various fields ranging from agriculture and excavation to more sophisticated products. Compared to cotton production or regenerated fibres this yarn production method has a smaller ecological impact. This research describes device that can reliably produce conical structures out of biodegradable wood fibre. These structures could be utilized as, e.g., planting bags in agriculture. This method can be used as a basis for the development of other similar wood fibre products.*
Key words: Cellulose, wood fibre, pulp, nonwoven, textile, biodegradable

1. INTRODUCTION

The current production chain of natural fibre fabrics is a long and complicated process. The keystone of this process is yarn production which includes multiple steps beginning from the harvesting, preparation and spinning to the rolling that prepares yarn for the actual fabric manufacturing. The fabric production itself is also a multiphase process, usually performed by knitting or weaving. Using shorter and simpler processes to produce textiles could lead to lower production costs and more sustainable products [1].
The technical Research Centre of Finland (VTT) has developed a new wet spinning method to produce yarn directly from papermaking grade pulp without dissolution of the cellulose fibres [2]. The method has been researched thoroughly and pilot prototypes have been developed that are capable of continuous production of yarn [3]. The method has several advantages. It is both resource and cost-efficient since the main raw material, pulp, is both affordable and readily available. The process does not require harsh chemicals and is scalable due to its simplicity.
Use of native wood fibres as the basic element in different kind of fabrics can lead to a whole new area in forest industry and has commercial potential in various market-areas [4].
The main goal of this research was to investigate the possibility of a new technique of producing a textile-like structure based on the same chemical process as the wet spinning method. The research hypothesis was that different structures can be achieved by extruding the suspension directly into different shapes without an intermediate yarn production phase. As the wet yarn has a tendency to adhere to itself, larger structures can be produced.

The achieved wood fibre products could be categorized as wet-laid nonwovens. Nonwovens are widely used across different industries. The term "nonwoven" refers to fabric-like materials that are composed of fibres that are bonded together. The binding is in most cases achieved with chemical additives such as latex. Papermaking grade pulp can be used to produce wet- and air-laid nonwovens. The structure produced in this research is however, composed of individual yarn layers and has different properties than traditional nonwovens.
The goal was to produce a product that exhibits more characteristics of a traditional textile than nonwovens in general. This paper describes how the basic functions of the manufacturing process are discovered and how the technical solutions are implemented in the form of a prototype.

2. METHODS

The main interest of this research was to examine the production of wood fibre yarn based products. The process used is based on a wet spinning method developed at VTT. In this method a cross-linking reaction with alginate is used to create an initial yarn-like structure.
When alginate reacts with certain multi- or divalent cations it forms a stable and strong hydrogel. The reaction is sufficiently fast to form a supportive outer layer for the pulp in a fraction of a second. [[5]]
In this research sodium alginate and calcium chloride (Ca2+ cations) were used. The concentration of calcium chloride solution was kept at 15% throughout the different tests. This research examines a process where the yarn is extruded straight to the final shape of a textile.
The main raw material used is a wood fibre based suspension which has a total dry matter consistency of 1.5%. Of the dry matter 66.2% are wood fibres, 33.1% is sodium alginate and 0.7% is a rheological modifier. The rheological modifier decreases flocculation and improves the flow behaviour of the suspension. In this study an ultrahigh molecular weight anionic polyacrylamide (A-PAM), AN 923 PG0, was used. The pulp was refined with a valley beater to achieve an even distribution of the wood fibres in the suspension. [[5]]
To evaluate the approach of direct forming, simple nets were manufactured (Fig. 1). The nets were produced with a 3-axis, stepper motor driven machine. Samples of approximately 150 mm by 150 mm were produced. Additionally, different types of paper machine wires were compared to discover the best platform for the production process.

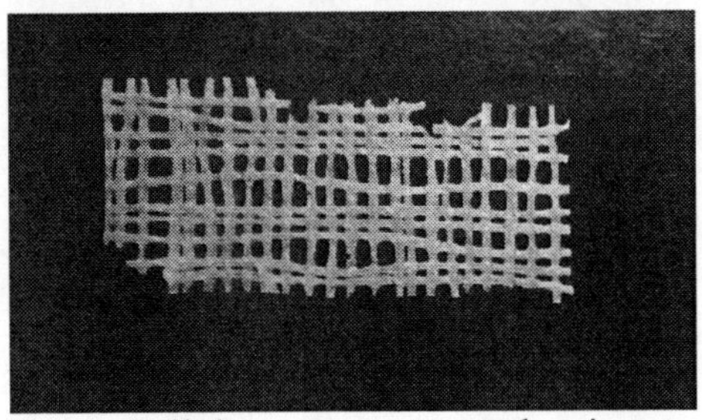

Fig. 1. Cellulose yarn net sample; size 70 x 25 mm

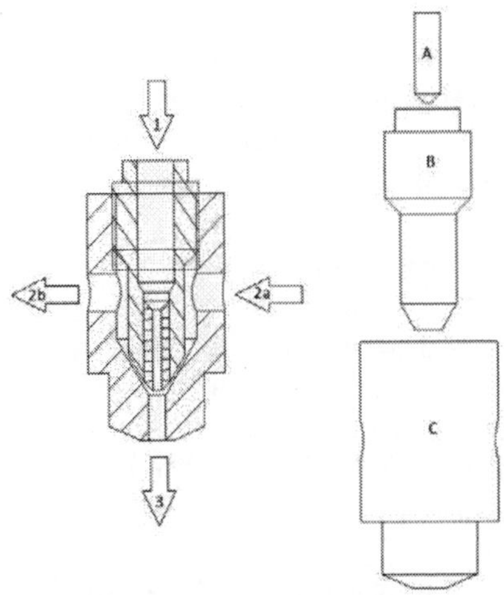

Fig. 2. Structure and flow directions of the extruder nozzle; 1 suspension, 2a $CaCl_2(aq)$, 2b unused $CaCl_2(aq)$, and 3 wet yarn.

To achieve the cross-linking reaction, the wood fibre suspension and the calcium chloride solution are fed through a nozzle. As the suspension and solution come in contact with each other a hydrogel is formed around the suspension's outer layer. Figure 2 presents the structure of the nozzle and the flows through it.

The yarn extrusion process has many critical parameters that need to be controlled, such as the flow rates and pressures of the wood fibre suspension and the calcium chloride solution. The diameter of the yarn is largely determined by the nozzle diameter and has a strong impact on the final structure. The diameter of the nozzle can be adjusted with interchangeable brass sleeves. The extrusion speed of the suspension and the linear motion of the nozzle have to be synchronized to produce good distribution of the cross-linked wet yarn.

It is a well-known phenomenon of fibre suspensions that they tend to form flocs. Flocs are tiny groups of fibres that are attached to each other that act as larger solids in the suspension. Flocculation needs to be taken into consideration during the extrusion. Due to the thixotropy of the fibre suspension, sufficient flow speeds are required to achieve a consistent flow. A syringe pump was used to pump the wood fibre suspension with precise control over the flow rate.

2.1 Design of the prototype

To test the possibilities of the production method further a prototype was developed. The prototype produces conical structures that can be used as planting bags in agriculture. The basic layout of the prototype is shown in figure 3.

The key component of the prototype is a rotating drum and the wire which is attached to it. Sparse stainless steel wires had produced the best results in previous tests and were therefore chosen. The stainless steel wire has a good resistance to the corrosive environment. To dewater the wet wood fibre structure the drum is rotated with a high angular velocity. The centrifugal force causes water to be expelled from the structure and holds the wet structure in place as it dries.

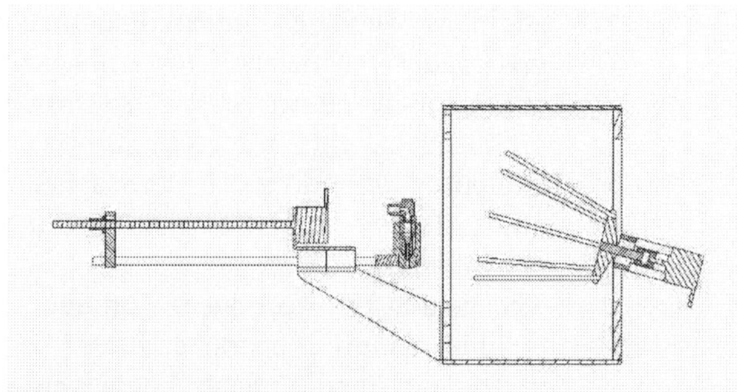

Fig. 3. Layout of the prototype

The nozzle is actuated along linear guides inside the drum and the yarn is extruded perpendicularly to the wire. By moving the nozzle back and forth a uniform structure is formed. The drum is enclosed in a polycarbonate frame. The frame acts as a reservoir for the calcium chloride solution which is expelled from the wood fibre structure.

The conical shape of the drum needs to be taken into consideration in the control of the linear movement of the nozzle. To achieve a uniform structure the flow speeds are kept at a constant but the linear velocity of the nozzle is increased as it enters further into the drum. The angular velocity of the drum is also decreased accordingly during the progression of the nozzle.

The designed nozzle, presented in Figure 2, allows precise control of flow and pressure of the calcium chloride solution and suspension. The flow of the calcium chloride solution can be controlled by adjusting the inner part of the nozzle (B) closer to the bottom of the nozzle frame (C). The pressure of calcium chloride solution can be controlled by resisting the return flow (2b). Furthermore, the yarn diameter can be determined by choosing a sleeve (A) with the desired nozzle

diameter. The calcium chloride solution is corrosive which is accounted for in the choice of materials.
The electro-mechanical implementation is based on three stepper motors, a DC vane pump and control electronics including motor drivers and a microcontroller. Stepper motors are used to rotate the drum, push the suspension out of the syringe pump and move the linear guides. A vane pump is used to pump calcium chloride solution to the nozzle. The prototype is controlled with a computer through a graphical user interface.

3. RESULTS AND CONCLUSIONS

The prototype was used to successfully produce conical wood fibre based structures. It was demonstrated that a 3-dimesional structure can be produced out of wood fibres with the alginate-based cross-linking reaction. The structures are uniform and can be consistently produced with the prototype.
However, in comparison with traditional knitted or woven textiles, the structure falls short in terms of strength and elasticity. Results indicate that this direct forming technique cannot be used to produce textile-like material suitable for, e.g., the clothing industry. The material does, however, have attributes which make it applicable in other fields such as agriculture.
Biodegradability was found to be one of the major assets of wood fibre yarn products during this research. The structures contain no harmful chemicals and can be left in nature with no harmful effects. Different speeds of decomposition can be attained by adjusting the physical dimensions of the structure.
The manufacturing process does not require any strong chemicals and is based on inexpensive raw materials. The bulk cost of alginate is high compared to the bulk cost of pulp and its use should therefore be minimized. The manufacturing process is simple and therefore scalable. The approach can be used to produce different products in a sustainable way.
The developed prototype for producing conical structures proved to be fully functional. The prototype was able to produce repeatedly uniform bags with different parameters. The stainless steel wire, which was chosen to be used as yarn extrusion surface, proved to have beneficial attributes related to drying process. Due to its high tolerance of heat, elevated temperatures can be used in the drying process. Higher temperatures decrease the drying time of the structure. Figures 4 and 5 present the final product that was produced with the prototype.

Fig. 4. Final product #1

Fig. 5. Final product #2

The final product fulfils all the requirements set for it; the structure is uniform and the yarns are bound together. After drying the product maintains its shape and can be used for the desired purpose. The height of the bag is approximately 11.5 cm, bottom diameter is approximately 10 cm and top diameter is 15 cm. Tests showed that a minimum of 8 layers are needed to produce bag that has sufficient wall thickness.

4. DISCUSSION

Even though the prototype built during this research was fully functional there are still some functions that require improvement. The linear speed of the extruder nozzle should be increased to improve the shape of the product.

With higher linear speeds, more net-like structures could be produced. With the current construction, the linear speed is set to its highest value but the yarns are not perpendicularly oriented with each other. The angular velocity of the drum cannot be decreased to remedy this issue since the structure folds and collapses on itself without an adequate centrifugal force holding it in place.

Use of calcium chloride as the cross-linking cation presents problems as well. After drying, calcium chloride remains on the surface of the structure. This may have a harmful effect when used in an agricultural context. The salinity of the final product is moderately low and can still be reduced by washing the structure with water. Nevertheless further investigations should be done to assess whether these steps are needed. To further improve the applicability in an agricultural context, the structures could be impregnated with nutrients by adding them to the suspension. The nutrients could be selected to suit specific plants.

During this research drying was performed manually, which proved problematic. In further development the drying process should be automated. The drying could be implemented with an even higher rotating speed of the drum. Hot air flow or even electrical heating elements could be used in drying. Compression can also be used to increase the efficiency of the drying process. Compression could be achieved by adding a roll that presses the net against to drum. Due to the high water content of the used suspension the drying is the most energy consuming stage of the manufacturing process. By improving the energy efficiency of the drying process the overall profitability of the whole product would increase.

Drying is also an issue regarding the scaling of the process. Currently the drying time constitutes a major part of the duration of the whole process. The manufacturing process of one plant bag takes approximately 10 minutes in total, with only the first three minutes consisting of extrusion.

5. REFERENCES

1. Hämmerle, F. M. The cellulose gap (The future of cellulose fibres). Lenzing - Nature Jeju Festival. Jeju, South Korea. June 2011.

2. WO2013034814, Method for the manufacture of fibrous yarn, fibrous yarn and use of the fibrous yarn, VTT, Jyväskylä, Finland, (Salmela, J. Kiiskinen, H. Oksanen, A.), WO2012FI50877, 10.9.2012. 14.03.2013, 18 p.

3. Koskinen, H.; Isomaa, T.; Lehto, J.; Stark, T.; Salmela, J.; Liukkonen, J.; Kiviluoma, P.; Widmaier, T. & Kuosmanen, P. DEVICE FOR CONTINUOUS CELLULOSE YARN DRYING AND FORMING. 9th International DAAAAM Baltic Conference "INDUSTRIAL ENGINEERING", 2009. (Cited 31.10.2014). Available: URL: "http://innomet.ttu.ee/daaam/proceedings/Design%20Engineering/Koskinen.pdf"

4. Finland to lead the way as a designer of cellulose-based products, VTT internet article, (Cited 22.4.2015). Available: URL: "http://www3.vtt.fi/news/2013/02102013_DDWoC.jsp?lang=en"

5. Klar, V. Development of wood fibre yarn machinery. Masters' Thesis. Aalto University, Espoo, 9.3.2015. [Unpublished work]

6. CORRESPONDING ADDRESS

Panu Kiviluoma, D.Sc. (Tech.), Senior University Lecturer
Aalto University School of Engineering
Department of Engineering Design and Production
P.O.Box 14100, 00076 Aalto, Finland
Phone: +358504338661
E-mail: panu.kiviluoma@aalto.fi
http://edp.aalto.fi/en/

7. ADDITIONAL DATA ABOUT AUTHORS

Laitinen, Heikki, B.Sc. (Tech)
E-mail: heikki.laitinen@aalto.fi

Vahtila, Vesa, B.Sc. (Tech)
E-mail: vesa.vahtila@aalto.fi

Liski, Timo, B.Sc. (Tech)
E-mail: timo.liski@aalto.fi

Komsi, Petri, B.Sc. (Tech)
E-mail: petri.komsi@aalto.fi

Klar, Ville, M.Sc. (Tech)
E-mail: ville.klar@gmail.com

Kuosmanen, Petri, D.Sc. (Tech.), Professor
Phone: +358 500 448 481
E-mail: petri.kuosmanen@aalto.fi

10th International DAAAM Baltic Conference
"INDUSTRIAL ENGINEERING -
12-13 May 2015, Tallinn, Estonia

OPERATING LEVERAGE AS A FACTOR OF PROMOTING INVESTMENT IN THE ENGINEERING INDUSTRY

Laskina, L.; Musalimov, V. & Musalimova, L.

Abstract: *Investment in new assets is the most common means of increasing a company's value. Further, improving the technical level of the company increases the proportion of non-current assets in the value of the total assets, resulting in an increase in fixed costs. This implies an increase in operating leverage. In order to evaluate operating leverage, the authors use the methods of mathematical statistics and regression analysis, and introduce a measure of evaluation which has not previously been used in the literature, namely depreciation-operating costs ratio. According to the regression analysis of time series, the degree of operating leverage in engineering averaged 0.835; when evaluated with the measure proposed by the authors it was 0.044. The latter can be interpreted to the effect that the average degree of depreciation in operating costs for the industry is 4.47%. Given that promoting investment in fixed assets increases enterprise value, we analyzed the effect of operating leverage on such value. In general, the industry demonstrates a positive correlation (correlation coefficient of 0.326), which means that an increase in operating leverage causes an increase in enterprise value.*

Key words: operating leverage, risk, investment, depreciation, firm value.

INTRODUCTION

The concept and measure of operating leverage – degree of operating leverage (DOL) – are widely used by experts in the field of finance. The DOL literature review, however, contains uncertainties and contradictions. Thus, J. Clark, M. Clark and P. Elgers say that "operating leverage measures the *sensitivity* (elasticity) of operating profit and changes directly with the share of fixed costs in the total cost." E. Solomon and J. Pringle make a similar statement: "The higher the ratio of the company's fixed operating costs to variable operating costs, the higher the operating leverage."

In the literature, there are two basic approaches to the evaluation of operating leverage: 1) regression analysis of time series; 2) multiplicative approach (point-to-point approach). Mandelker and Rhee (1984) pioneered the use of *regression analysis of time series* to evaluate the degree of operating leverage, followed by Ang and Peterson (1985) and DeYoung and Roland (2001). In this case, evaluating DOL requires using a relationship between earnings before interest and taxes (EBIT) and sales volume.

Mandelker and Rhee [1] undertook to empirically assess operating leverage with the following regression equation

$$\ln EBIT_t = D \ln S_t + A + \varepsilon \qquad (1)$$

where $EBIT_t$ – operating profit for the period t;

S_t – revenue from sales for the period t;

D – regression coefficient representing a degree of operating leverage;

A – regression coefficient.

Following the *multiplicative approach*, the degree of operating leverage (DOL) can be evaluated with a variety of measures: 1) as the ratio of the change in operating profit (per cent) to the change in sales volume (per cent), 2) various methods to measure, proxy-variables (static proxy measures): the ratio of fixed assets to total assets [1,2], or as the ratio of depreciation to sales revenue, or the ratio of depreciation to assets [3]. There are other measures of the degree of operating leverage. For example, Novy-Marx (2007) defines the degree of operating leverage as the ratio of operating costs (cost of goods sold plus selling, general and administrative expenses) to total assets.

The most *demonstrative* measure of operating leverage is the "proportion of fixed costs in the total operating costs". In this case, operating leverage is interpreted as a factor of variability in financial results. This measure reflects the effect of internal factors inherent in the company. Raising the technical level of the company will lead to an increase in the proportion of non-current assets in the value of the total assets, resulting in an increase in fixed costs. This in turn will increase operating leverage. The positive effect of operating leverage is manifested in the fact that raising the technical level of the company is profitable, as it provides additional income. However, there is an adverse effect expressed in the fact that investment in fixed assets may not be justified.

A more complex measure of the degree of operating leverage, which evaluates the impact of both internal and external factors on profit variability, is the so-called "classical model". We can identify such external factors as market factors associated with price fluctuations, volatility of demand for products, growth of prices for material resources, etc. The above parameters tend to be poorly controlled by the company. In this case, "operating leverage" is regarded as the coefficient of elasticity which allows evaluating the sensitivity of the firm's operating profit to changes in sales volumes:

$$DOL_r = \frac{T_{EBIT}}{T_S}, \qquad (2)$$

where T_{EBIT} – operating profit growth rate, %;

T_s – sales revenue growth rate, %.

The significance of operating leverage is particularly evident in resource-intensive industries. Examples are oil producers (regardless of the degree of utilization of a well, the maintenance of the well requires significant costs), metallurgy, railway, etc. Such companies need intensive investment in fixed assets. This generates high fixed costs and a significant proportion of fixed costs in the total cost. For example, companies with low operating leverage may include trading companies with lower fixed costs [4].

The correlation between DOL and the above "proxies", as well as the names of scientists suggesting "approximation", are presented chronologically in Table 1[1].

Ratio	Author, year	DOL correlation
Assets / Sales revenue	B. Rosenberg, W. McKibbon (1973)	– 0.059
Fixed assets / Assets	B. Rosenberg, W. McKibbon (1973) M. Ferri and W. Jones (1979) G. Mandelker and S. Rhee (1984)	– 0.043
Depreciation / Sales revenue	T.J. O'Brien and P.A.Vanderheiden (1987)	– 0.045
Depreciation / Assets	T.J. O'Brien and P.A.Vanderheiden (1987)	– 0.028

Table 1. Experimentally-obtained approximate ratios offered as DOL.

As follows from Table 1, none of the alternative parameters is related to the degree of operating leverage (correlation coefficient of about 0), which indicates that the "proxies" are not effective and may not be used as "substitutes" for DOL. Let's discuss the possible reasons for the lack of such correlation. For example, two firms with the same proportion of fixed assets in total assets may use different volumes of labor (variable costs), which would change the proportion of fixed costs. Another example is service companies with relatively small investment in fixed assets (consulting, auditing companies),

[1] The correlation coefficient was calculated by T.J. O'Brien and P.A. Vanderheiden on the basis of a sample of 62 companies.

which may have a relatively large proportion of current assets and relatively high costs in the form of salaries and bonuses to the qualified personnel and management. Such company may have a high DOL and a small proportion of fixed assets in the value of the total assets.
Building the company's technical capacity involves significant investment. Investing in new assets is the most common means of increasing enterprise value. An important consequence of the acquisition of assets must be the generation of extra income, increased cash flow or improved efficiency of the company. If this is achieved, the value of the company increases. So, we can say that operating leverage is one of the factors of the enterprise value creation.
A variety of approaches is used to estimate enterprise value. There are three basic approaches: income, cost and comparative (market) approaches. Within the market approach, estimating enterprise value involves different multipliers. The best-accepted indicators are based on the relationship of the share price, market capitalization, or the value of business and a particular index in the denominator (revenues, income, or book value of assets). If we know the value of the multiplier, we are able to determine the enterprise value by multiplying it by the corresponding index in the denominator.
In this paper, the enterprise value EV was determined by multiplying the value of the multiplier EV/S by sales revenue. Theoretically, using the multiplier EV/S is the most correct, since sales revenue is a source of income for shareholders and creditors, as well as a source of taxation. Moreover, the experts point out that one of the main advantages of the multiplier EV/S is that revenue is practically the only index to compare companies, including those using different accounting standards.
In this paper, we study the sample of Russian issuers in the engineering industry. Engineering became one of the leaders in manufacturing in terms of growth of production in 2014. Note a pretty good result achieved against the background of the investment slowdown in the country. Of the twelve sub-sectors into which the Russian engineering was divided by the Russian Federal State Statistics Service, production growth was observed in only two: "Production of electronic components, equipment for radio, television and communication" and "Production of ships, aircraft and spacecraft and other vehicles". The most significant contribution to the overall dynamics was made by the second sector where production growth in the reporting period was 28.5%, being the highest figure not only in engineering, but also in the entire industry. It should be taken into account that this sub-sector is the largest in the engineering industry in terms of production. According to Rosstat, its share in the total volume of engineering production as of the end of 2013 amounted to almost 29%. The sector includes railway engineering, shipbuilding and aerospace engineering. From the fact that a decline in production in railway engineering was registered in January-September, it can be concluded that the main growth was provided by shipbuilders and aircraft manufacturers. According to RIA Rating experts, such result was largely due to the implementation of a defense order. The second sector which showed an increase in production in the reporting period was the "production of electronic components, equipment for radio, television and communication". In this case, production growth was 15,3% compared to January-September 2013. This could be partly explained by import substitution, but since the share of this sub-sector in the total engineering production is comparatively low – about 5%, the contribution of this effect is negligible. It is also possible that communication producers implemented the orders of the Ministry of Defense.
Typically, the production dynamics of the Russian engineering industry correlates highly with the dynamics of investment. Of all the industries, engineering is the most sensitive to changes in the investment climate. While during previous challenging periods a decline in investment meant a simultaneous or even premature decline in engineering production, in 2014 this relation was violated. According

to official statistics, investments in fixed assets are decreasing, while engineering production volumes are growing, and this growth has been almost continuously gathered pace.

Enhancing investment results in the growth of the share of fixed assets in the value of the company's assets, which causes an increase in fixed costs (in the form of depreciation charges). It is known that depreciation is an integral part of fixed costs, and the amount of depreciation charges is indicated in a separate line in the profit and loss account drawn up in accordance with IFRS. So, we consider it reasonable to use an additional measure to evaluate operating leverage, which has not previously been applied in the literature – depreciation-operating costs ratio. To evaluate operating leverage, the authors also used the methods of mathematical statistics and regression analysis. Two random variables – operating profit and sales revenue – are the subject of our study in terms of establishing their correlation. A preliminary analysis of the corresponding dependency diagrams showed that they are characterized by pronounced exponential nonlinearity. We decided to take the logarithm of the sample data. Then we used MATLAB technology to develop a model of linear dependence of the sample data transformed. In our study, we took the opportunity of using the random variables where they appeared in a standard form with the parameters m and σ^2:

$$X=m_x+\sigma_x\varepsilon(\mu,\sigma), \qquad (3)$$

where $\varepsilon(\mu,\sigma)$ is random perturbation.

For simulation modeling, the values of $\mu = 0$ and σ as appropriate are normally used, although in terms of the "three sigma rule" $\sigma = 3$ is more preferable. At such values, random perturbation is called "white noise". Regression equations are similar to the model proposed by Mandelker and Rhee

$$\ln EBIT_t = D \ln S_t + A + \varepsilon(\mu,\sigma) \quad (4)$$

where $EBIT_t$ – operating profit for the period t;

S_t – revenue from sales for the period t;

D – regression coefficient representing a degree of operating leverage (DOL);

A – regression coefficient determining the average value of profit at the beginning of the review period;

$\varepsilon(\mu,\sigma)$ – random error.

According to the method, the degree of operating leverage in engineering averaged 0,835 on the basis of regression analysis of time series; in accordance with the measure proposed by the authors, it was 0,044. The latter can be interpreted to the effect that the average share of depreciation in operating costs in the industry is 4.47%. The practicability of the DOL measure proposed by the authors can be determined by evaluating the relationship between the values of DOL_1 calculated using regression analysis of time series and DOL_2 estimated using our measure of evaluation "Depreciation/ Operating costs". The correlation coefficient between the two DOL measures was estimated at 0,611, indicating that this measure can be used as an estimate of DOL.

Taking into account that enhancing investment in fixed assets causes the growth of enterprise value, we analyzed the effect of operating leverage on enterprise value. The correlation was established between the degree of operating leverage calculated with the measure proposed by the authors (DOL_2) and enterprise value calculated on the basis of the multiplier EV/S. The relationship between the indicators is calculated as follows

$$\rho_{xy}=COV_{xy}/(D_xD_y)^{1/2}, \qquad (5)$$

where D_x, D_y – sample variances;

COV_{xy} – covariance between X and Y.

In general, the industry shows a positive correlation – the correlation coefficient was 0,326, which means that technical capacity building increases enterprise value.

We note finally that operating leverage is an indicator used in financial management primarily as a characteristic of operational risk. An enterprise with a relatively high degree of operating leverage is at the same base volumes of sales considered more risky in terms of operational risk, and in this case the company runs the risk to the extent that at best it will make a lower-than-expected profit, at worst - will not be able to cover its own production costs.

REFERENCES

1. Mandelker, G. N., Rhee, S. C. *The Impact of Degrees of Operating and Financial Leverage on Systematic Risk of Common Stock.* Journal of Financial and Quantitative Analysis, 1984, 19, 45–57.
2. *Ferri, M. G., Jones, W. H.* Determinants of Financial Structure: a New Methodological Approach. The Journal of Finance, 1979, 34. June, 215-226.
3. O'Brien, T. J., Vanderheiden P. A. *Empirical Measurement of Operating Leverage for Growing Firms.* Financial Management, 1987, 16, 45-53.
4. Brigham, E., Gapenski, L. *Financial Management: A Full Course:* in 2 vol. / Translated from English and edited by V.V. Kovalev. – St. Petersburg: Ekonomicheskaya Shkola, 1997.

ADDITIONAL DATA

1) Lyubov Laskina, Victor Musalimov, Liudmila Musalimova,

2) Operating leverage as a factor of promoting investment in the engineering industry

1. Lyubov Laskina, *cand. econ. sci., associate professor,* ITMO University, 191002, Russia, St. Petersburg, Lomonosov str., 9
+7-952-235-32-26., risk05@mail.ru

2. Victor Musalimov
Professor Department of Mechatronics, ITMO University,197101,St.Pb, Kronverkskiy,49 .tel +79214217917, musvm@ya.ru

3. Liudmila Musalimova
Corresponding Author,
Professor of ITMO University, 197101,St.Pb,
Kronverkskiy,49.tel +79214217917
musvm@ya.ru

10th International DAAAM Baltic Conference
"INDUSTRIAL ENGINEERING
12-13 May 2015, Tallinn, Estonia

SHIP MODEL BASIN CARRIAGE CONTROL SYSTEM

Liyanage, D.C., Aasmäe, E., Sutt, K.O., Tamre, M. Hiiemaa, M.

Abstract: *Modern ship model basin towing carriage control systems demand advance features which are not possible with ordinary motion control systems. When it comes to relatively shorter length towing tanks which are designed to use with belt driven carriages the control system development is more challenging task. This paper concerns about the design of towing carriage control system for ship model towing tank built in Kuressaare College of Tallinn University of Technology, Saaremaa Estonia. The following study covered requirement analysis, system design, control equipment selection and concept validation of the proposed towing carriage control system. And also it includes proposals for future tasks regarding robust positioning system for belt-driven servo mechanism.*

Key words: Ship model basin, towing tank, towing carriage control

1.INTRODUCTION

There are number of ship model tow tanks that are used by various research institutes to investigate hydrodynamic performance of ship hull designs. Majority of them are longer than 100m [1] lengthwise and are capable of different simulations from calm water resistance tests to vessel maneuvering characteristics predictions. Most of the towing carriages are powered by direct drive mechanisms. Due to longer length of the tank it is possible to achieve higher velocity with enough acceleration distance.

The size of the basins used for ship model testing varies greatly depending on the tests desired. A basin with bigger dimensions can be used for deep water testing or maneuvering testing. The paddles used in the basin also define its use. In modern testing facilities two kinds of paddles are used. Flap paddles that are mounted to the bottom of the tank for generating deep water waves where there is negligible motion at the bottom and the orbital particle motion decays with depth. And piston paddles that for generating shallow water motion. Neither of these types of paddles achieve the motion that accords to real waves. For that a combination of paddle types are used.

In our case calm water resistance tests and head or following sea resistance tests were required. For that the measurements of the pool are 60 m x 6m. and flap paddles are used for wave generation. After receiving the instructions from the client and getting a final overview of the objective a clear plan of action was made. Firstly it was necessary to select suitable equipment for the carriage control system. As a next step we had to work out the concept of the human machine interface and the graphical user interface for the system.

2. SYSTEM DESIGN

System design was started with requirement analysis of the ship model basin. Then the design concept has been developed. After that the equipment selection, field bus selection and concept validation were carried out.

2.1 REQUIREMENT ANALYSIS

The requirements for the carriage movement parameters and the motor selections were set by the client in the documentation presented to us. Based on that information ..

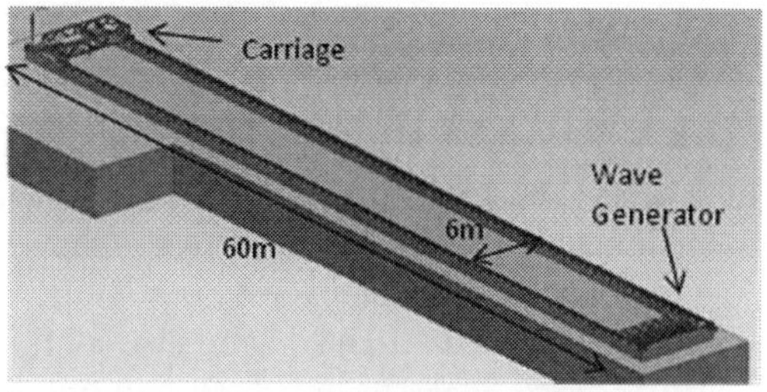

Fig 1. Dimensions of the ship model basin

The requirements given for the carriage movements showed the maximum speed of the model to be at least 5.5 m/s and the acceleration distance nor the deceleration distance to exceed 6 m. The required level of speed has to be kept with the accuracy of ± 0.005 m/s. Also the accuracy tolerance of the carriage trajectory must not exceed ± 01 mm/m in relation to water surface.

2.1 SYSTEM CONCEPT

The towing carriage has been designed to use belt drives to pull the from both sides. So that drive control system comprised of two synchronous servo motors to drive the carriage.

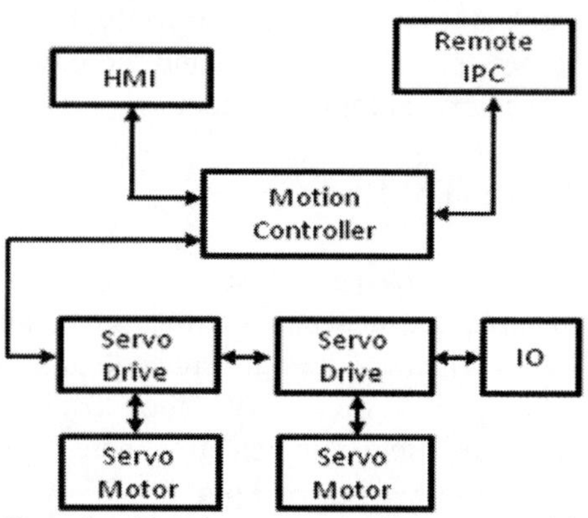

Fig 2. Carriage control system block diagram

According to the conceptual diagram Fig 2, the motion controller is controlling two servo drives and IO modules. And also HMI and remote IPC connected to the same motion controller.

The motion profile, which governs the carriage movement will be built desktop application running in the IPC. This profile defines the position of the carriage, velocity, acceleration or deceleration values. And the status information and basic system controls are presented in HMI.

Line topology used in servo drives and IO modules sub network. Since multiple drives driving the carriage, they are need to synchronize with fast industrial fieldbus.

When it comes to locating the components, motion controllers need to be installed in the operator room area which is starting side of the pool and servo drives and IO modules need to place on the other end which is close to servo motors and sensors. This concept makes the controllers and drives apart more than 60m in distance.

2.2 CONTROLLER SELECTION

It was proposed that Lenze synchronous servo motors and servo drives would be most suitable for the application. This selection imposed constraint to select suitable controller by maintaining interoperability. Other than that, there were several key requirements to fulfill by the controller which are listed below.

- Modular system - To be expand based on future requirements.
- Multiple motion axes synchronization
- MATLAB/Simulink integration
- Communication interface to desktop application software development (Visual C#, VB .etc.)
- Multiple programming language support
- Web server availability for remote data visualization.
- Multiple fieldbus options availability - Interoperability with other devices.

Based on above mentioned requirements, several motion control systems has been investigated. As a result Beckhoff CX2030 series controller has been selected. It is an embedded PC which allows to connect up to 20 servo axes and connect many IO modules in different network configurations.

2.3 SELECTION OF FIELD BUS

Choose a suitable field bus protocol was influenced by the high speed data communication requirement of the project. Since each servo motor will be driven by individual servo drives respectively, the time delay between velocity commands for each axis should be less than 20ms. Fig 3 shows the potential failure by de-railing the carriage in case of delay in velocity command execution. Below are the list of requirements imposed by the application to select suitable field bus.

- Fast data rate to communicate messages between synchronous axes.
- Synchronized clocks to trigger motion in order to minimize lag.
- Interoperability - Open standard protocol which is compatible with multiple vendor motion control devices.
- Topology - There is a need to have several network topologies in the system.
- Distance of the communication bus more than 60m.
- Ethernet based communication protocol.

EtherCAT which is based on ethernet protocol, and known to be fastest fieldbus in the market, priority dependant, deterministic communication and multiple topology support are fulfill all requirements of the application[2]. Therefore EtherCAT protocol was selected as the field bus for the towing carriage motion control application.

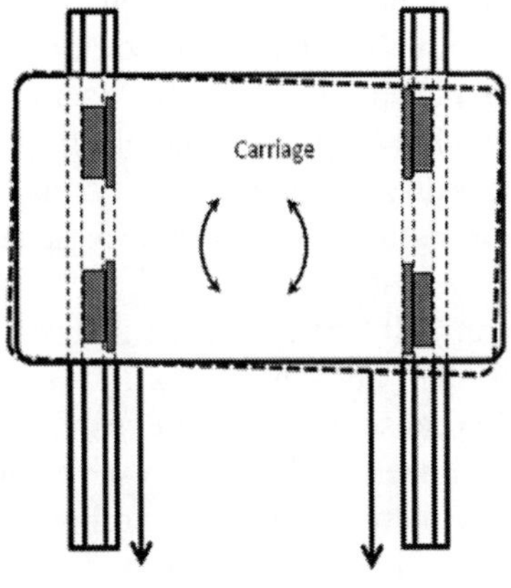

Fig 3. Carriage movement in case of belt tension difference.

2.4 CONCEPT VALIDATION

After selecting the controllers, IO modules and interface devices, it was investigated the functionality of different motion concepts and different software integration techniques in a simulated environment. This was done in order to check whether the controller can be used for the carriage control system application.

By using the built in motion and communication functions of the controller software development tools (TwinCAT 3) several test programs has been done. They were to synchronize two motors and move them according to different motion profiles which is similar to the end user requirements. And also communicate with desktop applications written in Visual C#.

3. DISCUSSION, FUTURE WORK

During the design of the system it has been completed system design, component selection and testing various motion functions available with the controller software development kit (TwinCAT 3). This can ensure the functionality of the system. But still there are technical challenges imposed by mechanics of the system.

Since the towing carriage uses tooth belt driven servomechanism to pull it from both sides, this can be induce vibrations while it is in motion with different motion profiles. There are several contributory factors to vibrations. One of them is the elasticity of the belt drive, which demands to incorporate robust position control scheme overcome belt-stretch issues. In a similar study about belt drive systems it has been identified that non-linear friction of the system also cause positioning errors[3].And also gear reducers add backlash which may result in vibrations during motion. The resonance in the system also accumulate positioning errors [4]. Therefore all these cases demands advance control schemes for positioning.

As part of the study it was investigated possible control solutions to overcome above mentioned positioning errors. There are some studies regarding accurate positioning of servo mechanisms using sliding mode control theories[5]. And also in a later study same authors has incorporated sliding mode control in conjunction with asymptotic disturbance observer[6]. And similar accurate positioning efforts has made by other authors using fuzzy-logic control systems[7].

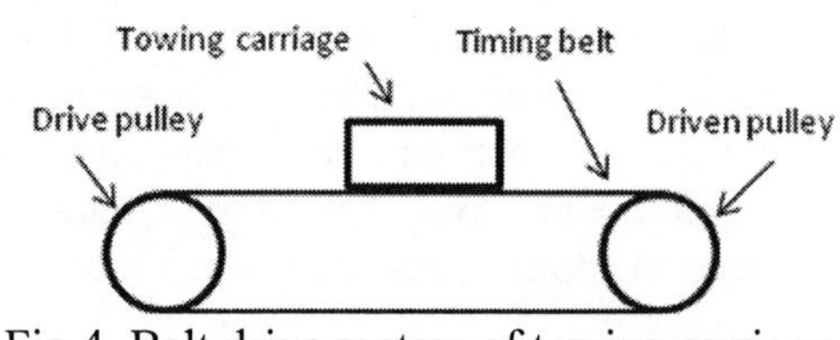

Fig 4. Belt drive system of towing carriage.

During controller selection, it was considered the possibilities of including MATLAB model into the motion controller. This is to try out different motion control models and optimize it.

4. CONCLUSION

The development of control system is currently underway to its completion. At present the conceptual system design has been done according to the requirements of the facility. Based on evaluation of the motion functions, Visual C# application integration it has been proven that the controller is capable of fulfilling demanding control requirements.

5. REFERENCES

[1].http://ittc.info/downloads/Catalogue%20of%20facilities/Index/index.pdf, retrieved on 9-May-2015.

[2].EtherCAT technology group, EtherCAT - the Ethernet Fieldbus, Nov-2014.

[3].Jokinen,M.; Saarakkala, S.; Niemela, M.; Pollanen, R.; Pyrhonen, J.; Physical Drawbacks of Linear High-Speed Tooth Belt-Drives, *SPEEDAM 2008 IEEE Conf*, June-2008, *pp 872-877.*

[4].Jayawardena, T.S.S.; Nakamura, M.; Goto, S.; Accurate Control Position of Belt Drives Under Acceleration and Velocity Constraints, *Int'l Jnl of Automation & Systems,* 2003, **1**, *pp 474-483.*

[5].Hace, A.; Jezernik, K.; Sabanovic, A.; A new robust position control algorithm for a linear belt-drive, *Proceedings of IEEE Int'l Conf Mechatronics*, June-2004, *pp 358-363.*

[6].Hace, A.; Jezernik, K.; Sabanovic, A.; SMC with Disturbance Observer for a Linear Belt-Drive, *IEEE*, Dec-2007, **54**, *pp 3402 - 3412.*

[7].Kulkarni,S.A.; El-Sharkawi, M.A.; Intelligent Position Control of Elastic Drive Systems, *IEEE Energy Conv*, Mar-2001, **16**, *pp 26-31.*

6. CORRESPONDING ADDRESS

Dhanushka Chamara Liyanage

Department of Mechatronics.

Tallinn University of Technology,

Ehitajate Tee 5,

Tallinn 19086, Estonia.

Phone: 372+620 3269,

E-mail: liyanagedc@gmail.com
http://innomet.ttu.ee/daaam

7. ADDITIONAL DATA ABOUT AUTHORS

1. Dhanushka Chamara Liyanage , MSc Student
+372 5636 2345
liyanagedc@gmail.com

2. Eva Aasmae, MSc Student
+372 5197 4506
eva.aasmae@gmail.com

3. Kristofer Ott Sutt, MSc Student
+372 5805 3789
kristofer.o.sutt@gmail.com

4. Mart Tamre, PhD, professor
+372 6203202
Mart.tamre@ttu.ee

5. Maido Hiiemaa, PhD, researcher
Maido.hiiemaa@ttu.ee

10th International DAAAM Baltic Conference
"INDUSTRIAL ENGINEERING"
12-13May 2015, Tallinn, Estonia

DEVELOPMENT AND DESIGN OF A CYLINDRICAL 3D PRINTER

Munigala, V.; Kemppainen, M.; Kärki, P.; Tudose, C.; Kiviluoma, P. & Kuosmanen, P.

Abstract: *3D printers offer many possibilities in manufacturing and prototyping. While being able to print complex structures, the most common Cartesian 3D printers lack the ability to print mechanically durable rotationally symmetric parts, such as pipes with fine surface quality. In this study a solution for this problem is presented by designing and building a 3D printer where traditional planar printing surface is replaced by a rotating cylindrical surface (mandrel). Challenges brought by this change included proper alignment of the axes, removal of printed pipe from the mandrel and designing a proper structure for the printer. Consideration was also required in programming the printer and printing different patterns for each printed layer. The resulting pipes were strong and they had better surface quality than similar Cartesian printers.*

Key words: non-cartesian, cylindrical, 3D printer

1. INTRODUCTION

There is a lot of excitement surrounding 3D printing, and the possibilities of 3D printing are huge. 3D printers are already used for a wide variety of tasks, ranging from prototypes to custom parts. While the technology is already in use, there is potential for further development. Many say that 3D printing will become even more widely adopted as the technology develops further.

While current 3D printers vary in both price and print quality, the more widely used cheap Cartesian 3D printers on the market have trouble printing plastic pipes. Currently, the pipes are printed in horizontal layers, one by one. As the number or layers increases, so does the probability of the structure collapsing on itself, especially if the walls are thin. Furthermore, horizontal layers are not an ideal structure for handling mechanical stress. A cylindrical 3D printer would avoid this problem by printing on a rotating mandrel. In this case, the print layers are along the walls of the pipe, this allows the 3D printer to print a mesh structure, which should be more durable than a traditionally printed equivalent structure.

There are articles and patents that might be useful in designing the printer [1–4]. Among the patents were several patents relating to mandrels (the pipe, which the part will be printed on), and how to remove the work piece intact. Removing the finished 3D printed pipe might turn out to be problematic.

There is also a patent for an inflatable mandrel. During printing it's inflated and when the printing is done, the mandrel can be deflated, allowing for easy removal of the finished product. While this kind of mandrel might not seem to be an accurate enough base for printing, it turns out that by applying sufficient pressure and then machining the mandrel, it conforms to tolerances very well. There was also a pneumatic release system for this mandrel, where compressed air is conducted into a sheath which then expands, locking the mandrel in place [2]. While useful, this kind of system might prove too complicated to be viable, as the mandrel is complicated to manufacture and it would need a pneumatics system for the mandrel release. There was also a separate article for a mechanical expandable and collapsible mandrel. While it too is relevant to this project, again, it is very difficult to manufacture since it consists of so many interlocking tightly fit parts.

There exists a patent for a cylindrical 3D printer [3]. However the patent is very general and doesn't cover any specific design. One working cylindrical 3D printer prototype has been made [4]. It's based on fused deposition modelling. This is similar to the aim of this research, with some exceptions. This research will be focused more on printing pipe structures. Nevertheless, since a cylindrical 3D printer like this has already been patented [3], there may not be commercial opportunities for it.

2. DESIGN AND METHODS

2.1 General layout

In comparison to a normal Cartesian 3D printer which has x, y and z movement axes, cylindrical 3D printer has x, θ and z movement axes. These three movement axes are the minimum number of axes required for printing a multi-layered pipe. In addition during preliminary tests it was noticed that manual y-axis adjustment could be useful, since location of the extruder along the mandrel surface can affect printing results.

Several ideas of how to implement printer's movement axes were considered. After consideration following design was chosen: x-axis is next to the mandrels rotational axis and parallel to it. Along the x-axis moves a slide carrier which houses the height movement mechanism for z-axis. Orientation of these axes along with the finished design of the printer is shown in Fig. 1.

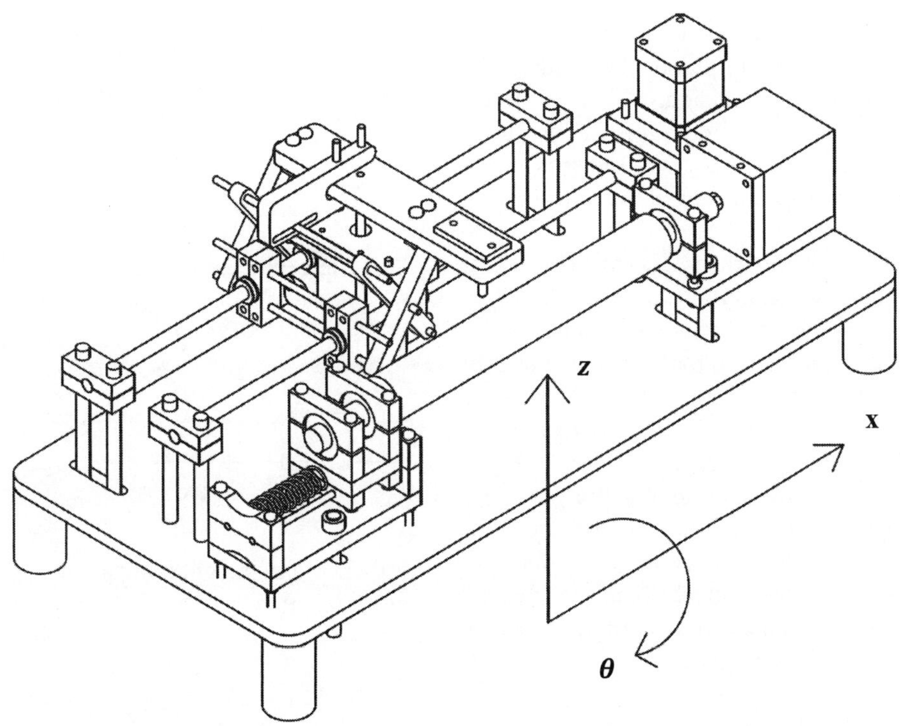

Fig. 1. The layout of the printer.

2.2 The Mandrel System

When designing the mandrel rotation mechanism, several issues had to be taken into consideration: possibility to adjust the alignment of the mandrel, torque and the resolution of the motor and easiness of removing the mandrel.

The mandrel is supported by two aluminium shafts, which are supported by bearings. In order to eliminate the torsion created by the weight of the mandrel, a motor supports the other end and the other shaft is supported by two rotational bearings.

The adjusting feature is implemented by using threaded rods to support the mandrel mechanism. The threaded rods will allow the height adjustment for the mandrels both ends. The rods will be attached into the baseplate, which has grooves that allow moving the mechanism along y-axis.

The used motor is a NEMA23 sized stepper motor, which has 200 steps per revolution (1.8 degrees per step). With micro stepping, this is considered to be precise enough, therefore no gear mechanism is required to widen the resolution of precision. With ¼ micro stepping, this configuration allows for a resolution of 0.24 millimetres. The resolution can be varied by altering the micro stepping, but the goal was to achieve resolution better than the extruder nozzle diameter, which in this case is 0.4 mm.

An easy mandrel removal mechanism is required for calibration. This is done by installing the left support mechanism on rails, thus it can be moved. The spring pushes the left side of the support mechanism against the mandrel, creating enough tension to hold the mandrel still. When removing the mandrel, the left side of the support mechanism is pushed manually against the spring, and the mandrel will be released.

2.3 X-Axis Movement

The x-axis movement is done with a timing belt. A timing belt, while simple and cost-efficient, also offers sufficient accuracy. The stepper motor controlling the x-axis movement is NEMA 17. With micro stepping this motor gives sufficient accuracy and torque. With ¼ micro stepping, this provides a resolution of 0.08 millimetres which is more than enough.

2.4 Z-Axis Movement

For accurate control the x-axis mechanism requires that the z-axis movement mechanism should be as light as possible to minimize inertia. The solution used in the printer is based on a scissor lift mechanism. This mechanism, while relatively lightweight, offers good accuracy and a good movement range. The mechanism with this design is shown in Fig. 2. In this design the slide carrier consists of two halves. There is a stepper motor in the other half and it controls the height of the extruder with a lead screw. The timing belt is attached to this half so it stays in the same x-axis position during the height movement. The other half moves along the rails of the x-axis when the height changes because of the scissor mechanism, as shown in Fig. 3. Benefits of this design are lightness and simple radial loads to linear bearings.

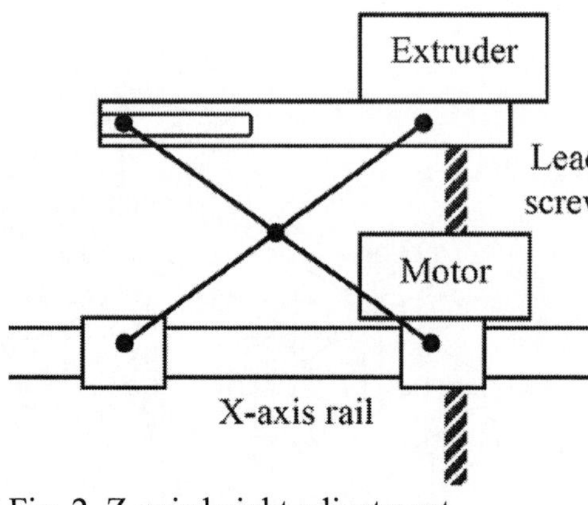

Fig. 2. Z-axis height adjustment mechanism.

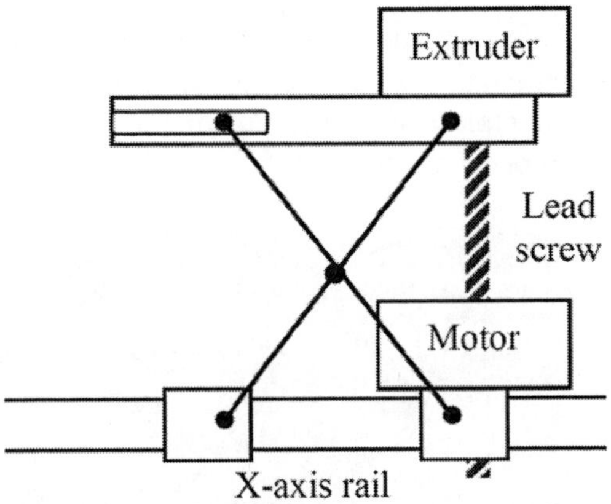

Fig. 3. Z-axis height adjustment mechanism raised.

The stepper motor controlling the height movement isn't a standard stepper motor but its size is close to a NEMA 14 sized motor. It has a special hollow shaft with internal threads. When the rotation of the lead screw inserted through the motor is prevented the screw will move linearly and since it is attached to the part which holds the extruder, the extruder will change its height. Advantage of this compared to a normal stepper motor with a lead screw coupled to its shaft is compactness and lightness: much less space and parts are needed between the motor and the object attached to the screw.

2.5 Extruder

When the idea of printing on a cylindrical surface was tested using traditional 3D printer, it was found out that extruder jamming was a real problem. While reviews were scarce, Wanhao MK9 was the best received, therefore this extruder was chosen for the printer. It comes with a standard stepper motor.

2.6 Electronics and Control System

The hardware for electronics and control system is based on an open source RepRap model. The stepper motors are controlled by Arduino Mega microcontroller with RAMPS (RepRap Mega Pololu Shield). The stepper drivers to power the stepper motors are based on Allegro A4982 Ice Blue Stepsticks. The RAMPS is powered by an external power supply with 12 VDC and 5 A. The advantage of the RAMPS is modularity which allows the stepper drivers to be replaced or upgraded easily in the case of a breakdown.

2.7 Programming

For programming the printer, two options were considered: use of existing open source software like Sprinter, Pronterface and Slicer as they were. Sprinter is the software controlling the microcontroller. Pronterface sends commands to the microcontroller from user's PC. Slicer is used to generate G-code from 3D models. Another option was to modify these programs so that they would be suitable for the axis layout of the printer. However modifying the printer software would have required modifying all the programs, so it was simpler to use existing software, as it is and take the axis layout of the printer into account when creating individual G-codes of the work pieces.

All G-code was coded manually. Since every move the printer makes has to be coded separately, the coding required numerous of iterations, especially since there were several parameters which required adjusting. These parameters include extruder speed, and feed rates.

3. RESULTS

First printed pieces had trouble sticking to the mandrel. Solution for this was at the beginning of the printing process print a short thick starting line along the x-axis. The line stuck well to the surface and prevented the pieces from detaching from the mandrel during the printing process.

During the first pieces it was also noticed that if the extruding rate isn't high enough compared to the movement speed of axes the material, the material winded along the mandrel like a string, instead of sticking to the mandrel or other layers. This was resolved by increasing the extruding rate.

Two types of pipes were printed. The difference was in the surface patterns. In the first type two different patterns alternated between the layers and in the

second type three different patterns were used. The first layer in both types was always a spiral layer, which was simply a line printed helically along perimeter of the mandrel.
In the first type of pipe the second layer consisted of adjacent lines along the rotational axis. These lines were in an approximate 90° angle to previous layer as shown in Fig. 4. The number of layers was always uneven and varied between three to nine layers so that the last layer could always be a spiral. Reason for this was that the spiral layers had better surface quality than the line layers.

Fig. 4. Layer patterns in the first pipe type.

In the second type the first spiral layer was much more spacious than in the first type. The reason for this was to make pipe easier to remove from the mandrel. Second layer was consisted of adjacent lines printed in a 45° angle to the rotational axis. In the third layer the lines were perpendicular to the previous layer as in Fig. 5. Only three layered pipes were printed with this technique.
The surface quality in the first type of pipe was finer than in the second. Generally the spiral layers had the finest and smoothest surface quality. However the quality varied according to the printing parameters: mainly how dense the spiral was. For these reasons most printed pipes had spiral layer as the inner and outer surface. These kinds of pipes were also very rigid. The five layered pipe was able to withstand a torque of 65 Nm. Same type of pipes with even more layers were also printed but their strength wasn't tested due to lack of suitable testing equipment for this size of pipe.

Fig. 5. Layer patterns in the second pipe type.

The second type of pipes with more than three layers could theoretically be printed but subsequent layers caused some challenges since the ends of the pipes bent when printing the diagonal pattern. This resulted in even worse surface quality in the ends of the next layer. Also generally the surface quality in the outmost surface was considerably coarser than in the first type of pipe since the outer layer in the second type consisted of diagonal lines.

4. CONCLUSION

Using the cylindrical 3D printing method quite lightweight and surprisingly flexible pipes were able to be printed. The printed pipes also turned out to be quite strong for their size.
To further develop this method, adding the capacity to print composite materials would be worth researching. With such setup, it should be possible to add metal wire or carbon fibre in between the layers to create even stronger structural mesh, while still retaining relatively lightweight and a good degree of flexibility.

5. CORRESPONDING ADDRESS

Panu Kiviluoma, D.Sc. (Tech.), Senior University Lecturer
Aalto University School of Engineering
Department of Engineering Design and Production
P.O.Box 14100, 00076 Aalto, Finland
Phone: +358504338661
E-mail: panu.kiviluoma@aalto.fi
http://edp.aalto.fi/en/

6. ADDITIONAL DATA ABOUT AUTHORS

Munigala, Vikram, B.Sc
E-mail: vikram.munigala@aalto.fi

Kemppainen, Mauno, B.Sc
E-mail: mauno.kemppainen@aalto.fi

Kärki, Pyry, B.Sc
E-mail: pyry.karki@aalto.fi

Tudose, Costin, B.Sc
E-mail: costin.tudose@aalto.fi

Kuosmanen, Petri, D.Sc. (Tech.), Professor
Phone: +358 500 448 481
E-mail: petri.kuosmanen@aalto.fi

7. REFERENCES

[1] Patent: A method for constructing an inflatable mandrel. Davis, Kevin; Yorgason, James A. (1993). US Patent no 5259901. Available: URL: http://www.google.com/patents/US5259901 (23.10.2014)

[2] Pneumatic Release Mandrel, United States Patent 4794858 A. Available: URL: http://www.google.com/patents/US4794858 (23.10.2014)

[3] Patent: 3d printing on a rotating cylindrical surface. Elsey, Justin (2012). US patent no 20120165969 A1. Available: URL: http://www.google.com/patents/EP2459361A1?cl=en (23.10.2014)

[4] Kenich, Alexandros, et al. Lathe type 3D printer. A design report. Available: URL: https://www.scribd.com/doc/147351838/Lathe-Type-3D Printer (23.10.2015)

10th International DAAAM Baltic Conference
"INDUSTRIAL ENGINEERING"
12-13 May 2015, Tallinn, Estonia

APPLICATION OF MEMRISTOR IN SIMULATION OF FRICTION PROCESSES

Nuzhdin, K.; Musalimov, V. & Sivitski, A.

Abstract: *Phenomena and processes occurring in the interaction zone of friction surfaces in the presence of various liquids and gases are multiple and complex. The purpose of this project is investigation electromagnetic phenomena accompanying the effects of friction, which take place in each tribosystem and under all methods of lubrication. Due to the constant changes in the electromagnetic field to record the values of the characteristics of the field it is possible to use the memristor. Numerical effects of electromagnetic phenomena are compared with mechanical effects, by estimating the coefficients of action. Also in this study was a comparative analysis of the electrical phenomena occurring in friction and characteristics of a memristor.*

Key words: memristor, friction phenomena, electromagnetic processes in tribosystem, model of friction.

1. INTRODUCTION

The electrical phenomena in the friction process have the significant interest for scientific rationale of the control of friction process. It is also necessary to study in greater detail the effect of external sources of electromotive forces and magnetic fields on the friction and wear processes. Currently, electromagnetic phenomena occurring in the process of friction not fully explored due to its complexity and features of origination of these phenomena.

2. PROBLEM STATEMENT

In order to better understand the nature of origination of electromagnetic phenomena in the process of frictional interaction of bodies (especially for friction pair consisting of a metal or a semiconductor) is necessary to explore in details the area tribocontact, in particular the parameters of the electric type: the contact resistance, friction electricity (triboelectricity) and thermo-emf. Another important aspect is the analysis of the effect of the tribological characteristics, such as the force of friction, the wear, the friction coefficient and others on electromagnetic processes occurring in friction. Also, there are grounds for believing that there is a regularity which connecting the tribological characteristics tribosystem with electromagnetic phenomena occurring in friction. The investigation of all these aspects makes it possible to extend the understanding the influence of the electric component on the laws of friction, and in further detail explore the process of frictional interaction.

3. TRIBOELECTRICITY

During the friction of dielectrics, semiconductors or conductors (having different or identical composition), during the mutual friction of liquid dielectrics, etc. is always electrifying the both bodies which involved in the process of friction. Their charges have the same value and different signs.

The contact electrification of the two metals or metal and semiconductor was generated in consequence of the transition of electrons across the interface from the body with less energy exit of electrons to the body with bigger energy exit of electrons (contact voltage). In the case of contacting the metal with a dielectric the electrification occurs in consequence of the transition of electrons from a metal to an insulator and positive ions of the dielectric to the metal surface.

In the study of tribological phenomena occurring in dry friction process of metals under the reciprocating movement were observed the formation and accumulation of significant electrical potentials, as shown in Fig. 1.

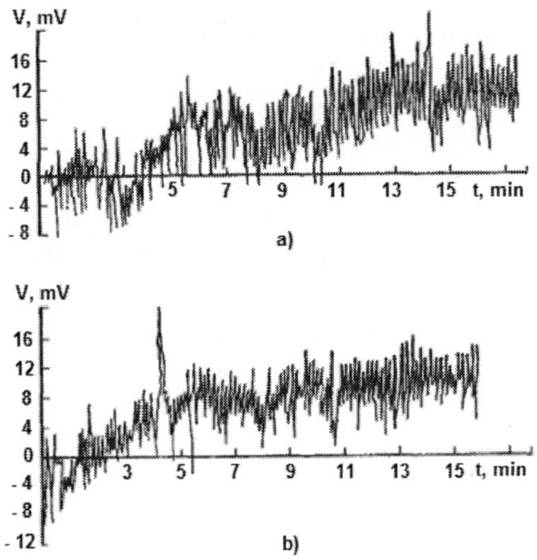

Fig. 1. Voltage was created during solid friction processes:
a) F_{normal} = 66 N, b) F_{normal} = 96 N; materials: Steel 45 – Teflon;
speed: v = 1,42 m/s

Introduction of the dielectric (oil or lubricant) between the two mutually moving conductors leads to the formation of a kind of capacitor in which electric charges are accumulated. Also other electrokinetic phenomena can have the specific influence. During friction, the lubricant film may be broken because of short-wavelength pattern, consequently it may origination of electrical phenomena, especially under high loads. They are similar to phenomena under dry friction: the emission of electrons, thermoelectric effects, contact-potential difference, etc.

As a result of rapprochement microroughnesses of the electrified surfaces, on the some distance may occur a local dielectric breakdown and a transport of electric charges from one surface to another, this is similar to atmospheric discharges. The moving electrical charges is becoming a source of magnetic field. The magnetic field, in turn, is a form of electromagnetic field which acting on the body with an electric charge or the body is having a magnetic moment, regardless of whether they are moving or are dormant [1].

4. MEMRISTOR

4.1 Definition of the memristor

In 1971, american professor Leon Chua proposed that there should be a fourth fundamental circuit element to set up the relation between charge and magnetic flux and complete the symmetry as shown in Fig. 2 [2].

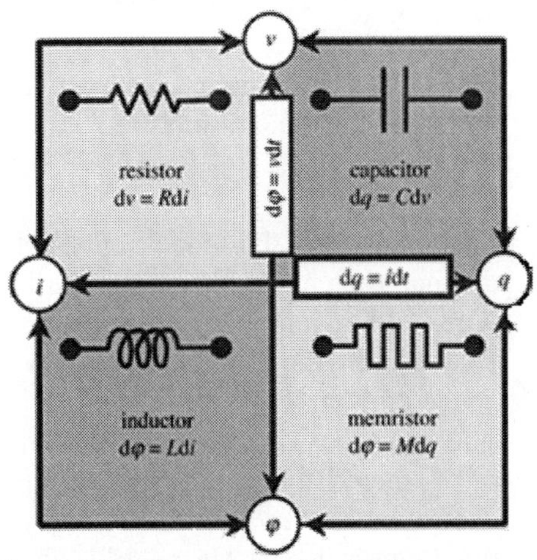

Fig. 2. The four basic circuit elements.

Prof. L. Chua named this the memristor. It is defined as a two terminal circuit element in which the flux between the two terminals is a function of the amount of electric charge that has passed through the device.

4.2 Properties of the memristor

The memristor devices we are primarily concerned with are based on metal-oxides the device models considered here are based on the theoretical underpinnings proposed by Hewlett-Packard Labs [4]. For modeling purposes, a thin film metal-oxide material of thickness D sandwiched between two conductors is modeled as two variable resistors connected in series, as shown in Fig. 7.

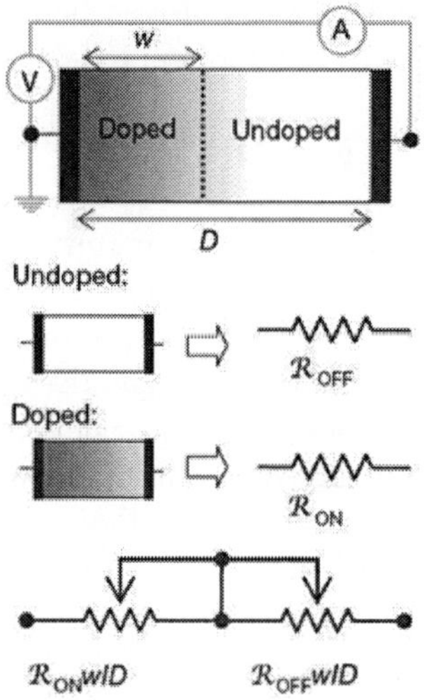

Fig. 7. A schematic diagram of the memristor with a simplified equivalent circuit

Each resistor in the model represents two distinct regions of the metal-oxide, one with a high ionic dopant concentration, R_{on}, and the other with a low concentration, R_{off}.

The most basic mathematical definition of a memristor is that of a current-controlled device for circuit analysis in the generalized class of nonlinear dynamical systems called memristive systems described by the equations:

$$v(t) = \left(R_{ON}\frac{w(t)}{D} + R_{OFF}\left(1 - \frac{w(t)}{D}\right)\right) i(t) \quad (1)$$

$$\frac{dw(t)}{dt} = \mu_V \frac{R_{ON}}{D} i(t) \quad (2)$$

where w can be a set of state variables. For simplicity and ease of simulation the memristor resistance or memristance definition has been reduced to that of a current-controlled time-invariant one-port device given by:

$$M(q) = R_{OFF}\left(1 - \mu_V \frac{R_{ON}}{D^2}\right) q(t) \quad (3)$$

where w represents the doped region of the memristor, D the total device length and R_{on}/R_{off} are the lowest and highest resistance states. Thus, there can be different values for the memristance M as changes in the electric current cause atomic rearrangement and x is varied between 0 and D [5].

5. THE COURSE OF RESEARCH

5.1 Model of the memristor

For a more detailed study of the characteristics of the memristor, there was built his model in an environment Simulink / Matlab (shown in Fig. 10). As a basis, it was used the model developed by Thang Hoang [6]. The subsystem i_v_flux (see Fig. 10) implements mathematical description of the memristor based on the equations (1) - (3).

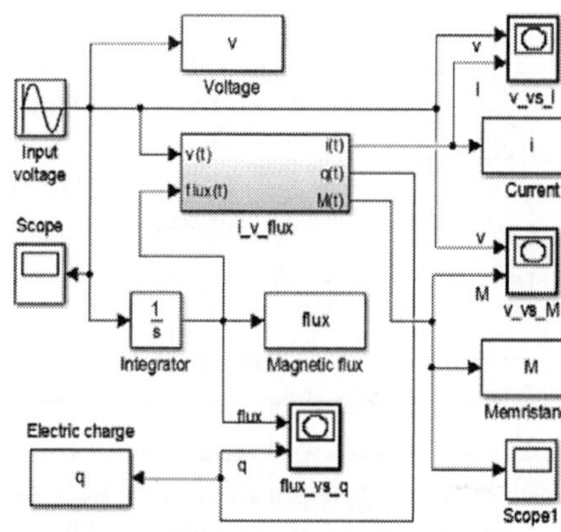

Fig. 10. Model of the memristor

During simulation were used parameters of the real memristor (Table 1), which was manufactured by HP-company [7].

Name title	Magnitude
Length of the memristor, [D]	10 (nm)
Average ion mobility, [μ_v]	10^{-14} ($m^2 s^{-1} V^{-1}$)
Resistance when on, [R_{ON}]	100 (Ω)
Resistance when off, [R_{OFF}]	10 (kΩ)

Table 1. Parameters of the memristor by HP

Results of the memristor modeling are represented in Fig. 11 - 15.

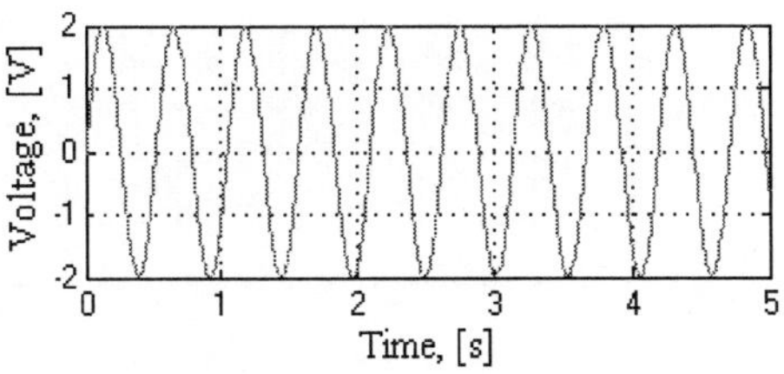

Fig. 11. The memristor input voltage

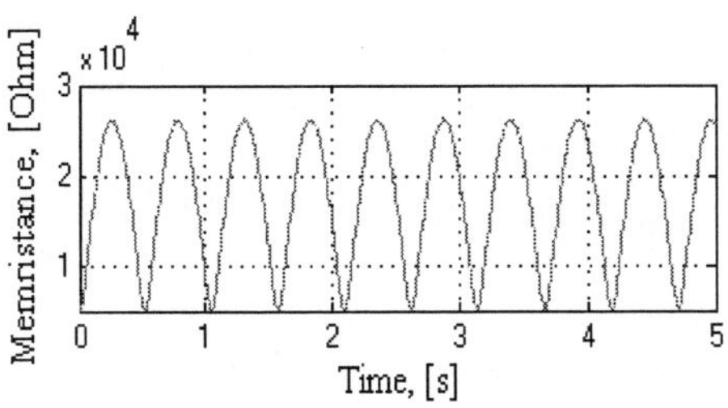

Fig. 12. The change of memristance

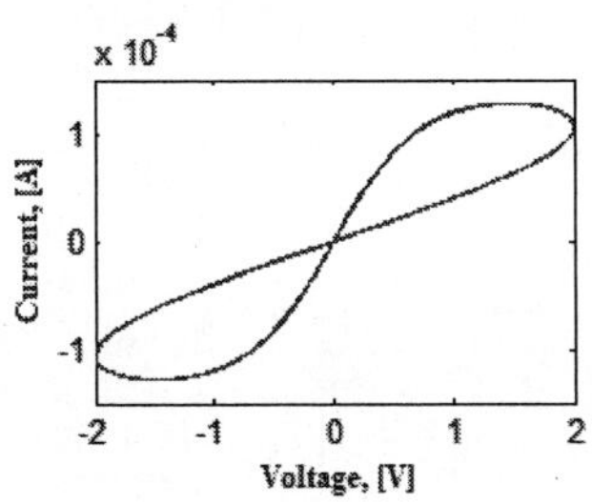

Fig. 13. The characteristic curve of voltage vs current

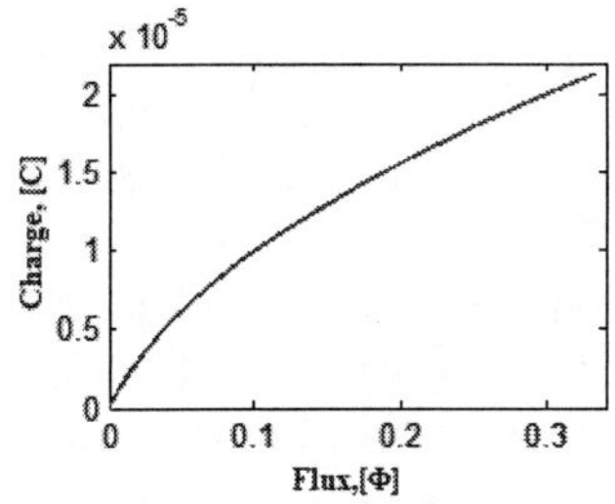

Fig. 14. The characteristic curve of magnetic flux vs charge

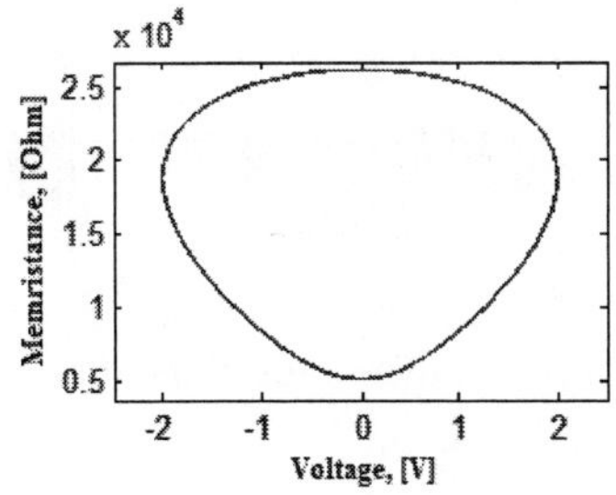

Fig. 15. The characteristic curve of voltage vs memristance

These results allow us to see the main characteristics of the memristor such as hysteresis loop (see Fig. 13) of electrical parameters. It means that current value depends up previous state.

5.2 Experimental data from the friction process

The experiments for the study of friction were conducted using tribometer. As specimens were used TiN and AlTiN materials. The scheme of friction process is reciprocating movement ball on plate. The experimental conditions are presented in Table 2.

Name title	Magnitude	
friction pair	Al_2O_3 - TiN	Al_2O_3 - AlTiN
movement amplitude	0.01 m	
frequency	5 Hz	
normal force	22 N	
Top specimen - ball		
diameter	0.01 m	
material	AL_2O_3	

Table 2. The experimental conditions

During the experiment the different characteristics of friction pair were recorded: the normal force, friction force, coefficient of friction, wear, etc. But of particular interest is the change of electrical resistance of contact friction pair, presented in Fig. 16 and Fig. 17.

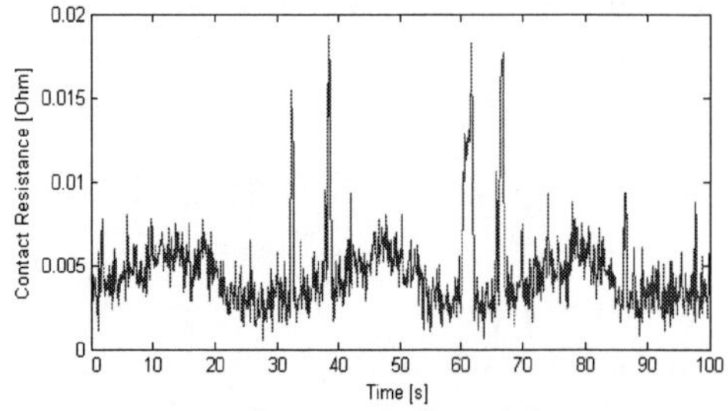

Fig. 16. The contact resistance of friction pair Al_2O_3 - TiN

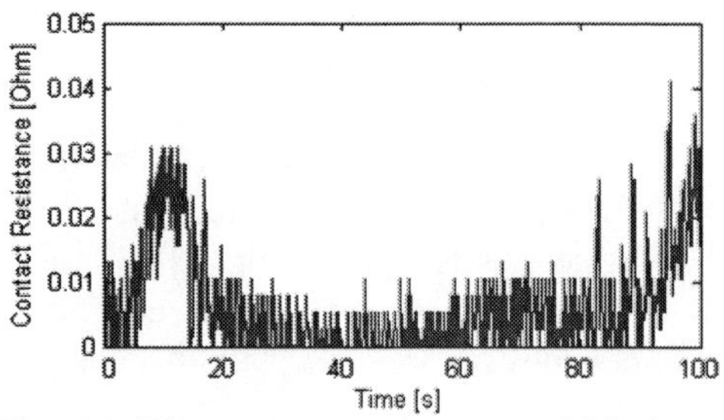

Fig. 16. The contact resistance of friction pair Al_2O_3 - AlTiN

5.3 Comparing the electrical characteristics

Further, the dependence was found between the electrical characteristics tribopair during friction process and parameters of the memristor. For this purpose was used the software package Matlab called System Identification Toolbox. As a result, there was obtained the transfer function, which describes the relationship between the electrical resistance of contact tribopair and resistance of the memristor (see Fig. 17 and Fig. 18).

Transfer function for friction pair Al_2O_3 - TiN:

$$y(u(r)) = 60 \cdot \left(\frac{-101 \cdot r^2 + 189 \cdot r + 10}{r^3 + 0.2 \cdot r^2 + 0.01 \cdot r} + 250\right) \quad (4)$$

where *y(u(r))* – output function; *u(r)* – input function.

Transfer function for friction pair Al_2O_3 - AlTiN:

$$y(u(r)) = 26 \cdot \left(\frac{669.4 \cdot r - 33.3}{r^2 + 0.000006 \cdot r + 0.0144} + 650\right) \quad (5)$$

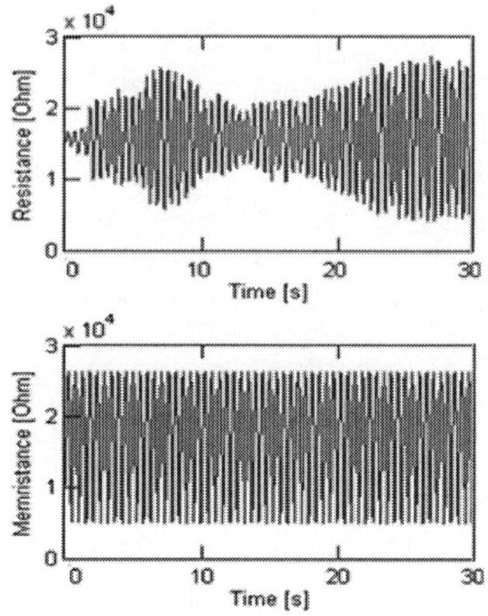

Fig. 17. The result of using transfer function for friction pair Al_2O_3 - TiN: the top graph – modeling of contact resistance change during the friction process;
the lower graph - change of memristance.

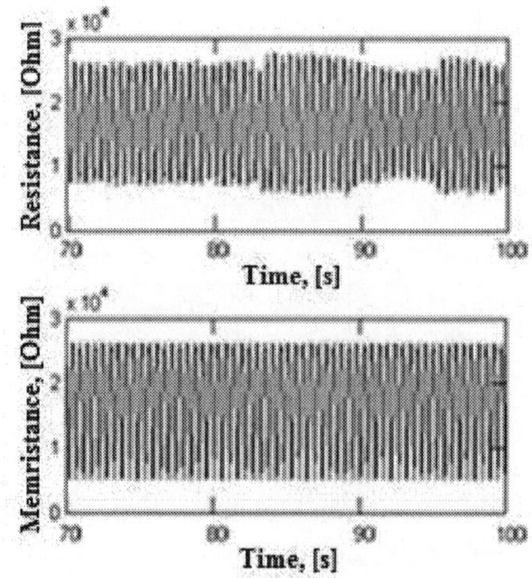

Fig. 18. The result of using transfer function for friction pair Al_2O_3 - AlTiN: the top graph – modeling of contact resistance change during the friction process;
the lower graph - change of memristance.

6. CONCLUSION

In the course of the works there was found the dependence of electrical parameters of the friction process and characteristics of a memristor.
There was obtained the transfer function that relates these dependencies.
To sum it up, we can conclude at some point of time, the resistance of the contact area has variation law of electrical characteristics to a high accuracy repeating regularity of change memristance.
Thus, in the process of friction, the contact area of tribopair can be considered as kind of memristor, however, this topic requires further more detailed investigation.

7. FUTURE WORK

As further work there is necessary to investigate the influence of all parameters was measured during friction (friction force, wear, elastic modulus, hardness, etc.) on the electrical characteristics of the contact friction pair.
After that we will be able to evaluate in full the characteristics of the electrical processes in friction and compare the obtained data with the theory of the memristor.

8. REFERENCES

1. Khebda, M.; Chichinadze, A. *Tribotechnology reference guide,* Manufacturing engineering, Moscow, 1989, pp. 75 - 78.
2. James M. Tour and Tao He, *The fourth element*, NATURE, Vol. 453, May 2008, pp. 42 – 43.
3. Yang, J. J.; Pickett, M. D.; Xuema, L.; Ohlberg, D.; Duncan, R.; Williams, S. *Memristive switching mechanism for metal/oxide/metal nano-device*, Nature Nanotechnology 3, 2008, pp. 429 – 433.
4. Chua, L.O.; Kang, S. M. *Memristive devices and systems*, Proc. IEEE 64, 1976, pp. 209 – 223.
5. Eliseev, N. Memristors and crossbars: nanotechnology for processors, *ELECTRONICS: Science, Technology, Business,* 2010, pp. 84 - 89.
6. Memristor model http://cn.mathworks.com/matlabcentral/fileexchange/25082-memristor-model
7. Biolek, D.; Biolkova ,V. Spice model of memristor with nonlinear dopant drift, *Radioengineering,* vol. 18, 2009, pp. 210 - 214.

9. ADDITIONAL DATA ABOUT AUTHORS

Kirill Nuzhdin, student MSc Mechatronics Double Degree program (TUT and ITMO University)
Department of Mechatronics, Ehitajate Tee 5, Tallinn 19086, Estonia;
Department of Mechatronics, Kronverksky Pr. 49, St. Petersburg 197101, Russia;
nkirill74@gmail.com

Viktor Musalimov, Professor, Head of mechatronics department;
ITMO University;
Kronverksky Pr. 49, Saint-Petersburg 197101, Russia;
musVM@yandex.ru
+7 (812) 232-31-50

Alina Sivitski, PhD, Teaching Assistant;
Tallinn University of Technology;
Department of Mechatronics, Ehitajate Tee 5, Tallinn 19086, Estonia;
alina.sivitski@ttu.ee
+372 6203302

10th International DAAAM Baltic Conference
"INDUSTRIAL ENGINEERING"
12-13 May 2015, Tallinn, Estonia

IMPLEMENTATION OF A CONTOUR CRAFTING SYSTEM TO A 3-DIMENSIONAL CONCRETE PRINTER

Nylund, J.; Järf, A.; Kekäle, K.; Rönnskog, J.; Al-Neshawy, F.; Kiviluoma, P. & Kuosmanen, P.

Abstract: *This paper examines the benefits of implementing a contour crafting system to a 3-dimensional concrete printer in order to improve the surface finish and geometrical shape of the printed concrete. This system was supplemented by three additional axis of rotation: two of which set the angle for the trowels that shape and smoothen the edges and the last that rotates the nozzle head around the Z-axis. Additional software was developed to modify the G-code and provide instructions for the three additional axes.*
Tests were performed measuring the surface roughness and the geometrical shape of the printed concrete object. The results show that the geometry can be improved by this contour crafting system. Applications for this technology can be small scale concrete structures for architects, non-standard pre-manufactured concrete elements and other minor concrete objects.

Keywords: layered manufacturing, additive, rapid prototyping.

1. INTRODUCTION

3D printing has become one of the fundamental processes for rapid prototyping and is currently being applied to actual manufacturing of components [1]. 3D printing was originally used to make plastic objects, but lately many alternative materials are being suggested for 3D printing such as different metals and concrete. Concrete 3D printing could, if implemented correctly, reduce building costs of concrete structures such as houses as no mold is needed and introduce a fast and flexible way for building emergency shelters. But as this idea is quite new, research into the subject is still needed to make it feasible and correct the existing problems such as: finding a viable concrete mixture, adding reinforcement to the concrete, logistics of the printer, improving surface quality and finding a way for the concrete to be made into free forms. This study will focus on the last two of these problems, which is adding a free form system called contour crafting to a 3D concrete printer which improves surface quality.

Contour crafting is a system used in layered fabrication processes, where trowel like plates are used to smoothen the outer edges of the printed concrete from the concrete printer nozzle. This gives the printed concrete a more uniform shape, instead of the layered shape of an object printed without contour crafting. The trowels improve the surface quality of the concrete and the aesthetics of the finished product. [1][2]

Previous research in this area has been made primarily at the University of Southern California and at Loughborough University. The University of Southern California built a 3D printer for various materials, which uses contour crafting to make angled forms possible. The printer uses a six axis system, three for linear movement, one for extrusion, one for rotating the nozzle and one for the singular trowel [3]. Loughborough University has also done research in this field [4][5][6].

This paper will focus on the implementation of a contour crafting system to a 3D concrete printer. The goal is to create a system which gives the user a greater degree of freedom in designing different forms, and that improves the aesthetic of the finished product. This is achieved by using trowels that can be angled at the sides of the nozzle. The goal of this project is to increase the surface quality significantly and introduce the possibility of printing angled walls.

2. METHODS

The original 3D concrete printer moves the nozzle head a maximum of 475 mm in the X-direction and 650 mm in the Y-direction and the printing bed 750 mm in the Z-direction (Figure 1). This is done by 300 W servomotors which actuates linear screws. A 660 watt servo motor drives a screw-type extruder [7]. The concrete printer is controlled by LinuxCNC, which is an open source software made primarily for CNC milling machines. An open source slicer-software converts stereolithography files to the G-code instructions used by LinuxCNC.

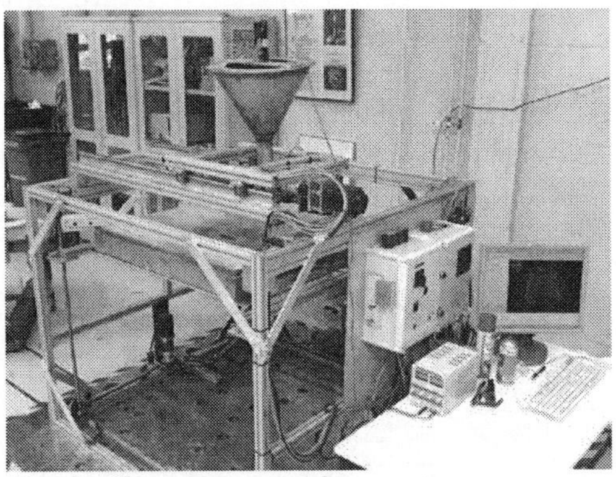

Fig. 1. The concrete printer setup.

2.1 Hardware

The purpose of contour crafting is to smoothen the surface of the concrete flowing out of the extruder. To implement this to the concrete printer, a new nozzle with trowels that follow the outer edge was needed. The trowels should also be able to be angled according to the shape of the printed object to allow angled walls to be printed. Additionally the trowels need to be completely lifted during the fill cycle.

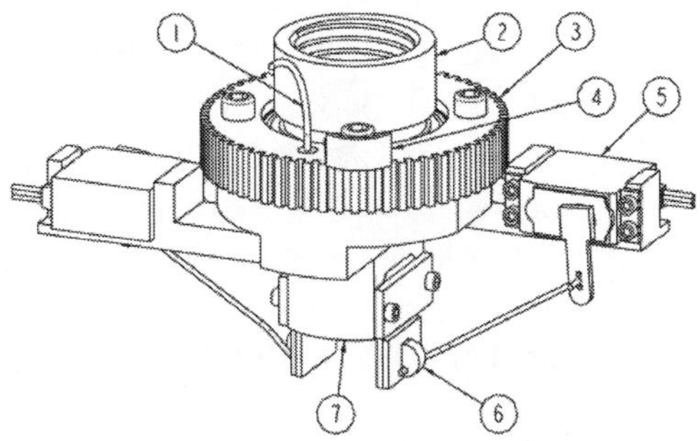

Fig. 2. The new nozzle assembly.

The new nozzle model is illustrated in Figure 2. The top of the nozzle (2) is threaded, to be screwed on to the extruder. The rest of the extruder assembly is fastened with a ball bearing and rotates around the top part. The rotational movement is controlled by a 63 watt servo motor, which is connected to the timing pulley (3) on the nozzle by a timing belt. The trowels (6) are made of 2 mm rubber and stabilizing aluminum pieces. The angle for each trowel is set separately by servo motors (5) that are mechanically connected to the trowels. Each trowel can be angled 135 degrees, as seen in Figure 3. Power and signals to the servo motors are supplied by a coiled cable (1), running through the nozzle assembly. This cable limits the rotational movement of the axis to ±540°. When zeroing the rotational angle, a cam (4) activates a limit switch on the frame of the concrete printer. Concrete is extruded through the 22 mm bottom part of the nozzle (7) and smoothened by the trowels. Between the upper and lower part of the nozzle assembly a sealing ring is used to prevent concrete from entering the nozzle assembly. This can be seen in the section view in Figure 3. The parts of the nozzle assembly that are in contact with concrete, are made of S355 steel, while other parts are made of aluminum.

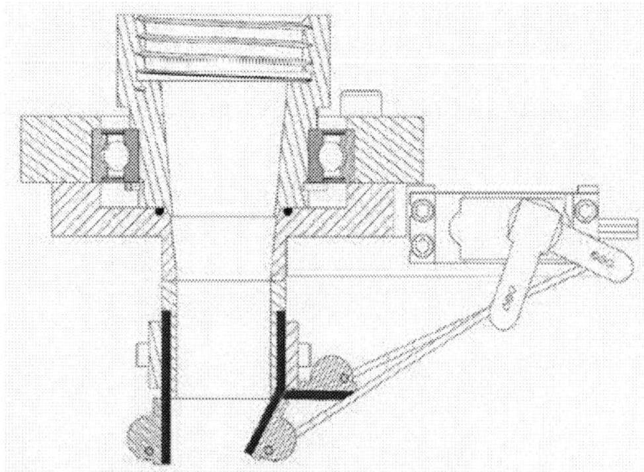

Fig. 3. Section view of the nozzle assembly. The trowel movement is illustrated on the right trowel.

2.2 Software

To implement contour crafting, G-code for the additional axis is also needed. In the original setup of the printer, G-code is generated from a CAD model by the open source software slic3r. Slic3r breaks the CAD model in stereolithography format down to points in the X-Y plane. LinuxCNC then moves the nozzle according to the path given by the G-code.
In order to control the additional axis, a program was written in C11 programming language. The program reads the coordinates, written in the G-code file and calculates the angle for the nozzle and the trowels. The rotation of the nozzle is limited to ±540°, so the spiral cable does not extend beyond its elastic limit.
The software has some considerations to avoid damaging the printed perimeter. If a sharp angle turn is detected, the program lifts the trowel that would cut into the printed concrete to avoid collision. This slightly worsens the surface quality near the sharp angle turn, but it is necessary to keep the form intact. The software also considers the movement of the trowels when returning to the starting point of the current layer, as in some shapes the trowel can collide with the printed concrete.

2.3 Concrete mixture

Contour crafting puts some additional requirements on the properties of the concrete being used. Measures of the printability of a concrete mix are defined as extrudability (the ability of the mass to easily flow through the extruder), buildability (the resistance against deformations by the subsequent layers) and formability [7]. The pre-existing recipe [7] is modified to accommodate the implementation of the contour crafting system. The water and plasticizer content of the mix was adjusted until the criteria listed above were deemed fulfilled.
A compression test was made on the resulting mixture, to test the properties of the concrete. To evaluate the contour crafting system two test shapes were printed with and without the contour crafting system. The shape was made in CAD software and is a cone with a base diameter of 155 mm and a slope of 78°.
The composition of the concrete mixture that was used for these tests is found in Table 1.

Ingredient	Composition [kg/m³]
Sand R 1-2 mm*	301.4
Sand R 0.5-1.2 mm	241.2
Sand R 0.1-0.6 mm	241.2
Filler	211.0
Cement**	464.0
Fly ash	133.8
Silica fume	22.4
Water	221.2
Super plasticizer	3.0

*) The aggregates used in the mix were 30% [R1-2], 24% [R0.5-1.2], 24% [R0.1-0.6] and 22% Filler.
**) Rapid cement [CEM II A 42.5R]

Table 1. The composition of the concrete mix.

Rapid cement [CEM II A 42.5R] type was used in the concrete mix. The cement was produced by Finnsementti's cement plant at Parainen. The aggregates used in the mix was natural granite gravel, which was washed, dried, and graded by sieving. The grading of the combined aggregates is shown in Figure 4. The water used was normal tap water provided by the water distribution system of the city of Espoo, Finland and its temperature was 20±2°C. The water-reducing agent was superplasticizer (Glenium 51), manufactured by Finnsementti Oy Finland. This agent was stored in a polyethylene bottle at room temperature.

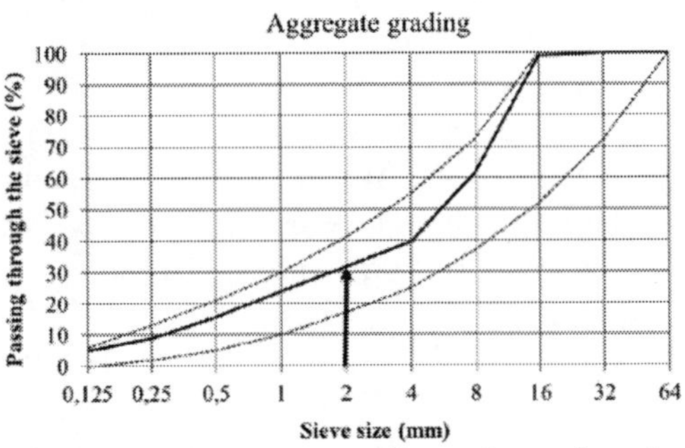

Fig. 4. Aggregate grading chart for the concrete mix.

The dry ingredients were hand-mixed for one minute after which it was poured into a container of wet ingredients for another 3 minutes of mixing. The batch was then immediately poured into the printer.

3. RESULTS

When comparing the pieces printed with the contour crafting method and the pieces printed without contour crafting it can be seen that the surfaces of the new printed pieces are smoother. The indentations appearing between each layer were measured on both types of printed pieces and were much deeper on the pieces printed without contour crafting. The indentations on the pieces without contour crafting were on average 7.9 mm, while the pieces with contour crafting had indentations with an average depth of 3.4 mm. Figure 5 illustrates the difference between a cone printed with and without contour crafting.

Fig. 5. Comparison of old and new result. The object on the right is printed with contour crafting.

To test the strength properties of the concrete, standard compression tests were performed on three cast cubic 100x100x100 mm samples [8]. The cubes were left to dry for five days before the tests were performed. The results can be seen in Table 2.

	f_c [MPa]	density [kg/dm^3]
Sample 1	46.4	2.199
Sample 2	44.5	2.209
Sample 3	43.3	2.139

Table 2. Results from compression tests of the concrete mix.

Cracks in the bend of the rubber trowels were formed after only a few printings. This was suspected to be due to either mechanical wear from the movement of the trowels or degraded elasticity due to chemical reaction with concrete. Another material should be used in this situation. Metal or rubber resistant to chemical degradation is proposed.

4. DISCUSSION

From the results, the conclusion can be drawn that with the new setup of the printer enabling contour crafting, smoother surfaces can be printed. The changes made to the concrete mixture have also helped improve the quality of the printed pieces.
Compared to normal concrete casting there is a fundamental problem with layer-based printing of concrete. This is the absence of

steel reinforcements inside the concrete structure. Since the printing is done layer by layer, steel reinforcements cannot be placed beforehand since they would collide with the printer. A solution for this could be to add steel fibers into the concrete mix. This could possibly strengthen the concrete enough to print larger scale structures.

There are multiple ways that this concrete printing system could be improved to achieve more precise forms with better surface quality. One of these is to implement a mass flow feedback system so that the extrusion feed can be adjusted according to the real output of the concrete through the nozzle. This would ensure that the right amount of concrete would be extruded. Further research to optimize the concrete mixture should be performed. The mixture used was designed to emulate the use of industrial construction concrete. Software that generates the needed G-code for all seven axes from the original stereolithography data could be developed. This would remove some of the limitations with the current setup, namely adding the possibility for discontinuous forms. An additional feature that could be added is to have a variable nozzle hole diameter that adjusts itself according to the required thickness of the printed piece. Since the properties of the concrete depends on multiple factors and is time dependent, an automatized way of mixing the concrete continuously while printing could improve the results. This could probably help reduce the differences between printed batches and even remove some of the differences in the same batch, that are caused by the unprinted cement drying while the printer runs.

It should be investigated if this technology could be scaled up to allow bigger objects to be printed. The hypothesis is that a bigger scale would allow the layers to harden more before the next layer is printed on top. This would allow higher objects to be printed. Contour crafting improves the aesthetics of the printed object and allows its forms to be freer.

5. REFERENCES

[1] Bogue, R. 3D printing: the dawn of a new era of manufacturing? Assembly Automation, 2013, **33,** 307-311

[2] Hwang, D. Experimental Study of Full Scale Concrete Wall Construction Using Contour Crafting. University of Southern California, 2005

[3] Khoshnevis, B. Automated construction by contour crafting - related robotics and information technologies. Automation in construction, 2004, **13**, 5-19.

[4] Le, T. T., Austin, S. A., Lim, S., Buswell, R. A., Gibb, A. G., & Thorpe, T. (2012). Mix design and fresh properties for high-performance printing concrete. Materials and structures, 45(8), 1221-1232.

[5] Lim, S., Buswell, R. A., Le, T. T., Wackrow, R., Austin, S. A., Gibb, A. G., & Thorpe, T. Development of a viable concrete printing process, 2011

[6] Lim, S., Buswell, R. A., Le, T. T., Austin, S. A., Gibb, A. G., & Thorpe, T. Developments in construction-scale additive manufacturing processes. Automation in construction, 2012, **21**, 262-268

[7] Kranz, C. Further development of the injection nozzle of a 3d concrete printer, Aalto university school of engineering, 2014

[8] SFS-EN 12390-3 Testing hardened concrete. Part 3: compressive strength of test specimens. Finnish Standards Association SFS

CORRESPONDING ADDRESS

Panu Kiviluoma, D.Sc. (Tech.), Senior University Lecturer
Aalto University School of Engineering
Department of Engineering Design and Production
P.O.Box 14100, 00076 Aalto, Finland
Phone: +358504338661
E-mail: panu.kiviluoma@aalto.fi
http://edp.aalto.fi/en/

ADDITIONAL DATA ABOUT AUTHORS

Nylund, Jimmy
Phone: +358 50 410 7302
E-mail: jimmy.nylund@aalto.fi

Järf, Alexander, B.Sc. (Tech.)
Phone: +358 50 492 8262
E-mail: alexander.jarf@aalto.fi

Kekäle, Kim, B.Sc. (Tech.)
Phone: +358 40 573 2532
E-mail: kim.kekale@aalto.fi

Rönnskog, Jan
Phone: +358 50 347 1498
E-mail: jan.ronnskog@aalto.fi

Al-Neshawy, Fahim, D.Sc. (Tech.), Researcher
Phone: +358 50 5649 372
E-mail: fahim.al-neshawy@aalto.fi

Kuosmanen, Petri, D.Sc. (Tech.), Professor
Phone: +358 500 448 481
E-mail: petri.kuosmanen@aalto.fi

10th International DAAAM Baltic Conference
"INDUSTRIAL ENGINEERING"
12-13th May 2015, Tallinn, Estonia

DEVELOPMENT OF A NURTURANCE EVOKING ROBOT

Peltonen, O.; Orhanen, S.; Venäläinen, J.; Auvinen, M.; Sepponen, R.; Kiviluoma, P. & Kuosmanen, P.

Abstract: *A robotic companion can have the same health benefits as pets, including alleviating loneliness, lowering stress and elevating mood. The non-allergenic and controllable nature of robots makes them preferable over real animals. This paper describes a nurturance evoking social robot for eldercare, which is able to express its emotions and act like a real animal, is robust and fits in user's lap. None of the existing social robots combine all these criteria. The robot expresses emotions with actuated neck, ears, nose and tail, vibration and sound. Flexible mechanisms ensure safe movements and durability. Touch, motion, temperature and sound affect the robot's emotional state. Preliminary studies show good results in evoking nurturance.*

Key words: social robot, companion robot, eldercare, artificial intelligence.

1. INTRODUCTION

In healthcare, a robotic companion can have the same health benefits as pets, including alleviating loneliness, lowering stress and mood elevation [1,2,3]. Although robotic and real pets have the same therapeutic effect, a robot is often preferred in healthcare, because it is not allergenic. Some patients might also be incapable of taking care of a real pet. Thus the need for social robots increases constantly, especially in the field of eldercare. A study shows that more than one third of elderly people suffer from loneliness [4], and the population in western countries is continuously ageing. It has been estimated that the proportion of people over 60 years old will be 37% of the whole population in Western Europe by 2050 [5].

There currently exists multiple commercial and non-commercial social robots with applications varying from eldercare to entertainment. These robots include Paro, Huggable and Probo.

Paro is a commercially available artificial pet seal. Under the soft hygienic artificial fur, Paro features a range of sensors, which allow it to sense touch and warmth, recognise its name, greetings and praise. Paro has actuators to move its eyelids, head and flippers and a speaker to play authentic seal sounds. Paro's behaviour alters to reflect user's interaction preferences. [6]

MIT's Media Lab is developing Huggable, a therapeutic robot companion with its whole body able to sense touch. To achieve touch sensing, Huggable utilises three sensor types all around the body: Quantum Tunnelling Composite (QTC) for force, thermistors for temperature and electric field sensing to determine for example if the object touching the robot is human. For motion, Huggable uses voice coil actuators, which provide smooth, soundless, lifelike motion. Huggable also utilises neural network to determine the nature of the touch. [1]

Probo is a mammoth-like robot developed to be a friend for hospitalised children. It features a fully actuated head with a moving trunk and natural, motion tracking eyes [7]. To achieve safe movement, Probo utilises flexible joints and Bowden cables between the motors and the actuated joints. In addition to safety, the use of Bowden cables allows the motors to be placed anywhere in the robot, thus keeping the head light-weighted. [8]

In this paper, the development of a Social Panda –robot (Fig. 1.) is presented. The goal of this research was to develop a robot that keeps company with elderly by evoking nurturance by the means of motion and voice. The robot had the following four requirements: it must be able to express its emotional state; it must contain an artificial intelligence (AI) to allow it to act like a lifelike animal; it must fit in the user's lap and be robust. All of these separate requirements have been met by the previously mentioned robots, but none of them combines all four criteria. This paper introduces a design of a social robot that combines all four aforementioned requirements.

Fig. 1. Appearance and features of the panda robot.

2. METHODS

2.1 Overview

The approach to make the robot fit one's lap, be expressive and to have lifelike artificial intelligence sets certain limitations and requirements for the design. For one, the hardware inside the robot has to be lightweight and small, selected actuators have to be able to accurately drive the used mechanisms to make the robot's expressions distinctive. Secondly the software design has to take human interpretations into account.

To meet the set requirements, the robot is able to move its head, ears, nose and tail, emit sound and vibrate to mimic purr of cats. The feeling states of the robot will be expressed with combinations of these outputs, and slight randomness is added to the AI to achieve a lifelike animal behaviour.

The robot was chosen to look like a panda for two reasons. Firstly, a study shows that it is psychologically beneficial to make the robot look like an animal that most people don't have experience on [6]. This choice of design is advantageous, because if users have no knowledge on how the robot should behave, they don't lose interest in it due to unsatisfied expectations. Secondly, pandas are commonly considered as sympathetic. A 19 cm x 24 cm x 30 cm WWF plush panda was selected as the exterior of the robot.

2.2 Mechanical Design

The first prototype of the robot features 6 degrees of freedom (DOF) to obtain a lifelike and nurturance evoking appearance. The robot is able to move its ears (2 DOF), move its nose (1 DOF), tilt its head (2 DOF) and wiggle its tail (1 DOF). These movements are used to express the feeling state of the robot. The prototype is powered with an external power supply.

To achieve safe movements and durability, flexible joints and mechanisms were used. The neck joint was manufactured from a coil spring and polycarbonate connecting plates. The joint is actuated by pulling monofil nylon fishing lines with RC servo motors (MG90S). Fishing lines operated with RC servo motors were also used to operate the mechanisms for the nose and ears. RC servo motors were chosen as main actuators for their affordable price, light weight, size and controllability.

The skeleton of the robot consists of two modules, the head and the body. Main parts of the skeleton are shown in the Figure 2. The function of the skeleton is to house and protect the electronics and serve as a contact surface for the touch sensors. The skeleton was 3D-printed, which allows great freedom of design and produces parts quickly. The skeleton was designed as a tight fit inside the plush panda so that it does not require any external attachment to the cloth exterior.

Fig. 2. Illustration of the two-module skeleton.

2.3 Electronics and Sensors

Arduino was chosen as the platform for the electronics, as it is well suited for prototyping: it features easy and fast coding, and the market is full of Arduino compatible modules. Arduino Nano V3 was chosen for its small size in order to fit in the robot.

The robot's touch sensing was achieved by placing sensors made of Quantum Tunnelling Composite (QTC, Peratech) pills on the surface of the skeleton. QTC material changes its electrical resistance when pressure is applied, thus making it suitable for touch sensors. QTC sensors were chosen for their affordable price and small size. Thermistors were used to detect the warmth of the holding human, other possible temperature changes and to detect possible electrical failures. A passive infrared (PIR, HC-SR501) sensor module was used for sensing motion in front of the panda. Motion and orientation of the robot is detected by an IMU module (GY-85).

Servo shield F08019 controls the 6 servos. The robot makes noises from a speaker, driven by a sound module (WTV020-SD) with audio stored on a SD-card. One eccentric rotating mass (ERM) motor (RF300) is used to create the purr vibration, purr effect is enhanced with corresponding sounds.

In order for the robot to be useful in eldercare or as a scientific tool for studies, one has to be able to monitor the functioning and usage of the robot. To achieve this, the monitoring data is sent by a serial-interface WiFi-module (ESP8266) to a monitoring computer.

2.4 Software Architecture

The architecture shown in Figure 3 has a hierarchical structure where high-level computation is executed on the top and, on the contrary, low-level computation on the bottom side. For example, software classes that represent different moods of the robot are on the top-side of the diagram, because they are more abstract concepts than, e.g., motor controller that lie on the bottom-side. This kind of hierarchical software structure is common in robotics and software design in general, because it offers two main advantages. Firstly, the designer of the software has strong understanding of ownerships of instances. Secondly, it is modular, as modifications made to classes in one hierarchical level do not affect the whole software. This is because each level works as an interface to the next level, which effectively hides all the modifications from further levels. In fact, modifications to one level will only affect the levels next to it.

The main loop that keeps the software running resides in the State machine –level of the hierarchy. As the name of the level

implies, it also contains a state machine that forms the AI of the panda robot. The transition logic between states is implemented with hard-coded control structures.
The mood-level includes the sequences of actions that create the characteristic behaviours of each mood. The mood-level uses the high-level hardware communication layer (HCL) that is an abstraction level of the hardware controllers. This high-level HCL provides simplified interface to the hardware. Finally, the low-level HCL communicates with the hardware, e.g., with motors and sensors via various communication protocols.
The software is implemented with C++ programming language, because our hardware is Arduino-based and there exist plenty of ready-to-use libraries to control various devices in C++.

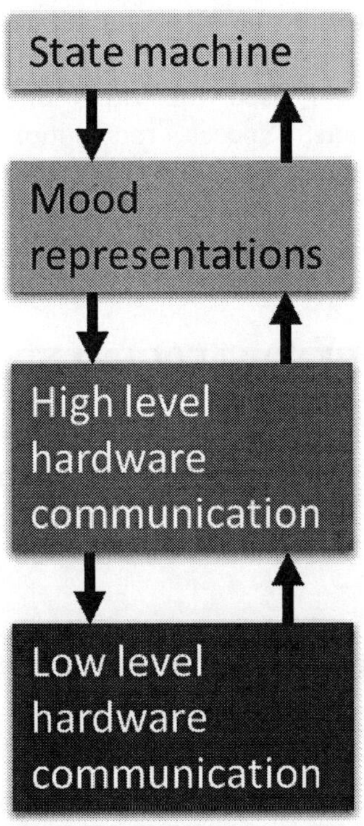

Fig. 3. Hierarchical software architecture used in the panda robot. High-level computation is executed at the top-side and low-level computation on the bottom side.

2.5 User Survey
The robot was tested with a user survey to investigate whether the robot fulfils the set goal, i.e., whether or not it can keep company with elderly. The participants of the survey used the robot for approximately 5 to 10 minutes each. Afterwards the participants took a questionnaire about the robot with 6 questions, which were answered on a three point scale: “Disagree”, “No opinion” and “Agree”. The questions asked in the user poll are described in Table 1.

Number of Question	*Question*
1	I understood if the robot was happy or sad.
2	The robot reacts to my care in a reasonable way.
3	The robot wanted my attention.
4	I enjoyed the company of the robot.
5	The robot was easy to hold in lap.
6	The robot felt alive.

Table 1. User survey questions.

3. RESULTS

The user survey was conducted with elderly people in a retirement home called Kustaankartano comprehensive service centre situated in Helsinki, Finland. A total of 8 elderly people participated in the survey and answered to the questionnaire shown in Table 1. The results of the questionnaire are presented in Table 2. The age of the participants varied between 70 and 97, the mean being 85.

Number of Question	*Agree [%]*	*Disagree [%]*	*No opinion [%]*
1	75.00	0.00	25.00
2	37.25	25.00	37.25
3	87.50	0.00	12.50
4	100.00	0.00	0.00
5	100.00	0.00	0.00
6	75.00	12.50	12.50
Number of participants			8

Table 2. User survey results. Question number is shown in the first column, see Table 1 for questions. In the second, third and fourth columns the distribution of users' answers is shown.

4. DISCUSSION

The following four requirements were set for the robot: it must be able to express its emotional state; it must contain an artificial intelligence to allow it to act like a real animal; it must fit in the user's lap and be robust. The results of the user survey indicate that most of the users were able to identify the emotional state (Question 1, Table 2) of the robot and the participants felt that the robot wanted their attention (Question 3, Table 2). The robot also fit in the lap of every user and was easy to hold (Question 5, Table 2), even though several of the survey participants had only one functional hand.

The results also show that although most of the users felt that the robot was alive (Question 6, Table 2), only few of the participants felt that the robot reacted to nurturing properly (Question 2, Table 2). This suggests that the AI of the robot does not meet the set requirement.

No specific experiment was conducted to determine the robustness of the robot. Nevertheless, tens of people have handled the robot using varying force without breaking it proving that the robot is robust.

The robot fulfils most of the set requirements and the survey indicates that the robot evokes nurturance. However, it must be noted that the sample size for the user survey was small and for more reliable results new survey with more participants should be arranged.

While developing the robot, fitting everything inside the robot was found challenging. The prototype was built mostly with separate easy to work sensor and drive modules. Building a custom PCB would save plenty of space. Fitting several motors and the speaker inside the head module increased the weight of the head, which forces the head to slouch.

5. FUTURE WORK

The robot should be designed battery-powered as a cordless robot feels more like a real animal. Batteries can be fitted inside the legs of the panda. The robot would sound less robotic, if the noisy servo motors used in our prototype would be changed with silent actuators, e.g. voice coil actuators as used in Huggable. The robot would seem more alive if it followed interesting objects with its eyes and head using machine vision. The cameras could be also used for video monitoring the patients. Speech recognition could be beneficial as real pets react to their name and other commands. For active use in health care, the robot should have an antimicrobial fur.

6. ACKNOWLEDGEMENTS

The authors would like to thank the personnel and the residents of Kustaankartano comprehensive service centre for participating in the user survey.

7. REFERENCES

1. Stiehl, W. D., Lieberman, J., Breazeal, C., Basel, L., Lalla, L., & Wolf, M. Design of a therapeutic robotic companion for relational, affective touch. In *Robot and Human Interactive Communication, 2005. ROMAN 2005. IEEE International Workshop on.* IEEE, 2005, 408-415.

2. Broekens, J., Heerink, M., & Rosendal, H. Assistive social robots in elderly care: a review. *Gerontechnology*, 2009, 8(2), 94-103.

3. Banks, M. R., Willoughby, L. M., & Banks, W. A. Animal-assisted therapy and loneliness in nursing homes: use of robotic versus living dogs. *Journal of the American Medical Directors Association*, 2008, 9(3), 173-177.

4. Savikko, N., Routasalo, P., Tilvis, R. S., Strandberg, T. E., & Pitkälä, K. H. Predictors and subjective causes of loneliness in an aged population. *Archives of gerontology and geriatrics*, 2005, 41(3), 223-233.

5. Lutz, W., Sanderson, W., & Scherbov, S. The coming acceleration of global population ageing. *Nature*, 2008, 451(7179), 716-719.

6. Shibata, T. Seal-Type Therapeutic Robot Paro [WWW] http://www.paro.jp (02.11.2014)

7. Saldien, J., Goris, K., Yilmazyildiz, S., Verhelst, W., & Lefeber, D. On the design of the huggable robot Probo. *Journal of Physical Agents*, 2008, 2(2), 3-12.

8. Goris, K., Saldien, J., Vanderborght, B., & Lefeber, D. How to achieve the huggable behavior of the social robot Probo? A reflection on the actuators. *Mechatronics*, 2011, 21(3), 490-500.

8. CORRESPONDING ADDRESS

Panu Kiviluoma, D.Sc. (Tech), Post-doc researcher
Aalto University School of Engineering
Department of Engineering Design and Production
P.O.Box 14100, 00076 Aalto, Finland
Phone: +358 9 470 23558,
E-mail: panu.kiviluoma@aalto.fi
http://edp.aalto.fi/en

9. ADDITIONAL DATA ABOUT AUTHORS

Peltonen, Olavi, B.Sc (Tech)
E-mail: olavi.peltonen@gmail.com

Orhanen, Samppa, B.Sc (Tech)
E-mail: smorhanen@gmail.com

Venäläinen, Janne, B.Sc (Tech)
E-mail: janne.venalainen@outlook.com

Auvinen, Matti, B.Sc (Tech)
E-mail: matti.a@outlook.com

Sepponen, Raimo, D.Sc (Tech), Prof
E-mail: raimo.sepponen@aalto.fi

Kuosmanen, Petri, D.Sc (Tech), Prof
E-mail: petri.kuosmanen@aalto.fi

10th International DAAAM Baltic Conference
"INDUSTRIAL ENGINEERING"
12-13th May 2015, Tallinn, Estonia

ADAPTIVE SELECTIVE LASER SINTERING TESTING DEVICE FOR PROCESS RESEARCH IN 3D PRINTING

Pomell, J.; Silvonen, A.; Lagus, N.; Kim, H.; Partanen, J.; Kiviluoma, P. & Kuosmanen, P.

Abstract:
Selective Laser Sintering (SLS) has been one of the most promising fields of study among different 3D printing techniques. The purpose of this study is to increase the knowledge of SLS printing and to give better understanding about the factors which affect the quality of printed parts. The article introduces a testing device which is designed especially for experimenting the capabilities of SLS. As a result, a device that enables high modifications from the user was built. The device includes an easily-replaceable and adjustable powder distributor mechanism, a large printing volume, powerful laser-scanner combination and durable building materials. With the construction, user is able to research various parameter effects, such as laser scanning speed, laser power, effect of preheating, printing material, layer thickness and layer building speed.
Keywords: Selective Laser Sintering, plastic, powder, parameter

INTRODUCTION

Rapid prototyping methods and additive manufacturing methods are technologies, which have been developing intensively during past decades. Most commonly these technologies concern 3D printing using polymers and plastics. Selective laser sintering (SLS) is among additive manufacturing methods enabling building components from powdered materials. Technique uses a laser as the power source to sinter powdered material layer by layer to form 3-dimensional objects.

Usually SLS involves the use of a high power laser, e.g. a carbon dioxide laser, to fuse small particles of plastic, metal, ceramic, or glass powders into a mass that has a desired three-dimensional shape. The laser selectively fuses powdered material by scanning cross-sections generated from a 3D digital description of the part from a CAD file or scan data on the surface of a powder bed. After each cross-section is scanned, the powder bed is lowered by one layer thickness, a new layer of material is applied on top, and the process is repeated until the part is completed.[[1]]
Because finished part density depends on peak laser power rather than laser duration, a SLS machine typically uses a pulsed laser [[2]]. The SLS machine preheats the bulk powder material in the powder bed somewhat below its melting point to make it easier for the laser to raise the temperature of the selected regions the rest of the way to the melting point. Unlike some other additive manufacturing processes, such as stereolithography (SLA) and fused deposition modeling (FDM), SLS does not require support structures due to the fact that the part being constructed is surrounded by unsintered powder at all times, this allows for the construction of previously impossible geometries.
The concept for process of selective laser sintering (SLS) was invented in mid-eighties and patented in 1986[[3]]. Over the last decades 3D printing has been a popular field of study and the scientific field of 3D printing is truly widespread because there are multiple different 3D printing technologies available. When considering SLS technique, remarkable work has been done especially in studies of

binding mechanisms of powder particles, mechanical and material properties of printed parts [4][5][6][7]. Kruth et al [4] give a comprehensive overview on different binding mechanisms in selective laser sintering (SLS) and selective laser melting (SLM). The paper classified SLS and SLM processes into four main binding mechanics categories: "solid state sintering", "chemically induced binding", "liquid phase sintering – partial melting" and "full melting". Most commercial applications can be classified into the latter two categories. The understanding of binding mechanism is a prerequisite for developing SLS materials with better mechanical properties. The most relevant mechanical properties studied are yield strength, elongation, Young's modulus, hardness, surface roughness, linewidth, layer thickness, shrinkage, porosity, wear rate, density, tensile strength and sintering depth [2]. Nowadays the range of printable materials is wide, ranging from metals to polymer and even on some ceramics, from single compound materials to powders that consist from multiple ingredients. [6]
The studies have also focused on other factors which affect the quality of final part. Studies discuss about the ways to spread the powder layer [8][9], importance of powder bed heating [10][11] and the effect of laser properties. When considering the laser used for SLS printing, the relevant factors have been laser power [6], scanning speed and spot size [12][5]. Relevant work has also be done in simulating the printing process and defining optimal parameter values with simulated models [13].
There have been only few companies that manufacture SLS machines and prices of these machines have been quite high, but more and more companies are moving in business. A recent development on commercial field has aimed to produce more available SLS printers for consumer and small business use. [1] The aim of this research is to develop an adaptive testing device to research process in SLS 3D printing method.

METHODS

In order to research the 3D-printing process in laser sintering method, the system needs to be highly adaptive. Three main functions and factors effecting to these functions in SLS printing process with polymers were recognized (Table 1). These were sintering with laser, layer building and material heating. To be able to research the process, the construction has to be adaptive in sense of sintering parameters, methods of building the layers and printing chamber heating system.
Sintering parameters concern the characteristics of the laser and the laser controlling system used. To discover optimal sintering parameters, research has to be done among the possible laser power, scanning/marking speed and laser spot size/focal length. Methods of building the layers are studied by implementing an easily replaceable powder distribution mechanism. This makes it possible to discover the most efficient way (roller, blade etc.) to spread a layer. The speed of the distribution mechanism has to be adjustable in order to research the impacts of spreading speed on layer quality. Due to the fact, that the sintering parameters also affect the depth of the sintered layer, the mechanism that defines the layer thickness has to be adjustable.

Sintering	Layer Building	Heating
heat capacity	layer area	heat capacity
laser wavelength	layer thickness	heat conduction
laser power	material flow	heater wavelength
layer thickness	particle size	layer volume
material IR absorption		material IR absorption
scanning speed		material melting point
spot distance		
spot size		

Table. 1. Main functions and effecting factors

The laser sintering system shall be constructed of robust materials to achieve high heat resistance. With the system, it is conceivable that printing material is changed and the influence of preheating the material is discovered.

RESULTS

As a result, a testing device was built. The device is a combination of a high power laser, adaptive powder distribution mechanism and spacious and durable design (Fig. 1). The laser used is a 100 W CO_2 laser combined with a scanning head. With the controlling software provided by the manufacturer, laser power and speed are controllable. The spot size of the laser depends on the lens that is used and laser's raw beam diameter. In our configuration, the lens has a focal length of 591 mm and a spot size of 614 μm.

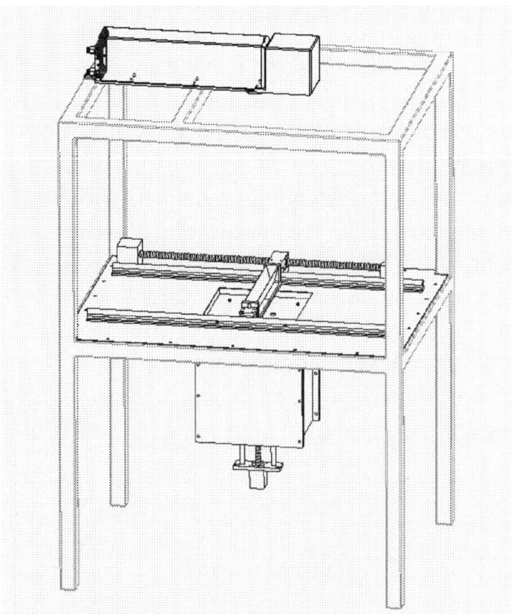

Fig. 1. The test construction

The powder distribution mechanism was constructed to be easily replaced if found out inoperative. Whereas commercial SLS stations usually spread the new layer by swiping with a blade or a roller, our construction implements a moving container (Fig. 2) that simultaneously provides the needed amount of powder and sweeps the layer. The gap between the two smoothing blades on the bottom of the construction is variable (between 0 and 10 mm) in order to find out the optimal value for different printing materials used. The minimum value to achieve the required material flow to produce good layer quality for nylon powder PA12, which was used in the first tests, was around 5 mm. The construction also enables the moving container to be replaced with different mechanisms so the most efficient spreading mechanism can be discovered.

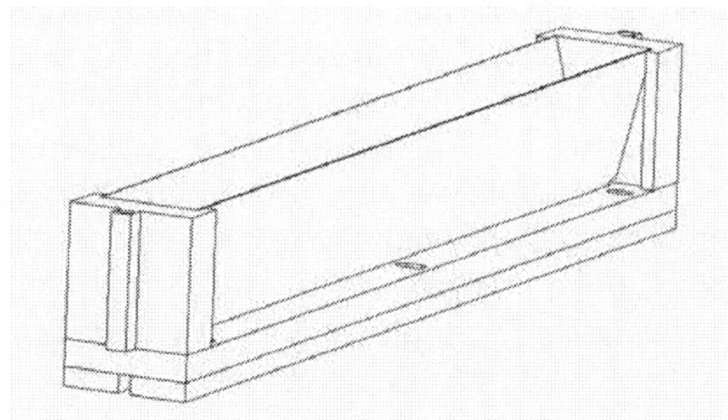

Fig. 2. Moving powder container

The layer thickness is defined as the distance of the building platform travelled in one cycle. The building platform is lowered with a stepper motor via a ball screw. With the combination selected, the layer thickness can be 100 μm or more.

Laser sintering of plastic materials requires preheating due to the strong heat expansion. Every layer is preheated somewhat below the sintering point with four 375 W infrared lamps. With the controlling system of the lamps the optimal value of preheating can be examined for different materials.

The control system of SLS printer can be implemented in many different ways. In this testing device (Fig. 3), the slicing of 3D models is executed with free slicing tools. The slicing tools were not in the scope of this study and the properties or functioning of these tools were not closely evaluated. Used slicing software can import 3D models in *stl* file format, export layers of the model in *png* format and let user to define the layer thickness. These images are moved to a specific directory, which works as a source directory for another software that was programmed for purposes of testing device. This program reads the image files from source directory and converts them to drawing files, which can be operated by the scanning head. The software can be manipulated with Visual Basic code using the ready-made functions of the scanner head, so it is easily adjustable for different needs of SLS research.

After the software convert the images and user starts a printing process, the software sends the first layer drawing to the scanning head, which generates the scanning track. Before the first scanning starts, the scanning head sends a signal to external control unit, which set the piston to the starting position and sweeps the first starting layer. The unit controls stepper motors though stepper motor controllers. Control unit also regulates IR heating, which raises the temperature of powder layer. When the first layer is built and heated, the scanning head turns the laser on and melts the powder according to the layer drawing. In the next phase the software sends the following layer to the scanner head and the cycle repeats until the part is finished.

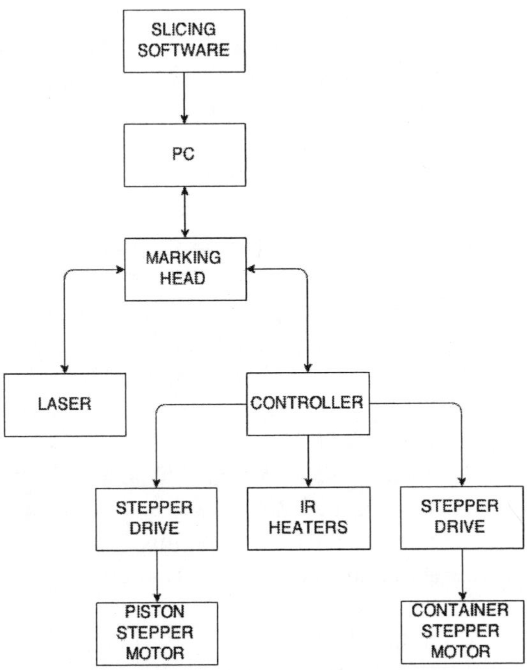

Fig. 3. Process block diagram

DISCUSSION

The design and building process of SLS printer pointed out extremely well the main challenges in SLS printers. Even though the SLS process is studied a lot and there are commercial solutions available, there are still many open design questions concerning laser type, laser positioning, beam sizes, powder distribution, heating and so on. Generally SLS devices are complex because of the challenging fine powder material and precise control and heating requirements.

The design process especially underlined the challenges arising from the heating. When sintering PA12 material without heating, the shapes experience significant deformation due to heat expansion, which leads to curved layers. The deformation happens after the laser has swept particular area and when the heated particles start to cool down to the room temperature. This deformation is especially problematic with SLS printers, where the new layer is swept above the previous layer. If the piston moves downwards one layer thickness, 100μm, there cannot be much curving upwards or the container will hit the printed layer and sweep it away. In this study too little attention was focused on heating. It can be stated that the main observation and outcome of this SLS project was that when designing a SLS device, considerable efforts has to be allocated to the design of the heating system.

IR-heaters should be chosen so that printing material is absorbent to their wavelength; hence required temperatures can be achieved with smaller power consumption. Heaters should also be fast, meaning that they achieve their maximum output in short time and that their power output can be changed quickly. This should make heating process faster and more refined. In our case IR lamp bulbs were chosen because they have short ramp up time, they were easy to use and install for prototyping purposes and they were accessible. Since the material is absorbent for wide range of IR wavelength, this wasn't considered as main constraint with heater selection.

The heating requirements lead also to other design challenges. When the temperature of the device should be around 180°C, also the structure, linear rails, electronics and other mechanical parts has to be designed so that they can withstand the heat without changing their properties. The temperature is especially difficult when considering moving parts like linear rails, motors and ball screws where proper lubrication is needed. Additionally, 3D printers have long operating cycles, which affects heat generation and wearing of the components.

Powder is a challenging material in 3D printers. Small grain size and good flowing properties are highly valued in SLS materials, but these properties also create design problems. The insulation of the piston is difficult to manufacture in a way that it keeps powder inside in high temperatures without leaking. Another problem was that the container can easily start to plow powder in front of it. Assumed reasons for plowing are the fact that the container is never perfectly aligned and that the powder bed is flexible. The powder bed compresses under the container and expands after the container has swept over it. Because of the plowing the powder piles up on the edges of containers operating range. The solution to this problem could be overflow bins, as presented in EOSINT P700 LS commercial device [[14]].

The design task was greatly simplified by using commercial laser and scanning head system. The scanning head has built-in tools and interface to change many properties, eg. laser power, scanning speed and area filling patterns. These functions are relatively difficult to implement in a testing system with reliable accuracy and this is why using commercial solutions is highly recommended. It was also possible to acquire extra accessories to the system, like in this study a new lens with a longer focal length. However, when using commercial devices it is extremely important to get familiar with all the properties, functions and restrictions of the device. For example in this study, the scanning head had a hard coded maximum value for lens's focal length, which made it difficult to operate scanning in the right scale.

REFERENCES

1. Gibson, I. & Rosen, D.W. & Stucker, B. 2010. Additive Manufacturing Technologies, Rapid Prototyping to Direct Digital Manufacturing. New York, New York, USA. Springer Science+Business Media. 447 s. ISBN: 978-1-4419-1119-3. e-ISBN: 978-1-4419-1120-9. DOI: 10.1007/978-1-4419-1120-9
2. Kumar, S. 2003. Selective laser sintering: A qualitative and objective approach. JOM. Vol 55. no 10. pp 43
3. US5597589. 1997. Apparatus for producing parts by selective sintering. Deckard, C.R. Patent Application US 08/251,609. Board Of Regents, The University Of Texas System.
4. Kruth, J.P. & Mercelis, P. & Van Vaerenbergh, J. & Froyen M. & Rombouts, L. 2005. Binding mechanisms in selective laser sintering and selective laser melting. Rapid Prototyping Journal. Vol 11. Iss 1. pp 26 - 36.
5. Caulfield, B & McHugh,P.E. & Lohfeld, S. 2007. Dependence of mechanical properties of polyamide components on build parameters in the SLS process. Journal of Materials Processing Technology. Vol 182. Iss 1–3. Pages 477-488. ISSN 0924-0136. http://dx.doi.org/10.1016/j.jmatprotec.2006.09.007.
6. Kruth, J.P. & Wang, X. & Laoui, T. & Froyen, L. 2003. Lasers and materials in selective laser sintering. Assembly Automation. Vol 23. Iss 4. pp 357 - 371.
7. Salmoria, G.V. & Leite, J.L. & Vieira, L.F. & Pires, A.T.N. & Roesler, C.R.M. 2012. Mechanical properties of PA6/PA12 blend specimens prepared by selective laser sintering. Polymer Testing. Vol 31. Iss 3. pp 411-416. ISSN 0142-9418.

8. Stichel, T. & Laumer, T. & Baumüller, T. & Amend, P. & Roth S. 2014. Powder Layer Preparation Using Vibration-controlled Capillary Steel Nozzles for Additive Manufacturing. Physics Procedia. Vol 56. pp 157-166. ISSN 1875-3892.
9. Van der Schueren, B. & Kruth, J.1995. Powder deposition in selective metal powder sintering. Rapid Prototyping Journal. Vol 1. Iss 3. pp 23 - 31.
10. Tontowi, A. E. & Childs T. H. 2001. Density prediction of crystalline polymer sintered parts at various powder bed temperatures. Rapid Prototyping Journal. Vol 7. Iss 3. pp 180 - 184.
11. Gibson, I & Shi, D. 1997. Material Properties and Fabrication Parameters in Selective Laser Sintering Process. Rapid Prototyping Journal Vol 3. Iss 4. pp 129–136.
12. Miller, D. & Deckard C.R. & William J. 1997. Variable Beam Size SLS Workstation and Enhanced SLS Model. Rapid Prototyping Journal. Vol 3. Iss 1. pp 4–11.
13. Ganeriwalaa, R. & Zohdib, R. I. 2014. Multiphysics modeling and simulation of selective laser sintering manufacturing processes. ScienceDirect. 6th CIRP International Conference on High Performance Cutting.
14. Pham, D. T., K. D. Dotchev, and W. A. Y. Yusoff. 2008. Deterioration of polyamide powder properties in the laser sintering process. Proceedings of the Institution of Mechanical Engineers, Part C: Journal of Mechanical Engineering Science. Vol 222. Iss 11. pp 2163-2176.

CORRESPONDING ADDRESS

Panu Kiviluoma, D.Sc. (Tech.), Senior University Lecturer
Aalto University School of Engineering
Department of Engineering Design and Production
P.O.Box 14100, 00076 Aalto, Finland
Phone: +358504338661
E-mail: panu.kiviluoma@aalto.fi
http://edp.aalto.fi/en/

ADDITIONAL DATA ABOUT AUTHORS

Pomell, Janne
E-mail: janne.pomell@aalto.fi

Silvonen, Atte
E-mail: atte.silvonen@aalto.fi

Lagus, Niko
E-mail: niko.lagus@aalto.fi

Kim, Hyunsoo
E-mail: hyunsoo.kim@aalto.fi

Partanen, Jouni, D.Sc. (Tech.), Professor
+358 50 5769 804
E-mail: jouni.partanen@aalto.fi

Kuosmanen, Petri, D.Sc. (Tech.), Professor
Phone: +358 500 448 481
E-mail: petri.kuosmanen@aalto.fi

10th International DAAAM Baltic Conference
"INDUSTRIAL ENGINEERING"
12-13 May 2015, Tallinn, Estonia

THE EFFECT OF FILTERS ON THE ACCURACY OF STEREOLITHOGRAPHY

Syvänen, J.; Karlén, N.; Salminen, P.; Kivekäs, K.; Kiviluoma, P.; Kuosmanen, P. & Partanen, J.

Abstract: *Stereolithography is a 3D printing technology in which photo-reactive resin is cured one layer at a time using a light source. The purpose of this study was to examine the effects of filtering of the exposuring light on the accuracy of stereolithography. Filtering only allows a certain bandwidth of wavelength to pass through, thus reducing undesired print-through errors caused by different wavelengths penetrating the resin at different depths.*

In this study, a 3D SLA (stereolithographic apparatus) printer was developed using a DLP projector in order to test different filters. Parts were printed with and without a filter and with blue light and the accuracy of their dimensions was compared. The results do not consistently show that accuracy is increased when projecting a narrow range of wavelengths. Filtering could be beneficial in stereolithographic applications in which there is a demand for a high degree of accuracy but the manufacturing time of parts is not critical.
Keywords: SLA, photopolymer, 3D printing, wavelength, print-through, DLP

1. INTRODUCTION

Computer-aided design (CAD) models can be fabricated directly using layer-based additive manufacturing (AM) processes. These processes create the models layer by layer making it possible to fabricate complex three-dimensional shapes that would otherwise be challenging. AM processes have been widely adopted in industrial product development as a method of creating prototypes due to rapid fabrication speed. [1] In recent years, the capabilities of AM technologies have significantly improved. As a result, they are being increasingly employed for the production of end-use products or their parts, mainly for geometrically complex products. [2]

Stereolithography is an additive manufacturing process in which three-dimensional parts are created by curing a resin (liquid photopolymer) one layer at a time on a platform. The layers are stacked on top of one another by moving the platform. [3] The basic operating principle of a stereolithographic apparatus (SLA) is presented in Figure 1. The process, which takes advantage of photopolymerization, was invented and patented in 1986 by Charles W. Hull [4]. Polymerization is the process of linking small molecules (monomers) into larger molecules (polymers) comprised of multiple monomer units [5].

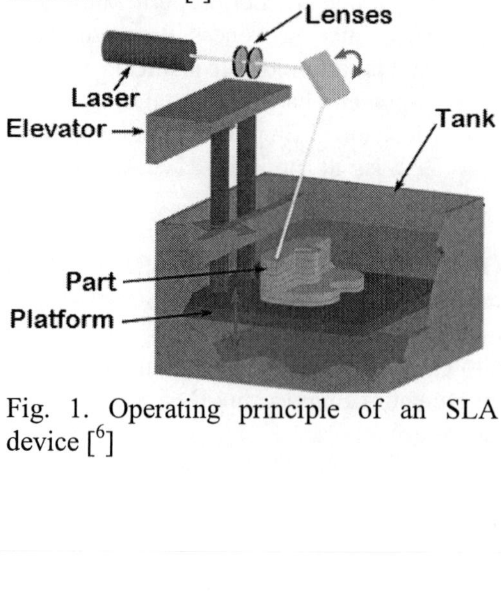

Fig. 1. Operating principle of an SLA device [6]

In stereolithography, cross-sectional patterns of the object are created at a selected surface of a resin capable of altering its physical state in response to appropriate synergistic stimulation [4]. The resin is usually cured with light rays of the ultraviolet region or by a high power laser [7]. Stereolithography has various uses ranging from biomedical engineering applications to producing architectural scale models for design evaluation and demonstration purposes [8] [9].

Projection stereolithography (PSLA) is a variation of the SL process. In PSLA, a projector is used as the source of light and the layers are formed on a platform one at a time by projecting a cross-sectional image of the model for that particular layer. Layer thickness is usually under 300 μm. [7]

The purpose of this study was to build a small and portable PSLA device and to test the effect of filters and color adjustment on the accuracy of the device. By default the projector produces white light which contains various wavelengths. The benefit of adjusting the output color according to the requirements of the resin is that the wavelength bandwidth output of the projector becomes narrower. A filter has the same effect. A narrower bandwidth of light should reduce unwanted curing of the resin as different wavelengths achieve different curing thicknesses. Thus, these print-through errors could be drastically reduced by limiting the bandwidth of light used to cure the resin, improving the accuracy of the created parts.

2. METHODS

The PSLA was aimed to be as compact as possible. The device was constructed with a bottom-up configuration (Fig. 2), which means that the light is projected from below and up through the transparent vat before radiating the resin. Two mirrors were included in the device in order to limit the size and to increase the length of the path which the light travels from the projector to the bottom of the resin tank. The light path length was 11.5 cm. The vat was coated with a silicone elastomer in order to prevent the cured layers from sticking to the vat.

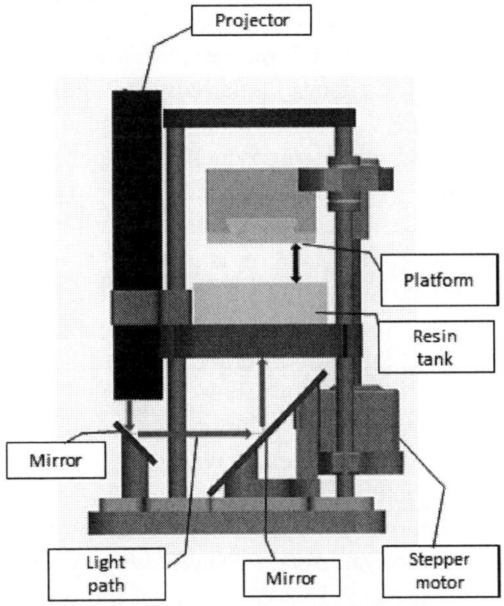

Fig. 2. Configuration of the device

The printing procedure is done in the following manner: first a 3D CAD model is brought in to a slicer software (Creation Workshop). This software slices the 3D model into layers and creates pictures, or slides, for each layer. Prior to curing of the first layer, resin is placed into the resin tank and the platform is lowered to a position where there is a gap between the silicone layer and the platform. The gap is equal to the set layer thickness, for example 100 μm. As the platform is lowered the resin is displaced leaving only a resin layer that fills the gap between the silicone layer and the platform. Next the first slide is projected on the resin layer which leads to curing of the first layer. After a layer is cured, the platform is lifted up in order to detach the cured layer from the coated vat and to let new resin flow in under the platform. Finally the platform is again lowered to the position where the appropriate gap between the previous layer and the platform is again created. This procedure is repeated until the complete height of the model is reached with the individual layers.

The projector used was a LED-based digital light processing (DLP) pocket projector with a native resolution of 640 x 480 Pixel (VGA) 4:3. DLP projectors utilize a digital micromirror device (DMD) that contains hundreds of thousands microscopic mirrors arranged in a rectangular array which correspond to the pixels in the projected image [10]. The output Peak Lumen of the projector was 50 Lumen. A band-pass filter was used to selectively transmit a portion of the projector spectrum in order to inspect the effect of light filtering on print accuracy. The filter had a center wavelength of 450 nm and a bandwidth of 10 nm. In addition to filtering, tests were also done by projecting blue light instead of white light. The mirrors were commercial grade first surface mirrors, coated with enhanced aluminum. They had a minimum reflection percentage of 90 % for wavelengths between 400 and 650 nm. The resin was Envision TEC® RCP30.

Spectrum measurements were conducted with a spectrometer. Tests were done with the following setups:

1. Projector, white light (R:255, G:255, B:255)
2. Projector + filter
3. Projector, blue light (R:0, G:0, B:255)

Layer thickness was set to 100 μm and the minimum curing for each setup was found through trial and error. Curing time was increased until successful prints were produced. For comparing the lateral accuracy between the different setups, a chessboard shape (Figure 3) was printed. For comparing the vertical accuracy a part with an overhanging roof (Figure 4) was printed with each setup. The results were visually inspected and compared with a microscope.

Fig. 3. Chessboard shaped test part

Fig. 4. Test part with overhanging roof

A summary of the test areas, their research questions and the methods and tools used can be seen in Table 1.

Test area	Research questions	Methods/tools
Curing time	What is the suitable radiation time for certain thicknesses?	Trial and error/PSLA device
Projector	Emitted wavelengths	Spectrum measurement/ spectrometer
Filter	Transmitted wavelengths	Spectrum measurement/ spectrometer
Accuracy	Which setup produces parts that represent the slide shapes best?	Visual inspection/ microscope

Table 1. Test areas

3. RESULTS

The results of the spectrometer measurements can be seen in Figures 5-7. Figure 5 shows the results for the projector when white light was projected. Figure 6 displays the results for the transmitted wavelengths when the filter was used. Figure 7 contains the results for the projector when blue light was projected.

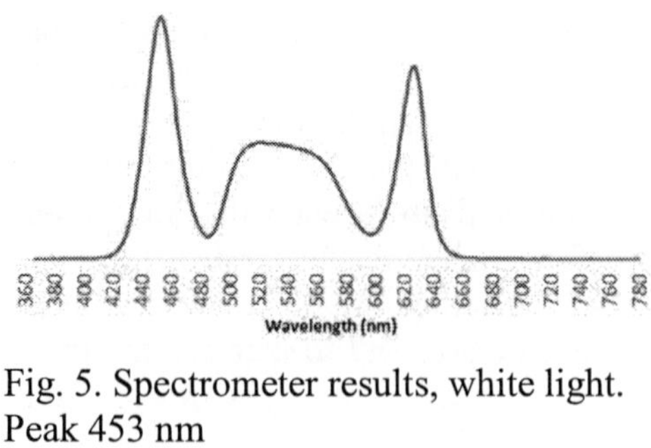

Fig. 5. Spectrometer results, white light. Peak 453 nm

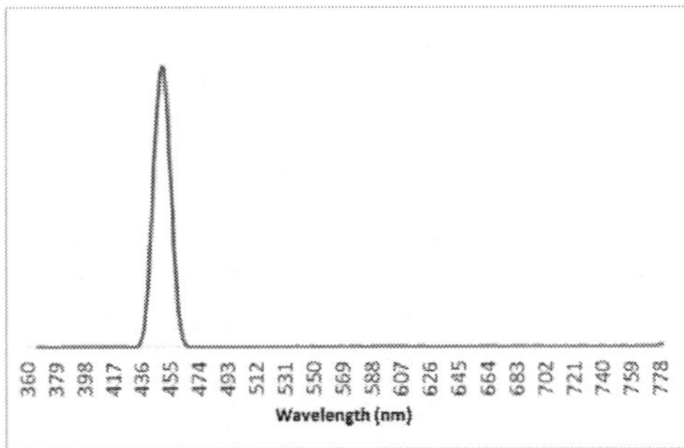

Fig. 6. Spectrometer results, with filter. Peak 450 nm

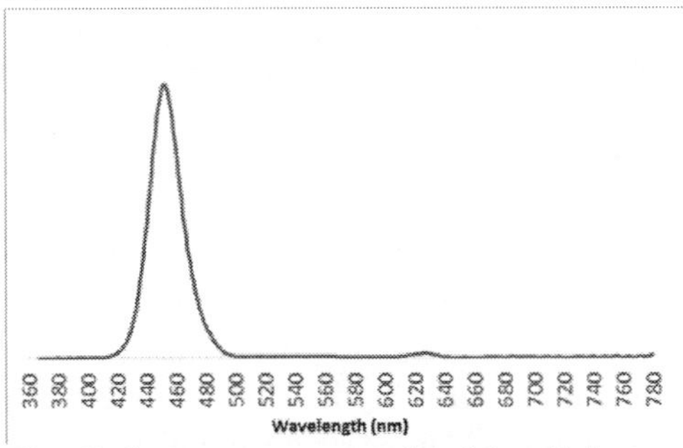

Fig. 7. Spectrometer results, blue light. Peak 453 nm

Attempts to cure the resin using the filter failed. Curing times up to 180 seconds were tried, but only very small amounts of the resin was cured. Therefore no printed parts were produced with the filter attached. The minimum curing time when projecting white light was 45 seconds. When blue light was projected the minimum curing time for producing successful parts was 90 seconds.

Figure 8 shows how the chessboard shape should have been printed, the width and height was 10 mm x 10 mm and the thickness was 1 mm. In Figure 9, the printed chessboard shaped parts can be seen.

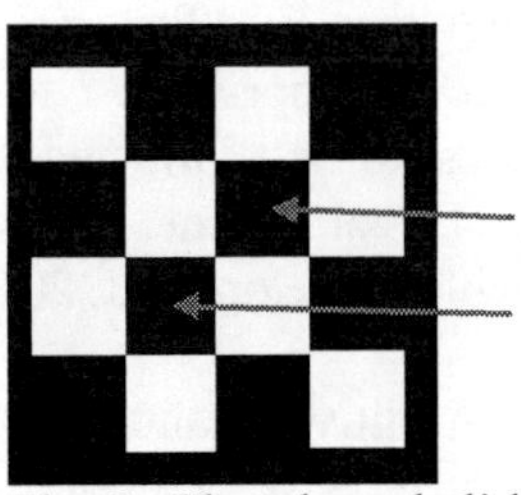

Fig. 8. Chessboard slide

Fig. 9. Part created with white light on the left, with blue light on the right

The chessboards looked fairly similar but the print produced with blue light does have sharper edges than the print produced with white light. This can be seen when looking at the two squares pointed at in Figure 8. These squares are slightly bigger in the part produced by blue light, which indicates that unwanted curing occurred more with the white light.

Figure 10 shows the part with overhanging roof printed with white light (upper part) and blue light (lower part). The height of the parts was 3.0 mm and the sides were 9.5 mm x 9.5 mm. With this part, more unwanted curing can be seen in the part printed with blue light. In the part produced with white light, the layers are more visible. These results might indicate that the curing time was slightly too low for the upper part and slightly too high for the lower part.

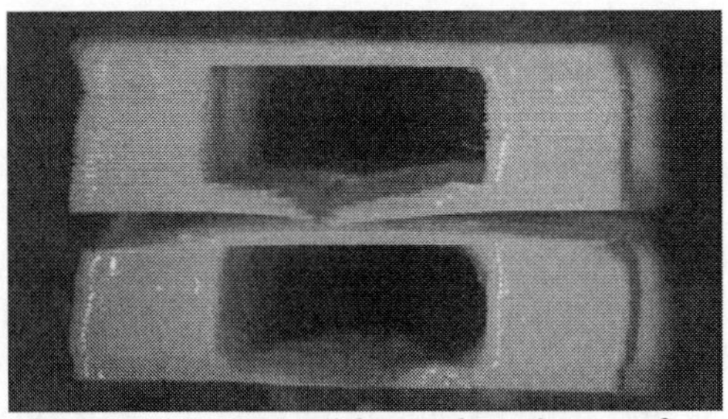

Fig. 10. Test part with overhanging roof. Upper part printed with white light, lower part printed with blue light

4. DISCUSSION

As the images show, there is not much perceivable difference between the parts created using white light compared to those created using blue light. The printed rectangular shapes of the chessboard patterns appear to be larger than intended and their edges are rounder than in the CAD model when cured with either white or blue light. These errors are, however, less pronounced on the part that was created using blue light. Even so, the inaccuracies are still clearly visible. This could be at least partly due to cross-talk effect which causes polymers near the light to also get cured [7]. Another explanation for why the part cured with white light is less accurate could be chromatic aberration which causes different wavelengths to reflect in slightly different angles from the mirrors used in the PSLA device [11].

Another difference between the usage of the two different colors of light in the printed parts is that the layers are more difficult to separate from each other in the overhanging part that was created using blue light than they are in the part created with white light (Fig. 11). The layers of the part created with white light have consistent protruding sections which could be indicative of print-through errors. They are not visible in the part that was cured with blue light as the layers of that part appear smoother and more consistent. It is possible, however, that the curing time was not high enough for white light. This would also explain why there is more unwanted cured mass visible on the overhanging roof of the part that was cured with blue light despite the layers being seemingly smoother.

White light proved to cure the layers faster, which was expected. This is due to the higher luminous flux. When limiting the wavelength bandwidth to the spectrum of blue light the luminous flux is also limited. Blue light required approximately 50% higher curing times per layer than white light. As mentioned earlier, applying the filter caused the curing to become very slow. Curing times were increased up to threefold for the filter in comparison to white light but even that was not enough. It appears that the filter reduces the luminous flux far more than simply changing the output color does.

The results have proven moderately inconclusive. This is a consequence of the inaccuracy of the created PSLA device. The device uses two mirrors to guide the light into the tank which can cause the image to get distorted unless the mirrors are lined up accurately. The mirrors were also not attached strongly enough to the base of the device, which may have allowed them to move slightly while the parts were being printed. The platform mechanism of the device also proved to be a source of inaccuracy as it was not strong enough and easily changed its orientation as its edges got stuck in the silicone of the tank when creating the first few layers of a part. These issues demonstrate that in order to have a great degree of accuracy when creating parts with a PSLA device, the device itself must also be constructed with great precision and of strong materials. The 3D-printed parts used in the construction of the device proved to be too weak and inaccurate. The next logical step of this research would be to build a more rigid device and to conduct more tests in order to obtain more reliable results.

5. REFERENCES

[1] Pan, Y.; Zhou, C. & Chen, Y. A Fast Mask Projection Stereolithography Process for Fabricating Digital Models in Minutes. Journal of Manufacturing Science and Engineering, 2012, **134:5**.

[2] Pearche, J. M.; Blair, C. M.; Laciak, K. J.; Andrews, R.; Nosrat, A. & Zelenika-Zovko, I. 3-D Printing of Open Source Appropriate Technologies for Self-Directed Sustainable Development. Journal of Sustainable Development, 2010, **3:4**, 17-29.

[3] Miller, R. Additive Manufacturing (3D Printing): Past, Present and

Future. Industrial Heating, 2014, **82:5**, 39-43.
[4] US4575330 A. Apparatus for production of three-dimensional objects by stereolithography. Uvp, Inc., San Gabriel, Calif. (Hull, C. W.) US 06/638,905, 08.08.1984. Publ. 11.03.1986. 16 p.
[5] Jacobs, P. F. Rapid Prototyping & Manufacturing: Fundamentals of Stereolithography. SME, Dearborn, 1992.
[6] Stereolithography [WWW] http://www.custompartnet.com/wu/stereolithography (25.10.2014)
[7] Rayat, H.; Pulkkinen, V.; Johansson, M.; Nyfors, V.; Partanen, J.; Kiviluoma, P. & Kuosmanen, P. The Effects of Minituarisation of Projection Stereolithography Equipment on Printing Quality. In *Proceedings of the 9th International Conference of DAAAM Baltic, INDUSTRIAL ENGINEERING*, DAAAM ESTONIA, Tallinn, 2014, 278-282.
[8] Giannatsis, J.; Dedoussis, V. & Karalekas, D. Architectural Scale Modelling Using Stereolithography. Rapid Prototyping Journal, 2002, **8:3**, 200-207.
[9] Melchels, F. P. W.; Feijen, J. & Grijpma, D. W. A review on stereolithography and its applications in biomedical engineering. Biomaterials, 2010, **31:24**, 6121-6130.
[10] Douglass, M. Lifetime Estimates and Unique Failure Mechanisms of the Digital Micromirror Device (DMD). 36th Annual Reliability Physics Symposium Proceedings, 1998. 31.3-2.4.1998. IEEE Reliability Society. 2004. P. 9-16. ISBN 0-7803-4400-6.
[11] Marimont, D. H.; Wandell, B. A. Matching color images: The effects of axial chromatic aberration. Journal of the Optical Society of America A, 1994, 11 (12):3113. doi:10.1364/JOSAA.11.003113

CORRESPONDING ADDRESS

Panu Kiviluoma, D.Sc. (Tech.), Senior University Lecturer
Aalto University School of Engineering
Department of Engineering Design and Production
P.O.Box 14100, 00076 Aalto, Finland
Phone: +358504338661
E-mail: panu.kiviluoma@aalto.fi
http://edp.aalto.fi/en/

ADDITIONAL DATA ABOUT AUTHORS

Kivekäs, Klaus
Phone: +358405262937
E-mail: klaus.kivekas@aalto.fi

Syvänen, Janne
Phone: +358407184606
E-mail: janne.syvanen@aalto.fi

Karlén, Niko
Phone: +358503579098
E-mail: niko.karlen@aalto.fi

Salminen, Pontus
Phone: +358408350189
E-mail: pontus.salminen@aalto.fi

Kuosmanen, Petri, D.Sc. (Tech.), Professor
Phone: +358500448481
E-mail: petri.kuosmanen@aalto.fi

Partanen, Jouni, D.Sc. (Tech.), Professor
Phone: +358505769804
E-mail: jouni.partanen@aalto.fi

10th International DAAAM Baltic Conference
"INDUSTRIAL ENGINEERING"
12-13 May 2015, Tallinn, Estonia

CALCULATION OF PLATE PLANE MOTION PARAMETERS USING INERTIAL MEASUREMENT SYSTEM

Zhigailov, S.; Verchenko, A.; Musalimov, V.; Aryassov, G.

Abstract: *This article is a logical continuation of the work [1], which deals with aspects of simulating the movement of the pelvis person in the vertical plane prototype design. More detailed attention has the measurement system used to obtain experimental data. The main elements of the proposed inertial measuring system are two accelerometers with gyroscopes and controller. An algorithm for processing the experimental data to calculate the parameters of motion of the plate simulating the pelvis is introduced in this article too. As a result the comparison of the found parameters of plate plane motion is done. Also the overall effectiveness of the proposed method of processing the experimental signals is evaluated.*

Key words: pelvic motion, IMU, signal processing, data filtering.

1. INTRODUCTION

In recent years for successful treatment diseases of musculoskeletal system (MSS) there have been many modifications of medical rehabilitation facilities and trainers, some of them have proved successful in practice. The most successful models of such systems make it possible to provide feedback to the patient, as well as take into account individual physiological and anthropometric characteristics of a particular patient in the preparation of the course of restoration programs [3]. However, talking about the possibility of using these trainers at each facility is inappropriate. It's primarily because of their high cost.

One of the main ways to reduce the high cost of medical rehabilitation systems is using of lower-cost measurement tools - inertial sensors (accelerometers primarily) for the study of personal human movements [2]. Nowadays there is no single model and complete description of human MSS using inertial sensors. It means that this area of biomechanics requires further study. The main task of this work is to show one of the ways how to process inertial sensors signals describing pelvis prototype motion and evaluate obtained results.

2. UNFILTERED DATA AND DIRECT INTEGRATION

After the design, assembly and test of a prototype, there were made experimental measurements of plane motion of the plate simulating the pelvis.

As a result, for the experiments 3 different angle of rotation of the support bearing, 3 combinations of load plate, simulating the pelvis and 3 modes of duration of discrete steps were selected. So, the total number of held experiments is 27 (3 angles x 3 loads x 3 durations of steps).

Figure 1 shows the data obtained with one of accelerometers for one of experiments.

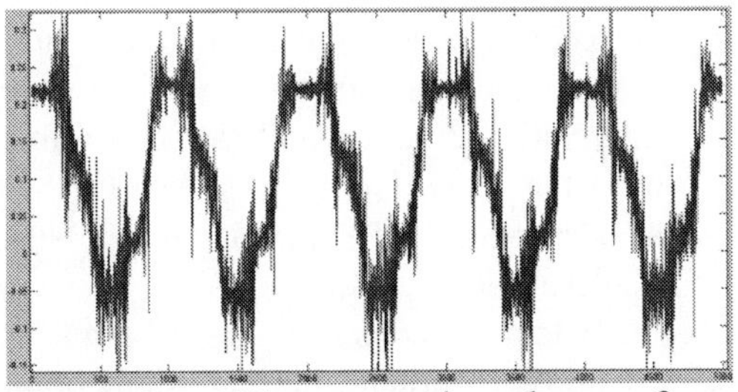

Fig. 1. Unprocessed data from accelerometers.

After the initial data was obtained, it was attempted to integrate the acceleration and angular velocities data along the axes Y and Z two times to get displacements along axes named axes.

As we can see in Figure 2 (axis Z, different colors mean different accelerometers), direct integration of unfiltered data can`t be used to get sufficient displacements.

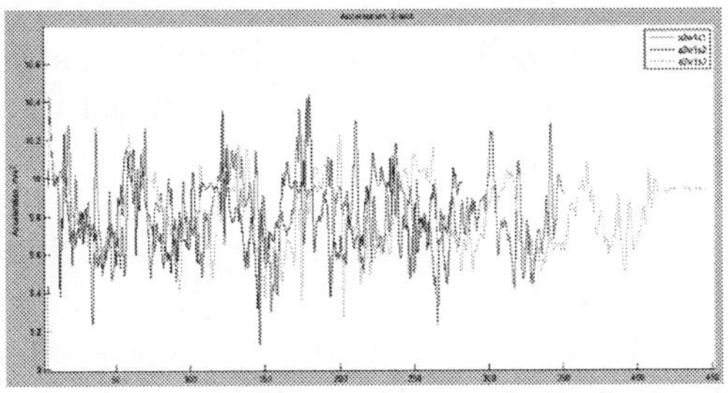

Fig. 2. Displacement along axis Z obtained from unfiltered data.

The main conclusions can be made from the visual analysis of experimental results:

1) The initial results obtained by Y axis are cyclical. It is possible to distinguish the individual steps of them.

2) There are too big noises on plots, which obliges us to filter obtained signals.

3) Some additional operations for signal processing for accelerometers data are necessary.

3. EXPERIMENTAL DATA PROCESSING

After the elimination of noise with low-frequency Butterworth filter, we got the following graphics accelerations for axes Y and Z in time (Figure 3).

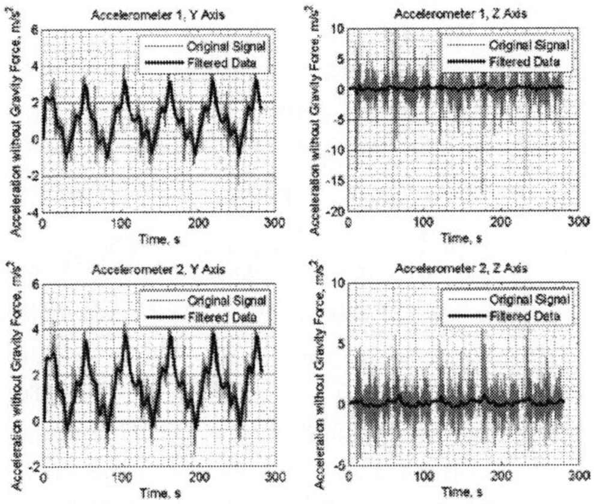

Fig. 3. Acceleration without Gravity Force.

In order to make the acceleration corresponding to planned work of prototype, it was decided to align the acceleration along the horizontal axis. To do this, all the elements of acceleration arrays were summed and the overall rate of displacement shifted from the horizontal axis was obtained. Then, the overall rate was divided on the number of measurements (elements of acceleration array) and the resulting quotient of the division is subtracted from all elements of the array. As a result, after the filtering operations were held, we got data shown in Figure 4.

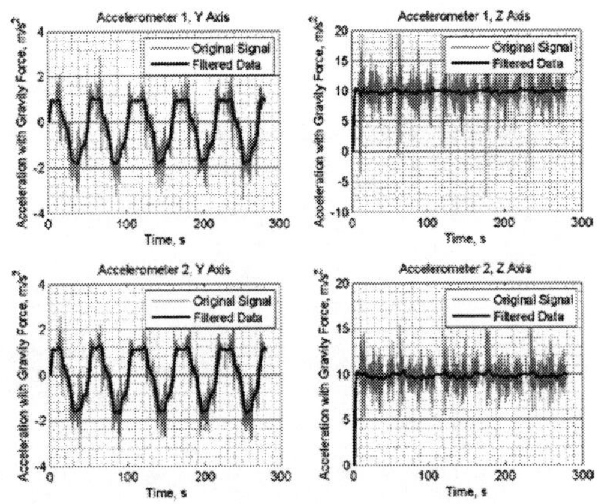

Fig. 4. Acceleration with Gravity Force.

After compilation and testing of various kinds of filters, the choice fell on the low-pass filter Butterworth, created in the DSP System Toolbox.

Butterworth filter designed so that its amplitude-frequency characteristic was as smooth as possible on the frequency bandwidth. At frequency suppression band, the frequency response in Butterworth filter reduces to zero.

As for frequency characteristics, the low pass filter we use for processing signals, passes the frequency spectrum efficiently below the cutoff frequency and suppresses signal frequencies above cutoff frequency. This option allows us to get rid of the noise appearing when we use instant discrete accelerations recorded by the accelerometer.

The order of the filter is equal to two, infinite impulse response.

After accelerations were aligned on an axis, it was necessary to calculate how good mechanical prototype has been assemblied and tested.

Since discrete accelerations were obtained it was difficult to evaluate accelerometers data. So it made sense to evaluate the curves of filtered data. Furthermore, as a comparative standard, was obtained additional middle curve whose coordinates received from adding the filtered first and second signals for the corresponding sensor axis for each mode. The result is shown in Figure 5.

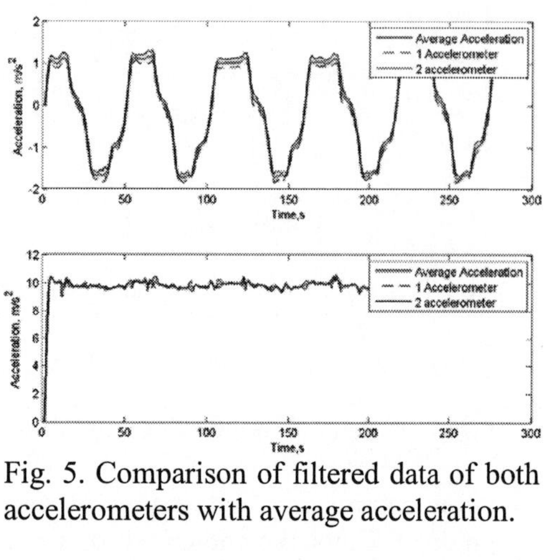

Fig. 5. Comparison of filtered data of both accelerometers with average acceleration.

Blue dotted line marks filtered acceleration curve for the first accelerometer, black solid line - filtered acceleration curve of the second accelerometer, the red solid line - the average curve of the filtered accelerations, yellow color - area referring to the difference between the measurements of the first and second accelerometers.

Since data of accelerations for both sensors should be identical, there are defects in the design and assembly of the prototype. What are some of these defects:

-Possible errors in the installation accelerometer on the plate,

-Incomplete relevant parts of the mechanical design of the prototype,

-Gaps in the joints,

-Incomplete fixation of plate in vertical plane. Full fixation of plate in vertical plane with bearings is not possible due to the occurrence of friction forces between the plate and bearings.

Analyzing it can be noticed that the biggest difference between the filtered acceleration sensors and the curve of averaged filtered acceleration is in those places where the angle between the filtered data and the horizontal axis (absolute zero) is the least.

For greater consistency and to reduce the amount of data to be processed we continue to process signals only for the curve of average values of the filtered acceleration.

After the average curve of the filtered accelerations was found, the values of gravitational constant g were subtracted of it (Figure 6).

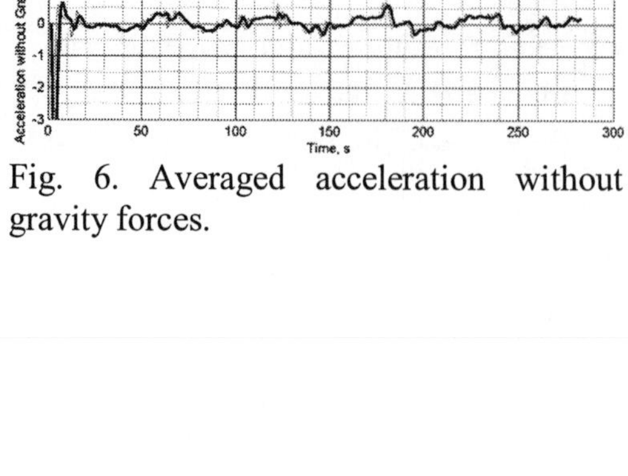

Fig. 6. Averaged acceleration without gravity forces.

Red color on the graph represents the original signal, black color means the signal after passing through the filter.

It is worth noting that coefficient of gravity would be impossible to find with only accelerometers. It is necessary to combine the accelerometers and gyroscopes for matching measured angles with accelerations to determine the tilt of the plate and see how gravity coefficient is distributed along the axes Y (horizontal axis) and Z (vertical axis). At a strainght plate position when the rails are on the same level the gravitational component affects only the axis Z. The figure 7 shows distribution of gravity forces depending on the time and position of the plate by axes Y and Z.

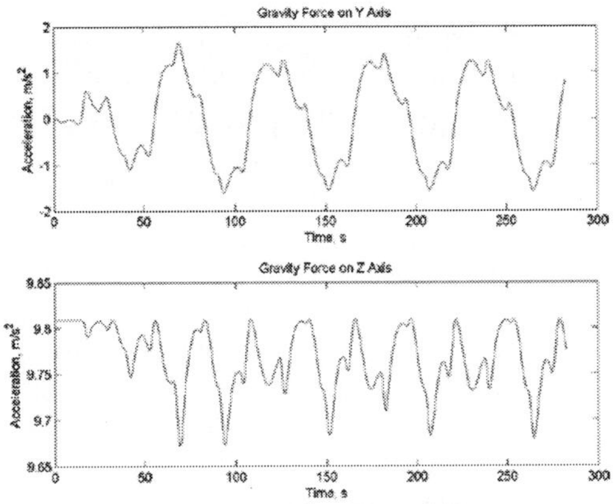

Fig. 7. Gravity Force by Y and Z axes.

The angular velocity obtained from gyroscopes is represented in Figure 8.

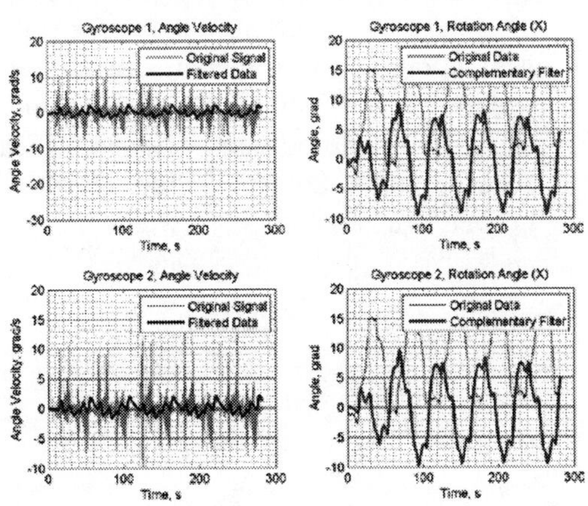

Fig. 8. Angular velocity and angles obtained from gyroscopes

To compensate drifts in the corners complementary filter is used. It plays a role of a compromise between the data collected from the accelerometer and gyroscope. The fact is that the accelerometer is under the influence of external forces that do not affect the gyroscope. Oppositely the gyro drift has no influence to accelerometer. Complementary filter eliminates the drift angle of up to 10 degrees as we can see from Figure 8.

Next we find the linear velocities by axes Y and Z (Figure 9).

The integration process used in the signal processing is always the same in essence. For example, if we want to integrate the acceleration to find a linear velocity, we use the expression:

$$V(i+1) = A(i) * (T(i+1) - T(i)) \quad (1)$$

where V – value of linear velocity , A – value of acceleration, T – time, i – number of elements in matrixes of accelerations, velocities, displacements of time.

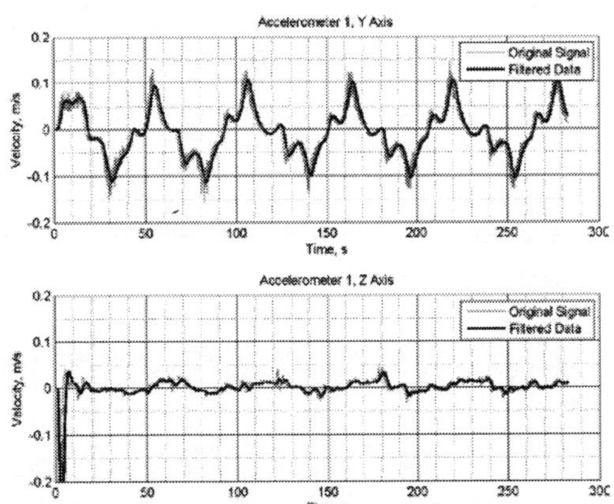

Fig. 9. Obtained linear velocity from filtered average acceleration curve

After we have found values of linear velocities, we integrate velocities to get real displacements by axes Y and Z (Figure 10).

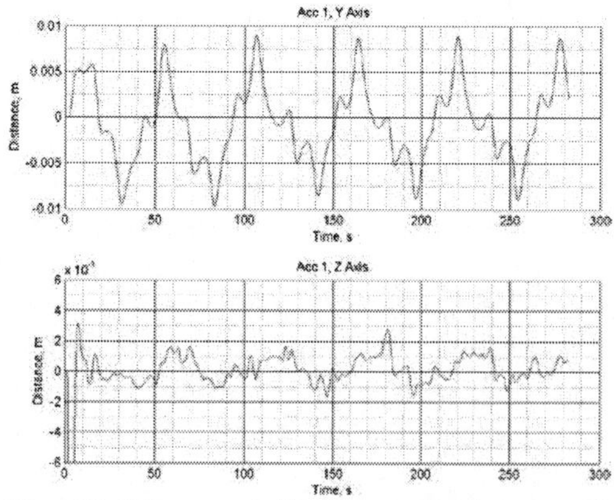

Fig. 10. Measured displacements by axes Y and Z.

In addition to the plots of the plate movements in the measured points by axes Y and Z, some interest has movement in vertical plane (Figure 11).

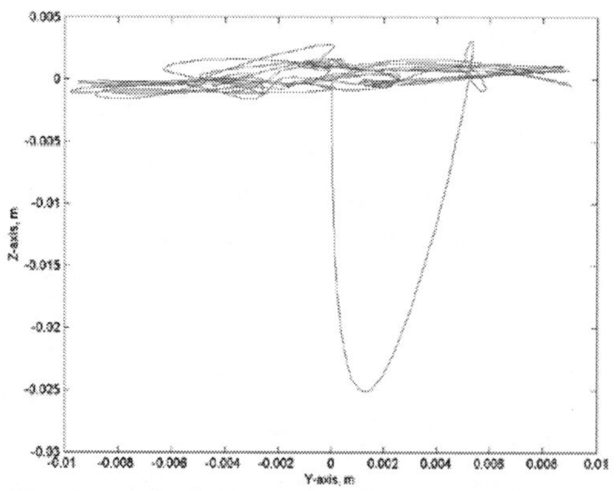

Fig. 11. Displacements of measured data in vertical plane

As we can see, the graph of movements looks pretty closely-grouped. The only loop standing out from the accuracy appears when we launch the prototype to make the first experiment.
Then we make the same operations for all 27 experiments and get displacements shown in Table 1.

Experiment	Axis Y	Axis Z
a1w1s1	0,016970	0,260571
a1w1s2	0,016627	0,254571
a1w1s3	0,015646	0,230385
a1w2s1	0,016136	0,273472
a1w2s2	0,016858	0,254695
a1w2s3	0,016897	0,239517
a1w3s1	0,018125	0,303382
a1w3s2	0,016271	0,272152
a1w3s3	0,015088	0,272789
a2w1s1	0,019467	0,269273
a2w1s2	0,017848	0,232846
a2w1s3	0,017139	0,262471
a2w2s1	0,016673	0,239904
a2w2s2	0,017796	0,255086
a2w2s3	0,020343	0,268898
a2w3s1	0,020731	0,325544
a2w3s2	0,019718	0,267148
a2w3s3	0,020979	0,266753
a3w1s1	0,020191	0,273343
a3w1s2	0,018971	0,280425
a3w1s3	0,019566	0,284237
a3w2s1	0,02154	0,293620
a3w2s2	0,023211	0,287130
a3w2s3	0,021504	0,299449
a3w3s1	0,020242	0,310109
a3w3s2	0,021004	0,290831
a3w3s3	0,021305	0,294809

Table 1. Displacements by Y and Z axes

Letter "a" means the angle value, ranging from 4,8 to 8°, letter "w" – the value of loads from 0,926 to 1,686 kg and letter "s" means duration of each step for stepper motors from 21 to 35 ms.

4. EXPERIMENTAL RESULTS EVALUATION

There are many different ways for statistic evaluations of experimental results. Let we use in this work Dixon`s Q-test, helping us to find possible outliers. The main idea of such Dixon`s Q-test is finding relations between gap and range, as well as:

$$Q = \frac{x_2 - x_1}{x_n - x_1} \quad (2)$$

The numbers $x_1 \dots x_n$ are ordered from smallest to largest, so that

$$x_1 \leq x_2 \dots \leq x_{n-1} \leq x_n \quad (3)$$

After calculation of coefficient Q, this coefficient needs to compare with theoretical coefficient given in [4]. If Q is larger than theoretical coefficient, experimental value is not outlier, if not – experimental value is outlier.

Analyzing visually displacements got in Table 1, the best way is to divide all 27 experiments on 3 parts: experiments 1-9 (angle 1), experiments 10-18 (angle 2), experiments 19-27 (angle 3).

As confidence level we choose 95%.

Experiment	Axis Y	Z-scores (Y)	Axis Z	Z-scores (Z)
a1w1s1	0,017	0,53	0,261	-0,52
a1w1s2	0,017	0,13	0,255	-0,77
a1w1s3	0,016	-1,00	0,230	-1,80
a1w2s1	0,016	-0,43	0,273	0,03
a1w2s2	0,017	0,40	0,255	-0,77
a1w2s3	0,017	0,44	0,240	-1,41
a1w3s1	0,018	1,85	0,303	1,30
a1w3s2	0,016	-0,28	0,272	-0,02
a1w3s3	0,015	-1,64	0,273	0,00
a2w1s1	0,019	0,31	0,269	-0,15
a2w1s2	0,018	-0,69	0,233	-1,70
a2w1s3	0,017	-1,13	0,262	-0,44
a2w2s1	0,017	-1,41	0,240	-1,40
a2w2s2	0,018	-0,72	0,255	-0,75
a2w2s3	0,020	0,85	0,269	-0,16
a2w3s1	0,021	1,09	0,326	**2,25**
a2w3s2	0,020	0,46	0,267	-0,24
a2w3s3	0,021	1,24	0,267	-0,25
a3w1s1	0,020	-0,51	0,273	0,03
a3w1s2	0,019	-1,48	0,280	0,33
a3w1s3	0,020	-1,01	0,284	0,49
a3w2s1	0,022	0,56	0,294	0,89
a3w2s2	0,023	1,88	0,287	0,61
a3w2s3	0,022	0,53	0,299	1,14
a3w3s1	0,020	-0,47	0,310	1,59
a3w3s2	0,021	0,13	0,291	0,77
a3w3s3	0,021	0,37	0,295	0,94

Table 2. Results of Dixon`s Q-test for axes Y and Z.

As result, we can see that the only coefficient Q for experiment a2w3s1 exceed possible limits. The most reason of this is strong vibrations on springs of experimental prototype appeared during this mode of experiment.

All remain result are tending to be normal.

5. CONCLUSION

Two ways of finding displacements of pelvis prototype during motion were analyzed. Direct integration did not have good results, so the need to use filters appeared. Low-pass filter was used to avoid noises. Moreover it allowed to obtain averaged accelerations from accelerometer and angular velocities from gyroscopes. To compensate floating angle complimentary filter was used. Finally, integrated displacements correspond to distances in real experiments.

As statistical basis for experiment result evaluation Dixon`s Q-test was chosen. One outlier was found.

To sum up, processed data corresponds with real motion of plate and can be used as the way for processing of inertial measurement units (ex. Accelerometers) signals.

6. REFERENCES

[1] Zhigailov, S., Musalimov, V., Aryassov, G., Design of experimental stand for human gait imitation. *In: Proceedings of the 9th International Conference of DAAAM Baltic, INDUSTRIAL ENGINEERING, (toim.)* Tallinn University of Technology, Tallinn, 2014, pp. 300-304.

[2] Zhigailov, S., Kuznetcov, A., Musalimov, V., Aryassov, G. Measurement and analysis of human lower limbs movement parameters during walking, *In: Solid State Phenomena* (Valiulis A.V., Černašėjus O., Mokšin V., eds.), Trans Tech Publications Ltd, 2015, 220-221, 538-543.

[3] Aryassov, G., Kuznetcov, A., V. Musalimov, V., Zhigailov, S. *Lower Limb Mathematical Modeling with Inertial Motion Capture during Normal Walking*, 4th International Conference on Integrity, Reliability & Failure, Funchal, (2013), (in press).

[4] Sebastian Rashka site http://sebastianraschka.com/Articles/2014_dixon_test.html

Sergei Zhigailov, Doctoral student, Tallinn University of Technology, Department of Mechatronics, Ehitajate tee 5, Tallinn, sergsil@gmail.com

Anton Verchenko, MSc student, ITMO, Kronverk prospekt 49, Saint-Petersburg, Russian
Federation, anton.verch@gmail.com

Viktor Musalimov, Professor, Head of mechatronics department, ITMO, Kronverk prospekt 49, Saint-Petersburg, Russian Federation,
musVM@yandex.ru

Gennady Aryassov, Ass.Professor, Tallinn University of Technology, Department of Mechatronics, Ehitajate tee 5, Tallinn, gennadi.arjassov@ttu.ee

10th International DAAAM Baltic Conference
"INDUSTRIAL ENGINEERING -
12-13 May 2015, Tallinn, Estonia

USE OF EXCELLENCE MODELS AS A TOOL FOR BENCHLEARNING AND KNOWLEDGE MANAGEMENT

Tammaru, T. & Kiitam, A.

Abstract: *In this paper some issues of using Excellence models as an efficient and innovative knowledge management and benchlearning tool, also an integrated approach to total quality management (TQM), knowledge management and benchmarking are discussed. The potential synergy resulting from the diversity of knowledge and experience of cross-functional self-assessment teams as well as external assessors of recognition schemes is outlined. Some results and conclusions based on the experience of recognition schemes in Estonia during past decades are presented. Need for supporting infrastructure for the promotion of performance excellence thinking and application, benchmarking and sharing best practices as well as alignment of different schemes is discussed.*
Key words: Excellence models, total quality management, benchmarking, benchlearning, knowledge management, self-assessment, recognition schemes

1. INTRODUCTION

Excellence models (EM) as management frameworks have been known and in use for a few decades. There have been several attempts to analyse the possible positive impact on organizational and financial performance with somehow contradictory results, depending on the research scope, focus and methodology [1]. Although benchmarking and knowledge management (KM) are intrinsic ideas embedded in EMs, there has been very limited research in the potential benefits of using the assessment processes based on the EMs as a tool for benchlearning and KM purposes. Firms, which are able to acquire and disseminate new knowledge and transform it quickly into processes, products and services, have a competitive advantage. It is discussed further whether and how EMs could be suitable instruments for gaining this goal.

2. ORIGINS AND LINKAGES

2.1 Total Quality Management (TQM) and Benchmarking

TQM and benchmarking concepts emerged and developed as approaches to continuous improvement (CI) in the 1980ies, gaining popularity in relation to the establishment of Malcolm Baldrige National Quality Award in the US [2]. The word *benchmark* originally means measurement against a reference point. In CI it is a 'best-in-class' achievement, which becomes a reference point against which similar results or process performance are measured. Benchmarking (BM) is a process of measuring and analysis that compares internal practices and processes with other organizations. The purpose is to identify best practices, which can be adapted and adopted and as a result increase business performance. Although BM has become a popular management tools and is not only related to TQM anymore, there is often limited understanding and application of BM, using it as a measurement instrument and neglecting the improvement aspect. This is the reason why sometimes *benchlearning* (BL) is being used instead, to emphasize the learning and improvement aspects. The term was already introduced in the early 2000ies [3].

TQM has been a well-known management concept for more than 30 years, but there is still a lot of ambiguity around it. There are different approaches, defining the main principles and/or critical success factors of TQM. One of the reasons behind the fuzziness and difficulties of implementing is that TQM is a culture change program, requiring transformation in organizational value system [4].

2.2 Knowledge Management

Knowledge management (KM) became a discipline later, in the 1990ies and has been gaining more popularity in this century [5]. The theory of organizational knowledge creation proposes that new knowledge is created through processes of conversion between tacit and explicit knowledge: socialization, externalization, combination, and internalization [6]. KM aims at creating sustainable competitive advantages by means of continuous organizational learning through internal formalization of diverse types of knowledge [7].

2.3 Links between TQM, BM and KM

There is no common understanding about the nature of relationship between TQM and KM. Some authors look at KM and TQM as completely different paradigms and independent systems of management practice [8], there are others claiming that KM is about to replace TQM as a quality management tool [9].

The majority of authors treat quality and knowledge management as close concepts that can be organically integrated [11-14]. Both have also Japanese origin – wide-spread quality circles were part of broader knowledge acquisition programs.

In 1997 EFQM and APQC carried out the first BM study project to search for good practice in KM, which was defined as 'all the necessary activities to orchestrate an environment in which people are invited and facilitated to apply, develop, share, combine and consolidate relevant knowledge in order to achieve their individual and collective needs.' [15]. The ultimate goal of KM was defined as to improve an organisation's effectiveness by three core-learning processes:

- Learning from success and failures, on individual, team or company level;
- Learning from each other, both from co-located colleagues as well as colleagues at further distance (geo-graphical as well as disciplinary-wise);
- Learning from 'outside-in', from partners, suppliers, customers and even competitors.

TQM, benchlearning and KM are closely related as they are based on common idea of organizational development. Learning involves accumulating of knowledge and it helps organizations to create new dynamic knowledge-related capabilities.

3. KM AND BENCHLEARNING AS AN INTRINSIC PART OF EM

Excellence models (EM) have been created to offer operational frameworks for TQM implementation as an integrated and holistic management tool. There are several variations of business excellence models (BEM) identified in the world, the basic three models being Deming Prize, Baldrige Performance Excellence Framework and EFQM EM [16].

Besides defining a set of core values or fundamental concepts they provide structured criteria frameworks and measurement instruments to enable scoring and benchmarking during assessment.

BEMs can also be seen as frameworks within which KM principles can be adopted as good management practices. Baldrige Model [17], EFQM EM 2013 [18] and CAF 2013 [19] are used as reference models in this article. The latter is an adaptation of the EFQM EM developed for introducing TQM and self-assessment in European public sector organizations.

Baldrige model has a separate category dedicated to KM – *Category 4 Measurement, Analysis and Knowledge Management.* In EFQM and CAF models there is no specific criteria for KM, but the topic has been embedded in different subcriteria of all enablers (criteria 1-5). *Learning, creativity and innovation* in the EFQM EM and *Innovation and learning* in CAF are the principles that are taken into account in all criteria, both enablers and results (criteria 6-9) when applying the dynamic RADAR (Results, Approach, Assess and Refine) in EFQM and PDCA (Plan, Do, Check, Act) assessment frameworks.

The following interrelated concepts have been defined in the three BEMs – knowledge (EFQM), KM and knowledge transfer (Baldrige), benchmarks and BM (Baldrige, EFQM, CAF), good/best practice (EFQM, CAF), learning (Baldrige, CAF), organizational learning (Baldrige, CAF), learning networks (EFQM) and benchlearning (CAF). In CAF the learning aspect of becnchmarking is emphasized by 'bench learning' - how to improve through sharing knowledge, information, and sometimes resources, as an effective way of introducing organizational change, reducing risks, increasing efficiency and saving time.

4. OPPORTUNTITIES FOR LEARNING AND KNOWLEDGE TRANSFER USING ASSESSMENTS

There are multiple ways how an organization can benefit from using BEMs. They are powerful future-oriented diagnostic tools, that can be used both internally (self-assessment - SA) and externally (recognition schemes, award processes). Combination of both gives the best results, providing valuable insights from different angles, helping to prioritize and focus improvement efforts.

The RADAR is a tool used to score organisations applying for the EFQM Excellence Award and most national awards in Europe [18]. It can be also used for SA and enabling internal and external benchmarking. SA is a comprehensive, systematic and regular assessment of organizational results against the criteria and fundamental concepts of a BEM, enabling to identify strengths and areas/opportunities for improvement. It is a process of organizational self-reflection. SA is a knowledge-generating and knowledge transmission process that ideally involves all the people in the organization through forming SA and improvement teams [20]. By discussions and consensus processes information becomes knowledge, which is transformed into improvements and innovations; the organization learns and improves its performance and capabilities. SA is also the first step and preliminary condition for external benchmarking with competitors.

While SA tools as well as external assessment methodologies have been quite thoroughly developed over the decades, relatively little attention has been given to the potential of using the knowledge and learning experience of assessor/examiner/ auditor team members in a more comprehensive way for organizational learning and KM purposes [21-22].

Knowledge creation cannot be separated from the context in which it is created. New tacit knowledge is socially constructed through the interactions amongst individuals or between individuals and their groups, rather than by an individual operating in isolation [23].

5. ROLE OF RECOGNITION SCHEMES

Different award and recognition schemes have been established around the world in parallel with the development of EMs to motivate organizations to use modern methods to improve management quality and raise overall quality awareness. Well-performing organizations are recognized, taking into account their specific features, eg size, sector, maturity level of management quality etc. [24].

Main benefits of recognition are increased reliability and improved image of organizations, enabling benchmarking, identification and sharing knowledge and best practices, encouraging learning (benchlearning). Long-term effect of such activities is better competitiveness of products and services, organizations and the society as a whole [25].

5. CASE OF ESTONIA

Recommendations for establishing a national quality award in Estonia as a quality promotion and awareness tool and means for raising competitiveness was made by experts in the document Quality Policy of Estonia already in 1996 [26]. The Estonian Quality Award was established in 2000, followed by the development of a series of sectorial recognition schemes and model adaptations (mostly in educational, also in tourism and public sector). Most of the initiatives were project based and partially or fully supported by ESF funding. A Strategy of Management Quality 2005-2008 was established by the Ministry of Economic Affairs, which was implemented mostly by Enterprise Estonia, using the expertise of Estonian Association for Quality (EAQ). Promotion of BEMs and development of recognition schemes was part of the strategy. As a result quite a remarkable a few hundred organizations and individuals attended different trainings (SA and external assessment), participated in SA and external assessment teams, acquiring practical knowledge and learning experience in the use of BEMs [27].

Based on the individuals' feedback from these projects during 2000-2012 the main value was found in profound learning experience, acquiring practical skills in using a systematic assessment tool (mostly RADAR-based), participating in cross-functional and cross-sectorial teamwork and opportunity to look at organizations (both their own and others) from a different perspective.

On organizational level the main value of participating in these programs was discovering a systematic diagnostic tool for organizational development and value-added feedback about improvement opportunities, also possibility to benchmark their maturity level of management quality.

As the project funding from ESF resources stopped, most of the projects also were finished (with the exception of Tallinn Quality Award for Educational Institutions, which is funded by the City Government). Although positive impact on business performance was identified among participating organizations during this period, it was too short time and limited number of organizations that were involved in these initiatives to have a bigger impact in society and economy [28].

A relatively low awareness about modern management techniques has been identified by different studies [28-30]. According to some studies the awareness about TQM is rather high [30] and the need for some TQM elements (customer focus, people involvement etal) is relatively well recognized, at the same time the awareness and level of implementation of TQM tools and BEMs is marginal [28,29].

Although the target groups of 2005 [28] (all sectors and sizes were represented, 540 respondents) and 2015 [29] studies where not the same (there were no public sector organizations in 2015, only companies that were older than 5 years, 111 respondents), some comparisons may still be made. In 2005 29% of respondents were not aware about BEMs; 51% were aware, but did not intend to use; 15% intended to use; 2% had started using and 3% were using. A major difference based on the size of organizations was identified in favor of larger organizations.

In 2015 58% of respondents were not aware of any BEMs; 30% do not intend to use BEMs, 3% were users. At the same time 60% of respondents expressed need to compare their current state with ideal, assessing the maturity level of different areas in realtions to full potential. 18% of respondents were not interested on this kind of approach and they are not aware of it.

4. CONCLUSION

In order to increase the positive impact of the use of EMs on performance, quality, sustainability and competitiveness of organizations as well as society as a whole, it is necessary to develop supportive infrastructure and facilitate schemes that promote benchmarking and benchlearning in a systematic and structured way, enabling efficient and quick knowledge transfer and transformation into new know-how, innovative and high quality products and services. Such schemes support identifying best practices, creating innovative approaches and opportunities for inter-sectorial learning and synergy.

Unified reference models are recommended to ensure comparability and benchmarking, though adaptations should be made for better usability in specific sectors, maintaining the same basic values and principles, overall model structure and assessment methodology of the EMs.

There is quite wide evidence that organizations can use the EFQM EM as the basis for designing, implementing and monitoring KM processes. Using the synergies of EMs and the critical factors of TQM makes the implementation faster and more successful [32]. EMs offer strategic framework for KM and innovation.

Combining SA and external recognition schemes enables multiple forms of learning and knowledge transfer:

- Intra-organizational – use of cross-functional self-assessment/submission document preparation teams
- Inter-organizational – external assessment teams with members from other organization, with different background and experience
- Inter-sectorial – external assessment teams with members from different sectors and professional background
- Two-way transfer – independent and fresh 'out-of –the-box' look and feedback for participating organization.

5. REFERENCES

1. Kiitam, A. and T. Tammaru. Impact of Application of Excellence Models on Organizational Performance. In: *Proc. of the 8th Int. Conf. of DAAAM Baltic Ind. Eng. 19-21st April 2012, Tallinn, Estonia*, (Otto, T. , ed.), 152 - 157.
2. APQC. *The Benchmarking Management Guide*, American Productivity & Quality Center (APQC), Productivity Press, Cambridge, Massachusetts, 1993.
3. Karlöf, B., K. Lundgren and M. Edenfelt Froment. *Benchlearning: Good Examples as a Lever for Development,* Wiley, 2001.
4. Kekäle, T. Et al. To Make it 'Total': Quality Management over Subcultures. *TQManagement*, **15**, 1093–1108, 2004.
5. Nonaka, I. and H. Takeuchi. *The Knowledge-Creating Company: How Japanese Companies Create the Dynamics of Innovation.* Oxford University Press, Oxford, 1995.
6. Nonaka, I., A dynamic theory of organizational knowledge creation. *Organ. Sci.* **5**, 14–37, 1994.
7. Davenport, T.H., L. Prusak. *Working Knowledge - How Organizations Manage What They Know.* Harvard Business School Press, Boston, 2000.
8. Ruzevicius, J. Integration of total quality management and knowledge management. *Informacijos Mokslai*, **37**, 30-38, 2006.
9. Adamson, I. Knowledge management - the next generation of TQM? *Total Qual. Manag.* **16**, 987–1000, 2005.
10. Hsu, S.H., H.P. Shen. Knowledge management and its relationship with TQM. *TQ Mag.,* **16** 351–361, 2005.
11. Ooi, K.B. TQM and knowledge management: literature review and proposed framework. *Afr. J. Bus. Manag.,* **3** , 633–643, 2009.
12. Linderman, K., etal. Integrating quality management practices with knowledge creation processes. *J. Oper. Manag,.* **22**, 589–607, 2004.

13. Ribière, V.M., R. Khorramshahgol . Integrating Total Quality Management and Knowledge Management *J. of Management Systems,* **16**, 39-5, 2004.
14. Choo, A.S., K.W. Linderman, R.G. Schroeder. Method and context perspective on learning and knowledge creation in quality management. *J. Oper. Manag.* **25**, 918–931, 2007.
15. Van der Spek, R., G. Carter. A survey on good practices in Knowledge Management in European companies. *In: Knowledge Management. Concepts and Best Practices* (Ed. K. Mertins, P. Heisig, J. Vorbeck). 2nd Ed; Springer-Verlag, Berlin, 2003.
16. Sampaio, P. etal. A comparison and usage overview of business excellence models, *TheTQM J*, **24**, 181-200, 2012.
17. NIST, *2015–2016 Baldrige Excellence Framework: A Systems Approach to Improving Your Organization's Performance*. Gaithersburg, MD: U.S. Dpt. of Commerce, National Institute of Standards and Technology, 2014.
18. EFQM, *EFQM Excellence Model 2013*, EFQM, Brussels, 2012.
19. EIPA, *Common Assessment Framework (CAF) 2013. Improving Public Organisations through Self-Assessment*, EIPA, 2012.
20. Martin-Castilla, J. I., O. Rodriguez-Ruiz. EFQM model: knowledge governance and competitive advantage. *J. of Intel. Capital*, **9**, 133-156, 2008.
21. Leonard, D. and R. McAdam, Impacting organizational learning: the training and experiences of quality award examiners and assessors, *J. of Eur. Ind. Training*, **27**, 16-21, 2003.
22. Beckett, R., P. Murray. Learning by auditing: a knowledge creating approach. *The TQM Mag.*, **12,** 125-136, 2000.
23. Erden, Z., G. von Krogh, I. Nonaka, The quality of group tacit knowledge, *J. of Strat. Inf. Systems*, **17,** 4–18, 2008.
24. Grigg, N., R. Mann. Rewarding Excellence: An International Study into Business Excellence Award Processes. *ASQ Qual. Man. J.,* **15**, 26-40, 2008.
25. Mann, R. *Report of the APO Survey on the Impact of Business Excellence/ Quality Awards on Enterprises.* Asian Productivity Organization, 2011.
26. Kiitam, A., Tammaru, T. e.a. *Quality Policy of Estonia.* Tallinn, Multico, 1996.
27. Tammaru, T., A. Kiitam. Excellence Models and National Quality Promotion in Estonia. In: *Proc. 7th Internat. DAAAM Baltic Conference Ind. Eng. 22-24 April 2010, Tallinn, Estonia*, TUT, 409-414, 2010.
28. Tammaru, T. *Preliminary Study for Identifying the Relevant Needs of Target Groups of a Quality Awareness Programme. Project Report. Vers. 2.1 (in Estonian).* TUT, Tallinn, 2005.
29. EAS. *Study of the Situation of Management in Estonia*, Enterprise Estonia (EAS), Tallinn, 2005.
30. Zernand-Vilson, M. *Adoption and Implementation of New Management Ideas in Estonian Business Organizations. Doc. Thesis in Man., no. 19.* Tallinn, EBS, 2014.
31. Laanemets, A. *Supply Chain Quality Management SCQM. Master Thesis (in Estonian).* TUT, Tallinn, 2015.
32. A. Calvo-Mora, et al. Project to improve knowledge management and key business results through the EFQM excellence model, *Int. J. Proj. Man.*, 2015.

6. ADDITIONAL DATA ABOUT AUTHORS

MSc.Eng. Tiia Tammaru
Phone: +372 56636679,
E-mail: tiia.tammaru@ttu.ee

PhD Andres Kiitam,
Phone +372 6203 204
andres.kiitam@ttu.ee
TUT, Department of Mechatronics
Ehitajate tee 5, 19086 Tallinn, Estonia

10th International DAAAM Baltic Conference
"INDUSTRIAL ENGINEERING"
12-13th May 2015, Tallinn, Estonia

IMPLEMENTATION OF A TORSION TESTING DEVICE FOR 3D-PRINTED PLASTIC TUBES

Vainikka, O.; Halinen, J.; Pajula, M.; Sarhaluoma, A.; Kiviluoma, P. & Kuosmanen, P.

Abstract: *Usage and popularity of 3d-printed parts is increasing but printed parts usually have different qualities compared to plastic parts made with traditional methods.*
This article describes the development and construction of a device for torsion testing of 3d-printed hollow plastic tubes. The device should measure the torque and twisting angle of the test specimen until it breaks. Strength values of ABS plastic tube with outer diameter of maximum 30 mm and wall thickness of maximum 5 mm were used.
A mechanically simple and robust testing device using incremental encoder and strain gages for measurement, was constructed. The device is calibrated to measure torque from 0 to 150 Nm while twisting angle can be infinite. The device uses four-jaw chucks for mounting the test specimen, which proved not to be an optimal mounting method for cylindrical items due to slippage.
Key words: shear strength, torque measuring, plastic tube, ABS plastic, cylindrical 3D-printing.

1. INTRODUCTION

The purpose of this study is to build a torsion testing device which can measure torque in function of twisting angle. The final use of the device will be the testing of the shear strength of plastic tubes 3D printed with a new cylindrical printing method. [1] Assumption is that cylindrical 3D-printing improves the shear strength behavior of the tube. The requirements for the geometry and forces of the device are defined so, that a tube made of ABS plastic which has an outer diameter of 30 mm, can be twisted until it breaks. Inner diameter of the 30 mm tube is defined to be no less than 20 mm so the thickness of the wall will be maximum 5 mm.
Testing speed of the device should be so slow that it will not interfere normal behavior of the material [2, 3, 4]. In this case maximum testing speed will be 0.22 rpm. Due to slow testing speed and fairly high torque required, high gear ratio is needed in power transmission. Device is built so that only shear stress is applied to the test specimen, meaning that the specimen is able to move without restrictions in axial direction. The fixing method of the specimen is designed so that the specimen should not slip during the test procedure. The fixing should be done so that the specimen measured is not damaged during the fixing process.
The torque-twist response of 3D-printed tubes is very important for determining the quality of the printed test specimen and also in testing the quality of the new cylindrical printing method. This device enables comparing the qualifications between traditionally manufactured tubes and 3D-printed tubes.
Torque produces shear stress into the material. Every material has specific amount of stress they can tolerate before breaking. Shear stress is used in measuring because 3D-printed objects usually cannot tolerate twisting. [5]
There are torsion testing devices on the markets, but they are mainly designed for

thin metallic materials and very expensive. They are usually made for testing solid specimens and do not cope well with hollow specimens such as tubes. The problem is that tubes tend to collapse under tension. Therefore fixing needs special attention so that only torque affects the specimens. [2]

2. METHODS

Several motors and drive systems were considered based on the required torque and specimen geometry. A single phase induction AC motor having an integral gear reduction was selected based on performance, size and cost. Because of the single phase supply voltage the motor was easy to get functional just plugging it to the wall socket. Nominal speed of the motor is 1200 rpm and with the integral gear box it is reduced to 7 rpm developing output torque of 7,8 Nm.

A second gearbox was selected to reduce the speed even more and increase developed torque. Based on initial test specimen geometry it was estimated that approximately 150 Nm of torque would be needed to overcome test specimens ultimate shear strength of 147 Nm. A worm gearbox with gear ratio of 30:1 was selected to provide sufficient nominal torque and high enough maximum output torque.

The torque developed by the machine was transferred to the specimen using self-centering four-jaw spindle chucks with outer diameter of 80 mm. Aluminum plugs were machined to eliminate a possible yielding of the plastic tubes while compressing with the chucks.

The torsion testing device was designed to test 3D-printed plastic tubes made from ABS. The shear strength of the tube can be calculated by using von Mises equation. [5] While twisting, the length of specimen will decrease and this would cause unwanted tension to the specimen. The frame of the device was implemented so that the only stress that loads the specimen during measurement, is the stress from the torque. This was done by using linear slide in the other end of the test specimen. Since the device is designed so that the axial stresses are prevented, the von Mises equation can now be written as [5]

$$\tau = \frac{\sigma_y}{\sqrt{3}} \qquad (1)$$

Where τ is the shearing stress and σ_y is the tensile strenght of the material.

The tensile strength of the ABS plastic is approximately 50 MPa [6]. The torque needed from the device can be calculated by using equation 2 [5]

$$T = \frac{\tau I_p}{r} \qquad (2)$$

Where T is the needed torque, I_p is the torsion constant and r is the radius of the tube. Equation 2 gives torque that is needed to break tube made from ABS plastic. By assuming that the printed tube will not be able to take as much load as traditional ABS tube, a low safety factor can be used. In this experiment safety factor of 1,3 was used and therefore the torque needed from the apparatus is approximately 150 Nm. [5]The schematic of the device is shown in Figure 1. The device consists of AC-motor, two gears, two fixing chucks and a linear slide. The AC-motor produces the needed movement and torque which are modified with the two gears to match requirements. The test specimen is attached to the chucks and the linear slide is used to prevent the axial stresses.

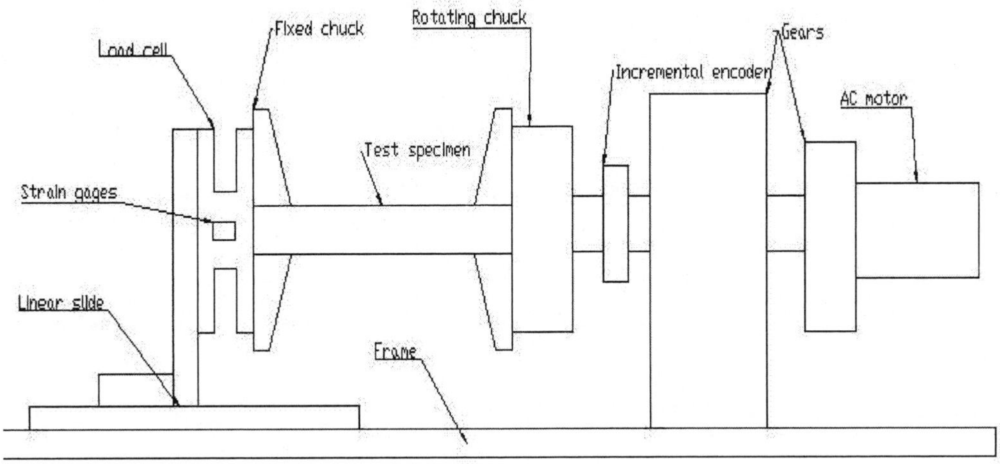

Fig. 1. The Schematic of the device.

The torque was measured with a custom made load cell and the actual measurement was done by using strain gages. Strain gages were attached to the neutral axis of load cell so that the bending stress is compensated. The gages were also aligned with main stress components. The measurement was conducted by using full Wheatstone bridge connection. In order to get precise results from the strain gages the dimensions of the load cell were designed so that relative strain in the load cell was between 0,0005 and 0,001. The relative strain applied to the load cell can be calculated by using equation 3. [5, 7, 8]

$$\varepsilon = \frac{Tr}{2GI_p} \qquad (3)$$

Where G is the Young's modulus. The load cell was attached between the linear slide and the fixing chuck as shown in Figure 1. The calibration of the load cell was conducted by using calibration circuit that was built in to the amplifier. The output signal was adjusted to zero when no load was applied to the device. After zero regulation the calibration equivalent load was applied and the output voltage was set with reference to the calibration equivalent load.The angle of twist was measured with an incremental encoder. The incremental encoder was placed on the rotating axle. The principal of the measurement system is shown in Figure 2. For data acquisition USB-powered data acquisition box was used. It can power the incremental encoder and the amplifier of the strain gage. The resolution of the encoder was 2000 pulses per revolution. The incremental encoder was connected to the digital input of the data acquisition. The measured data was analyzed by using LabView software.The temperature of the test specimen cannot change too much during the measurement, because properties of the ABS vary a lot depending on the temperature. The temperature changes of the test specimen can be prevented by using very low angular velocity during the measurement. As result the device will plot the torque as the function of the angle of twist. [1]

3. RESULTS

The accuracy of the device was determined by using 50 cm bar attached to the chuck that is attached to the load cell and calibration weights that were attached to the end of the bar. Due to installation of the chuck jaws and a square shaped fixing part of the calibration bar, the bar was not horizontal in the measurements. Due to that the angle of the bar was measured before each measurement with angle rule and a water level. Calibration arrangement is shown in Figure 3.

*** Formatting Issue - Best Available Paper/Graphic ***

Fig. 2. The principle of the measurement system.

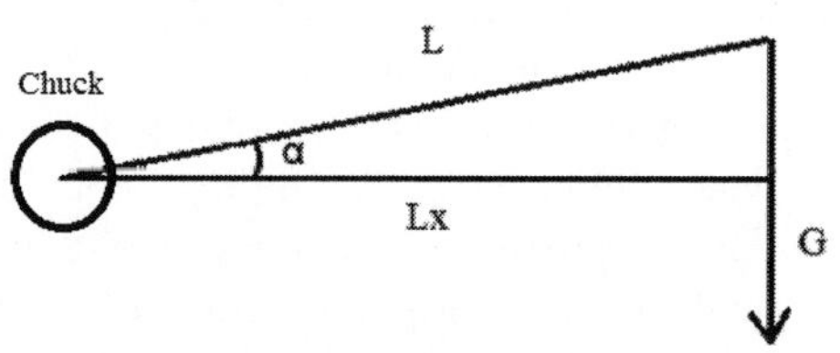

Fig. 3. Calibration arrangement.

Torque caused by calibration weight (G=mg) was calculated with formula 4:

$$M=mgcos(\alpha)Lx \quad (4)$$

Calibration weights, measured angles and calculated and measured torques are shown in the table 1. Difference in measured torque and calculated torque is shown Figure 4.

As a result of previous measurement, the measuring accuracy of the device between 0 and 90 Nm can be determined.

Average error between measured and calculated value was 1,3 % and largest individual error was 2,7 %.

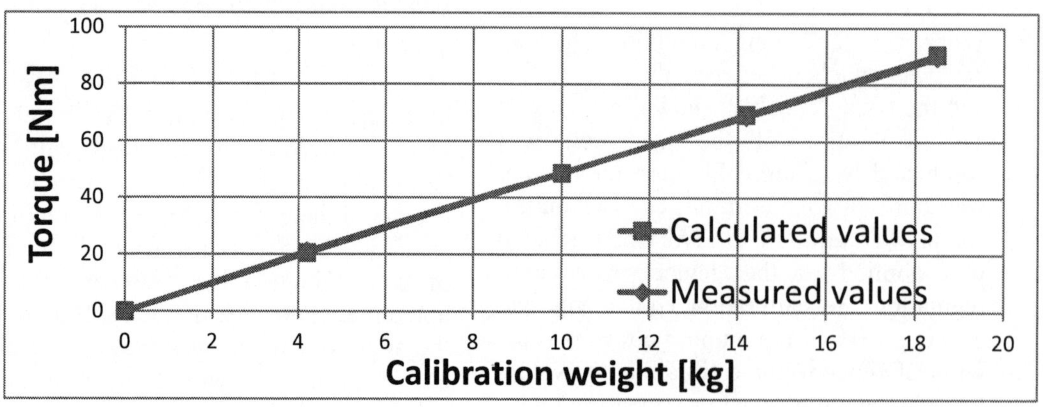

Fig. 4. Difference in measured torque and calculated torque.

measurement	m (kg)	angle (deg)	calculated torque	measured torque
1	0,0	0,0	0,0	0,0
2	4,2	6,0	20,5	21,1
3	10,0	6,5	48,7	48,9
4	14,2	4,0	69,3	68,6
5	18,5	2,0	90,4	89,5

Table 1. Measurements.

Reason for the error might be the slight inaccuracy in the angle measurement, which might cause error to the calculated values. By observing the values from Figure 4 and table 1 it can be seen that the error remains quite similar throughout the whole range.

A critical part of the torsion measurement is the fixing of test specimen. On our device the greatest concern was that does the fixing made by chucks create enough friction to prevent the slipping. It was investigated that by tightening the jaws against round surface and using a sleeve made of aluminium inside the tube, only about 65 Nm torque could be applied to the test specimen without slipping. Due to that it is recommended that all test specimen should have rectangular mounting surfaces on them.

Another concern was that will the machine be able to provide the required 150 Nm and will the structure be able to stand the stress caused by that. While testing specimen with rectangular mounting faces, we were able to achieve 95 Nm. In Figure 5 is the data plotted from that test. Y-axis is the torque and x-axis is the angle of twist in degrees.

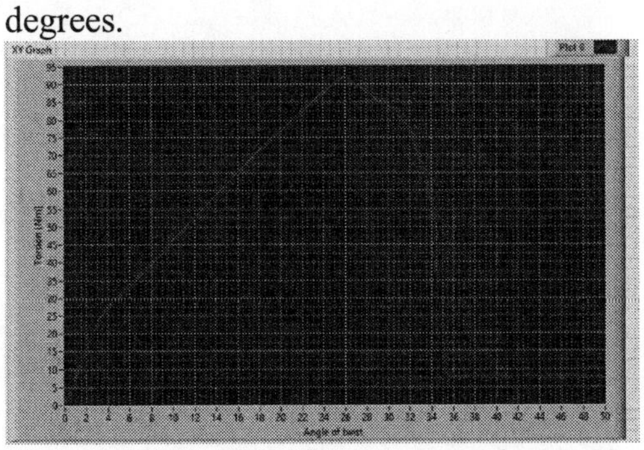

Fig. 5. Torque-angle curve with 95 Nm peak torque.

4. DISCUSSION

This torque testing device is able to twist and break the tubes up to a certain torque. The device is also able to measure the torque and angle accurately. Moreover this device was done with a low budget compared to commercial testing devices.

Based on the first tests the cylindrically printed tubes broke in a 45 degree angle, parallel to the main shear stress. Traditionally, layer-by-layer printed tubes ruptured between the printing layers. This supports the idea that the new method provides tubes that can endure bigger shear strengths [1]. More tests should be made to determine which method is better. The fixing points of the testing device should be upgraded before new tests. Also the quality of the new tubes is not yet constant enough to draw real conclusions.

The next step should be upgrading the fixing mechanism of the test device. The spindle chucks do not provide enough friction to create more than 40 Nm of torque. The current mechanism works well with tubes with square ends, but the cylindrical printer cannot produce these. Thus a new fixing mechanism has to be developed. Other improvements could include speed control for the rotation and a proper kill switch for the device.

5. REFERENCES

[1] Kemppainen, M., et al. *The Development and Design of a Cylindrical 3D Printer,* 2015 DAAAM Baltic Conference.

[2] SFS-EN ISO 6892-1. *Metallic materials. Tensile testing. Part 1: Method of test at room temperature.* Mechanical Engineering and Metals Industry Standardization in Finland. 2009.

[3] Callister, W.D. & Rethwisch, D.G. *Materials Science and Engineering.* 8th ed. John Wiley & Sons, 2011. ISBN: 978-0-470-50586-1.

[4] Du Bois, P.A. et al. *Material behaviour of polymers under impact loading.* International journal of impact engineering, 2006. Vol. 32. Issue 5. pp. 725-740. DOI:10.1016/j.ijimpeng.2005.02.007.

[5] Parnes, R. *Solid Mechanics in Engineering.* Wiley, 2001. ISBN: 0-471-49300-7

[6] Granta Design. Ces Edupack 2015. ABS material properties. A computer software on Windows. Commercial software available: http://www.grantadesign.com/education/edupack/index.htm

[7] National Instruments. *Top Five Considerations for Taking the Stress Out of Strain Measurements.* 2012. Available: http://www.ni.com/white-paper/11488/en/ (16.04.2015)

[8] Omega. *Positioning strain gages to monitor bending, axial, shear and torsional loads.* Available: http://www.omega.com/faq/pressure/pdf/positioning.pdf (16.04.2015)

6. CORRESPONDING ADDRESS

Panu Kiviluoma, D.Sc. (Tech.), Senior University Lecturer
Aalto University School of Engineering
Department of Engineering Design and Production
P.O.Box 14100, 00076 Aalto, Finland
Phone: +358504338661
E-mail: panu.kiviluoma@aalto.fi
http://edp.aalto.fi/en/

7. ADDITIONAL DATA ABOUT AUTHORS

Halinen, Juho
E-mail: juho.halinen@aalto.fi

Pajula, Matti
E-mail: matti.pajula@aalto.fi

Sarhaluoma, Anssi
E-mail: anssi.sarhaluoma@aalto.fi

Vainikka, Otto
E-mail: otto.vainikka@aalto.fi

Kuosmanen, Petri, D.Sc. (Tech.), Professor
Phone: +358 500 448 481
E-mail: petri.kuosmanen@aalto.fi

10th International DAAAM Baltic Conference
"INDUSTRIAL ENGINEERING"
12-13th May 2015, Tallinn, Estonia

EFFECTS OF THE ANGULAR VELOCITY OF A FLYWHEEL ON THE GYROSCOPIC STABILIZATION OF A BICYCLE

Vepsäläinen, J.; Peltola, M.; Nygren, T.; Mälkönen, J.; Heikkilä, E.; Kiviluoma, P.; Kuosmanen, P.; Socie, D.; Teerihalme, S.

Abstract: *Learning to ride a bicycle requires simultaneous development of various motoric and perceptual capabilities. During the learning process, training wheels are often used both as a training device and as a safety measure. Gyroscopic stabilization offers an alternative way for balancing the bicycle, enabling the possibility to control the amount of the assistive balancing force for more efficient learning. In this study, the effects of a gyroscope's flywheel velocity were observed on a small-sized bicycle. To measure the stabilizing effect, a bicycle was fitted with a control moment gyroscope and an inertial measurement unit. The resulting stabilizing forces were measured to show the correspondence between the stabilizing effect and the flywheel velocity.*

Key words: gyroscope, stability, learning, training

1. INTRODUCTION

Cycling is a great way to get daily exercise and an environmentally friendly way of transport. In many cities cycling is one of the fastest ways to get around. Cycling is also a very competitive sport. The main thing needed for cycling is the skill to balance the bike even at low speeds.

Children are often taught how to ride a bicycle when they are between ages of 3 and 8 years old, averaging just over 5 [1, 2]. Some children take to it naturally, others do not. Every child has a different physical and mental development. This can easily lead to frustration if siblings or other children learn faster. While learning how to ride a bicycle, kids are a risk to themselves and others in traffic.

The earliest sketches of a bicycle are said to be from 1493 by Leonardo da Vinci's pupil, Gian Giacomo Caprotti. The bicycle was later invented in the early 19th century. The first vehicle that was powered by a human and had only two wheels was the German Draisine dating back to 1817 by Karl von Drais. The bicycle has evolved quite a lot over the years. There were also trends of multiple wheels, big wheel on the front or no pedals at all. Nowadays the basic structure of a bike has been standardized. [3]

The goal of this study is to help children to learn how to ride a bicycle by designing a device that can be attached to a bicycle and self-stabilizes it. In this study the optimization of the balancing control system in the bicycle is peripheral to measuring the forces generated with the device. The main focus is on the self-stabilization and the training aspect is secondary.

There's a saying, "It's as easy as riding a bike". In the mathematical world, it definitely is not so. A bicycle has an intricate geometry and it has many degrees of freedom. Thus creating a comprehensive and accurate model of a bicycle is complicated. This is definitely a particular challenge when trying to design the stability control for a bicycle. This study simplified the situation by focusing only on the torque needed for stabilizing the bicycle in an upright position.

This article is structured as follows. First we introduce the methods that are used for stabilizing a bike. In this section the calculations, simulations and prototyping are showed and explained. The next part relates the results from the tests. The last part of the article is discussion concerning the project's success and thoughts about future plans.

2. METHODS

Lam, Yetkin and Ozguner have successfully constructed small bicycles that can autonomously stabilize themselves so that they stand upright [4, 5]. These bicycles were based on a control moment gyroscope (CMG) which consists of a flywheel spinning at an even speed and a gimbal that is used for rotating the spinning flywheel around the vertical axis. Rotating the spinning flywheel moderately around the vertical axis causes a moment that can be utilized for balancing the vehicle.
In this study a CMG was used for stabilizing a bicycle. The gimbal was attached to a child-size bicycle as show in Figure 1. The original frame was marginally lengthened to fit the CMG.

Fig. 1. Representation of the CMG attached to the bicycle.

The flywheel was driven using a pneumatic motor attached to the gimbal of the CMG. At first a DC-motor was used instead but the small pneumatic motor was more suitable for frequent testing. Accelerating the flywheel using the pneumatic motor was more practical and there was no risk of overloading the motor. The shaft of the flywheel was connected to the output shaft of the motor via a belt drive. The rotation around the vertical axis was achieved using a servomotor via a belt drive. The gear ratio in both belt drives was 1:1.
The mass and the inertial moment of the flywheel were optimized to the need. On the one hand the structure was to be as light as possible. At the same time the shape of the flywheel had to be designed for maximal inertial moment. Thus the mass of the flywheel concentrates on the perimeter.
As a safety measure the frame of the CMG was constructed using flat bar iron. The result was a sturdy casing that would hold the flywheel inside even if something would go wrong. Bearings were fitted to both ends of the shaft of the flywheel. The frame was supported by a bearing from the top and from the bottom so that it could be rotated around the vertical axis.
The tilt angle of the bicycle was measured using an inertial measurement unit (IMU) which is capable of measuring 6 degrees of freedom. The whole system was controlled using a microcontroller. When stabilizing the bicycle the angle of the flywheel was controlled based on the tilt angle of the bicycle. The system was powered by a lithium polymer battery.
The principle of using a CMG was first tested using Adams and Simulink simulations. The Adams simulations verified that both the system and the stabilizing phenomenon worked as expected. The simulations gave some advice for the magnitude of the parameters in the actual structure.
Simulink simulations were made to verify the results from Adams simulations. These

simulations are based on the basic physics formula for torque. The formula is

$$\boldsymbol{\tau} = \mathbf{I} \times \boldsymbol{\alpha} \quad (1)$$

where **I** is the inertia of the object, **α** is the acceleration and **τ** is the torque produced. This formula can be used to calculate the torque generated by a gyroscope. The formula is then added with the speed **ω** and the inertia **I** of the flywheel.

$$\boldsymbol{\tau} = \mathbf{I} \times \boldsymbol{\alpha} + \boldsymbol{\omega} \times \mathbf{I} \times \boldsymbol{\omega} \quad (2)$$

The formula is simplified because the speed of the flywheel is perpendicular to the acceleration off the gimbal motor. Then the formula takes the following form:

$$\boldsymbol{\tau} = \mathbf{I}(\boldsymbol{\alpha} + \boldsymbol{\omega}^2) \quad (3)$$

This formula is then used to run the simulations to estimate the final technical specifications. In addition to the simulations, the 3D-models created with Creo 2.0 were relevant for adjusting the dimensions of the structure. The final technical properties (Table 1) were based on the computer models.

Rotational speed of the flywheel	7,000 rpm (max)
Mass of the flywheel	7 kg
Flywheel's moment of inertia	0.059 $kg*m^2$
Material of the flywheel	steel
Range of the IMU	±16g and ±2000°/s

Table 1. The technical specifications.

The purpose of the study was to measure the useful torque created with the CMG. In order to remove one degree of freedom the rotation of the handle bars and the front wheel was prevented.
The torque generated with the CMG was measured at different angular speeds of the flywheel. The speed of the flywheel was monitored with a laser tachometer. The rotation around the vertical axis was executed by running the servomotor from one side to the other in the same way in each test. As mentioned later in the Results chapter the servo was not able to rotate the CMG at a similar rate at all flywheel speeds.
In the test setup the bicycle was virtually in an upright position though just leaning to one side. The bicycle was held in place by a force gauge attached to a wall from the other end. The initial situation was set as the neutral point where the force value was tared to zero. As the CMG was rotated around the vertical axis, the generated torque made the bicycle pull the force gauge. The bicycle staid in the initial position as the force gauge did not allow it to tilt. The force reading from the gauge was stored. The generated torque was calculated by multiplying the force with the vertical distance of the force gauge compared to ground. Measurements were carried out with the setup presented in Figure 2.

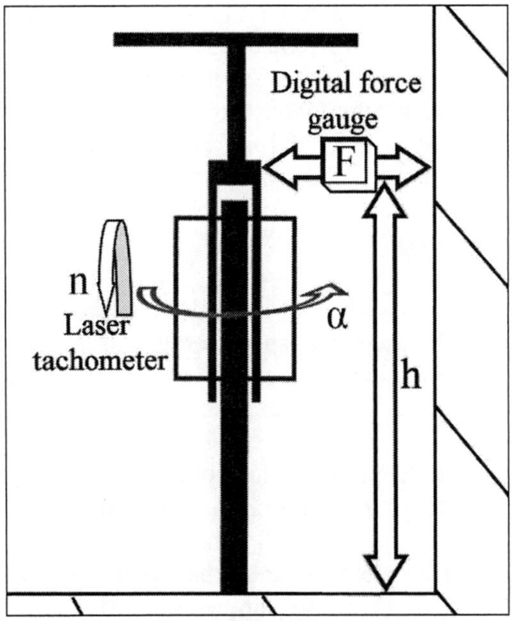

Fig. 2. Experimental setup where the CMG is attached to the frame of a bicycle. The generated force (F) at height (h) is measured as the CMG is rotated around the vertical axis. The force gauge is attached to the bicycle from one end and to the wall from the other.

3. RESULTS

Figure 3 presents the measured dependency between the rotational speed of the flywheel and the torque generated with the CMG.

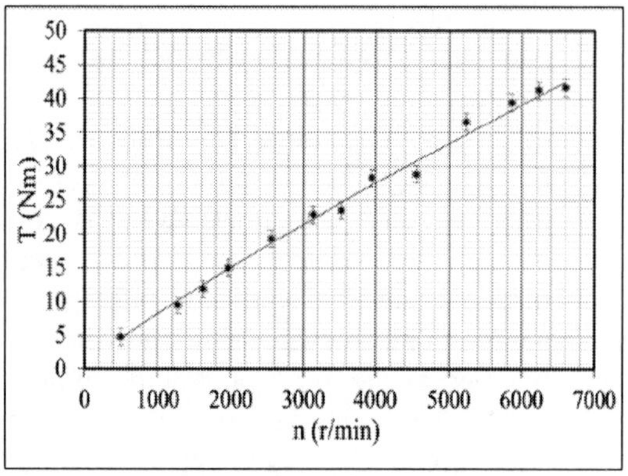

Fig. 3. Experimental results of the gyroscopic torque generated as a function of the rotational speed of the flywheel.

During the tests it was clearly visible that the servo motor was not able to perform the rotation around the vertical axle equally fast at all flywheel speeds. As the speed of the flywheel was increased the rate at which the CMG turned was reduced. The servo motor did not have the capability to maintain the same rotational speed as the torque of the flywheel increased with the speed.

Since the servo was not capable of rotating the flywheel at a similar rate at all speeds the ascending speed-torque curve in figure 3 gradually becomes less steep. The torque should grow exponentially because the angular speed vector is to the power of two in the presented formula (3).

The maximum torque was generated at the highest used flywheel speed. As the flywheel was run at the speed of 6600 rpm the generated torque was 41 Nm. The torque needed for bringing a bicycle with a rider back up from a given tilt angle can be calculated using formula (4).

$$\boldsymbol{\tau} = \mathbf{h} \times \mathbf{m} \times \mathbf{g} \times \sin(\boldsymbol{\beta}) \qquad (4)$$

where **τ** is the torque needed, **h** is the combined centre of gravity, **m** is the combined weight of the rider and the bicycle, **g** is the standard gravity, and **β** is the tilt angle.

If the combination of a rider and a bicycle ways 35 kg and the combined centre of gravity is located at the height of 0.45 meters 41 Nm is enough to bring the bicycle back to an upright position from the tilt angle of 15 degrees. This would be quite an extreme situation and even not desirable. According to Yetkin [6] it is possible to keep the tilt angle at a maximum of 1 degree with well-tuned control. The same control system can also bring the bicycle back from a tilt angle higher than 1 degree [6]. Taking this is into consideration gives a good perspective to the practical capabilities of the system. 41 Nm is enough to bring back up a combined mass of 100 kg at the height of 0.8 metres from a tilt angle of 3 degrees. Based on this the system could stabilize even adults riding a bicycle.

4. DISCUSSION

For now the study focused on confirming the torque that can be drawn from the CMG. The torque generated is easily enough for stabilizing the combination of a bicycle and a rider. The torque generated can be utilized for stabilizing or assisting riders and bicycles of various sizes. A CMG stabilization system could be used for teaching children to ride a bicycle but also for assisting adults who have difficulties is balancing a bicycle. The group of adults requiring assistance could consist of elderly people and persons who have difficulties because of medical issues.

Within the framework of this study the control system of the CMG was not developed so far that the bicycle would self-stabilize itself for a time span of much more than ten seconds. The bicycle would quickly get unstable and tip to either side.

Based on the study it is quite challenging to fit a control moment gyroscope to a bicycle. Through decades of testing a bicycle is a highly refined vehicle which does not essentially have room for a CMG. In this study the small bicycle frame had to be extended from the middle to fit the CMG. Thus the bicycle was no longer driveable for a child. To power the motor of the flywheel a CMG system requires either a very high capacity battery or a pneumatic compressor which both are difficult to implement to a mobile and light system. The support structure of the CMG was also slightly in the way of pedalling. In addition the CMG system weighs over 20 kg. The weight of the CMG is an enormous problem for the user and especially for a children's bicycle. If the CMG technology is used in the future for learning or other assisting purposes it almost inevitably requires a new frame design for the bicycle and better solutions for storing energy.

5. CONCLUSION

The test system demonstrates how torque can be generated with a CMG for stabilizing a bicycle. A clear dependency between the rotational speed of the flywheel and the torque can be seen in the results. However, the torque attained is not dependent only on the rotational speed of the flywheel. The non-linearity of the angular servomotor was one of the parameters affecting the results.

The test system can next be further developed so that the CMG makes the bicycle truly self-stabilizing even with a rider as a load. Then, gradually decreasing the amount of applied stabilizing assistance according to the improved skills of the cyclist could be beneficial. The system proved to generate enough torque for stabilization purposes. Thus making the system self-stabilizing even when manned is most of all down to refining the control system of the CMG.

The test system cannot be properly used in a bicycle in its present configuration as the bicycle is nearly undrivable. The control system does not yet take intentional tilting during cornering into consideration. The CMG could be made more feasible by reducing the weight of the flywheel and the whole system. The reduced weight of the flywheel could be then compensated by increasing the rotational speed of the flywheel.

6. REFERENCES

1. Iannelli, V. *Riding a Bike*. 2011. [WWW] (Cited on the 1st of April 2015), Available: http://pediatrics.about.com/od/learningtorideabike/a/riding-a-bike.htm.

2. Wood, T.D. *Teaching a Child How to Ride a Bike*. Rei.com. 2014. [WWW] (Cited on the 1st of April 2015), Available: http://www.rei.com/learn/expert-advice/teach-child-to-ride-a-bike.html.

3. Lessing, H-E. *The evidence against Leonardo's bicycle*, Cycle History 8, San Francisco 1998, pp. 49-56.

4. Pom Yuan Lam, *Gyroscopic stabilization of a kid-size bicycle*, 2011 IEEE 5th International Conference on Cybernetics and Intelligent Systems (CIS), 17-19 Sept. 2011, p.247 – 252, doi: 10.1109/ICCIS.2011.6070336.

5. Yetkin, H. and Ozguner, U. *Stabilizing control of an autonomous bicycle*, 2013, Control Conference (ASCC), 2013 9th Asian, Istanbul, Turkey, pp. 1 - 6.

6. Yetkin, H. *Stabilization of Autonomous Bicycle*. Thesis. The Ohio State University. 2013. [WWW] (Cited on the 28th of April 2015), Available: https://etd.ohiolink.edu/!etd.send_file?accession=osu1373947913&disposition=inline

CORRESPONDING ADDRESS

Panu Kiviluoma, D.Sc. (Tech.), Senior University Lecturer
Aalto University School of Engineering
Department of Engineering Design and Production
P.O.Box 14100, 00076 Aalto, Finland

Phone: +358504338661
E-mail: panu.kiviluoma@aalto.fi
http://edp.aalto.fi/en/

ADDITIONAL DATA ABOUT AUTHORS

Vepsäläinen, Jari
E-mail: jari.vepsalainen@aalto.fi

Peltola, Matti
E-mail: matti.2.peltola@aalto.fi

Nygren, Toni
E-mail: toni.nygren@aalto.fi

Mälkönen, Joonas
E-mail: joonas.malkonen@aalto.fi

Heikkilä, Eetu
E-mail: eetu.heikkila@aalto.fi

Teerihalme, Santtu
E-mail: santtu.teerihalme@aalto.fi

Socie, Darrell
E-mail: darrell.socie@aalto.fi

Kuosmanen, Petri, D.Sc. (Tech.), Professor
Phone: +358 500 448 481
E-mail: petri.kuosmanen@aalto.fi

10th International DAAAM Baltic Conference
"INDUSTRIAL ENGINEERING -
12-13 May 2015, Tallinn, Estonia

THE IMPACT OF TAXATION ON THE SUSTAINABILITY OF MACHINE-BUILDING ENTERPRISES: A CASE STUDY

Vlasova, M.; Laskina, L.; Musalimov V., Musalimova L.

Abstract: *In the current context, stabilizing the growth of enterprise development indicators is becoming urgent. However, there is still a problem of value maximizing, including through the optimization of taxation in order to promote and expand business activity.The study is aimed at investigating the impact of taxation on sustainability indicators. To assess this factor, the authors propose to pay attention to the identification and assessment of tax risks using specific indicators such as fiscal leverage tax and tax sustainability.This case study of machine-building enterprises assesses the relationship between tax leverage and tax sustainability. The study demonstrated the inverse relationship between tax leverage and tax sustainability of enterprises, where the correlation coefficient was -0.77. We discussed the practical application of this index for predicting the growth of productivity. Thus, tax leverage associated with the company's sustainability and determining the acceptable level of tax risk will further allow the authors to consider the company's performance in relation to taxes as one of the types of costs.*

Key words: risk, tax burden, tax leverage, tax sustainability.

INTRODUCTION

Stabilizing the growth of enterprise development indicators is currently becoming urgent. In this context, it is important that owners and top managers of the company pursue the same goal that is maximizing its value, including through performance management and optimization of taxation in order to promote and expand business activity. This is due to the well-known fact that one of the conditions for improving performance is cost reduction (or increase, if it is slower than the growth of production), while taxes as one of the types of costs have a significant impact on the company's activity and can destabilize the economic status of the corporation. Given the fact that, as a function of the combination of factors operating both in the external and internal environment, performance is constantly changing due to their dynamic effect, the authors propose to focus on the identification and assessment of tax risks when analyzing such effect.

Within the existing approaches to risk classification, many researchers never assign a separate category to tax risk, considering the latter to be part of other types of risk. Tax risks are typically identified with tax risks and losses, as taxes are part of the financial system.

We believe that this approach fails to reflect the actual current opinion in risk assessment, since tax risk is indirectly present within the regulatory framework.

Specifically, it is "recognized" by tax legislation. For example, Article 64 of the Tax Code of the Russian Federation regulates the terms of granting a deferred payment of income tax in the event of a threat of insolvency (bankruptcy) to the taxpayer resulting from a one-off tax payment (paragraph 3, Clause 2, Article

64 of the Tax Code of the Russian Federation). In this case, tax risk appears in the presence of any signs of bankruptcy after tax payment.

Tax regulations thereunder also specify the concept of "financial and economic activity associated with high tax risks". In accordance with Order of the Federal Tax Service of Russia of May 30, 2007 No. MM-3-06/33 @ "Concerning the Approval of the Conceptual Framework for the On-Site Tax Audit Planning System" (hereinafter – the Conceptual Framework), taxpayers are recommended to make a systematic independent assessment of tax risks. The Conceptual Framework lists publicly available assessment criteria the most interesting of which is Clause12 "Conducting financial and economic activity associated with a high tax risk".

Pursuant to Letter of the Federal Tax Service of Russia of December 28, 2012 No. AS-4-2/22619@ "On methods of conducting financial and economic activity associated with high tax risk", an unjustified tax benefit may be enjoyed:

- by using fly-by-night companies;
- when selling real property;
- when producing alcohol products;
- through the employment of disabled people.

As is evident, the presence of any signs of tax risk results in the increased focus of tax authorities on the taxpayer. However, the very concept of tax risk is not available. Therefore there is a need for the establishment of institutional risk identification [1]. The institution shall be set for paying its funds to multilevel budgets and extra-budgetary funds for taxes and obligatory payments in case of delayed payment of the goods sold, and the works and services rendered.

In the economic literature, the concept of tax risk has appeared relatively recently, but it is currently described in many ways (Table 1, Appendix A).

Thus, we have two polar approaches to the concept of "tax risk": for the taxpayer, as an opportunity to eliminate losses incurred as a result of poor decision-making by the company's management and changes in state tax policy; for the state, as possible financial losses due to insufficient and untimely payment of taxes by taxpayers.

There is no concept of tax risk in the current tax system, since its meaning has not yet been defined. At the same time, the question of defining the concept of "tax risk" is raised in the scientific literature, but mainly in terms of discussion.

Tax risk diagnosis is based on the analysis of tax burden, which provides the economic and financial security of the economic entity. To cover the entire spectrum of the company's operational and financial activity, the authors recommend discussing tax risks in terms of fiscal leverage. This approach is more than evident if we remember that taxes collected on the territory of the Russian Federation may be included in the cost of production (resource payments, payments to extra-budgetary funds), and charged to profit or loss and profit before tax.

The concept of "tax leverage" is practically nonexistent in the domestic and foreign literature. We can mention only a few researchers who addressed the issue: M.A. Limitovsky, L.Yu. Filobokova, L.V. Samkhanova, so the study of the latter seems quite interesting.

For the purposes of tax risk management, we propose to use " tax leverage" (DNL) indicator which is defined as "a lever to manage net profit through the variability of tax burden"[2,3]:

$$DNL = \frac{NP + T}{NP} = \frac{EBT}{NP}, \quad (1)$$

where EBT – earnings before tax,

NP – net profit.

This is the total amount of corporate profit tax. Given the high degree of tax burden-dependence of "net profit", the level of tax leverage is a coefficient that reflects the degree of the tax risk determined by income tax exemption, including the low optimality of tax policy [3].

Whereas, tax risks and tax burden indicate the *tax sustainability* of the company, which characterizes the level of tax burden in the company's own sources of working capital required to fulfill its tax obligations.

Tax sustainability is based on the retrospective analysis of tax burden and ensures:

- timely mandatory payments to the budget and extra-budgetary funds;
- calculation and monitoring of the accuracy of assessment of payments in order to prevent excessive or insufficient payments;
- identification and assessment of the level of tax risk;
- development of measures to reduce the tax burden in order to achieve a given level of economic sustainability.

To define tax sustainability, we recommend using the formula proposed by the authors, which shows how many tax liabilities are covered by the company's working capital[4]:

$$TSI = \frac{\Delta T}{WC}, \quad (2)$$

where FSI – *tax* sustainability index;

ΔT – the amount of tax deductions per year;

WC – the company's working capital.

The calculation is based on the company's budget drawn up for the planning period. Since the index implies that mandatory payments should be covered by the company's working capital, ie, TSI is 1.0, the estimated amount of the company's working capital will be equal to its tax obligations.

If TSI is less than 1.0, the company can be considered sustainable in terms of taxation. If TSI is more than 1.0, the company is directly dependent on the amount of taxes paid. However, to diagnose the company, provided TSI is less than 1.0, it is necessary to extend the economic boundaries so as to recognize that the amount of working capital or mandatory payments is near a critical level. The calculation takes into account the method of determining the burden of taxation, proposed by the Department of tax reforms of the Ministry of Finance of the Russian Federation.

On this basis, the estimated limits of tax sustainability are summarized in Table 2. This calculation takes into account the share of working capital in the planned revenue from sales, which is different among companies with different production technology and different production and economic conditions.

Level of tax burden (TB)	**Relevant limit of TSI**
TB less than 15%	TSI < 0.5
TB 20-35%	TSI = 0.5-0.8
TB 35-60%	TSI = 0.8-1.1
TB more than 60%	TSI > 1.1

Table 2. Relationship between tax burden and tax sustainability index

- If TSI <0.5, there are two possible situations: either mandatory payments are not paid in full, or there is an excess of the company's working capital, which may adversely affect the economic stability.
- If TSI is from 0.5 to 0.8, the company is sustainable in terms of taxation.
- If TSI is between 0.8-1.1, the company is not sustainable in terms of taxation.
- If TSI is more than 1.1, the company is directly dependent on the amount of taxes paid.

This index is calculated for each tax period and is adjustable for individual companies on the basis of the production cycle and need for working capital, as well as for other sectors of the economy. The advantage of tax sustainability index (TSI) over the indices calculated on another basis can be explained by the fact that by using this index one can manage mandatory payments and regulate the amount of the company's working capital. Further, this makes it possible to use tax sustainability index to calculate economic sustainability by adjusting the coefficients and determining the type of the company's tax sustainability. It is also important to increase productivity, since it remains the most important indicator of economic activity and must take into account the interrelation of production efficiency criteria.

Nowadays, the stable development of *machine-building enterprises* as leaders in boosting the entire Russian industry and providing food security is the most important condition of overcoming the challenging business environment by the Russian economy and is a strategic objective for the nearest future. Only through this will the machine-building enterprises be able to update worn-out and obsolescent equipment, introduce new technology, and improve their competitiveness, since according to statistics, 70% of domestic engineering equipment has an average life of 20 or more years, with a share in GDP of only 6.3% (while tax revenues of the consolidated budget of the Russian Federation are merely 4.4%). In the industrial structure of the country, the share of machine building is less than 15-20%, whereas in developed countries this share is near 35-50% (the share of engineering production in the USSR in 1990 was 40%). In terms of economic security, the threshold share of machine building in GDP is 30%.

The economic sustainability of the company means continuous and sustainable development during a specific calendar period of time. In this case, important is how to evaluate the development characteristics and their dynamic change. The most important factors of the economic sustainability of the company include not only taxation, but also investment, particularly investment in productive capacity and efficiency. It is investment that not only provides the long-term maintenance of economic sustainability, but also makes it possible to work out a sustainable development strategy. This investment property is based on the determination of capital renewal trends, where capital assets present a material basis for the long-term quality characteristics of the product. This approach is based on understanding the meaning of performance management as a process taking into account the variable parameters (taxes, in particular input legal and regulatory compliance and output optimization of taxation) of its targeted impact.

This investment property is based on the determination of capital renewal trends, where capital assets represent a material basis for the long-term quality characteristics of the product.

This case study of machine-building enterprises assesses the relationship between tax leverage and tax sustainability. The study confirmed the assumed inverse relationship between tax leverage and tax sustainability of the enterprises, where the correlation coefficient was -0.77. We discussed the practical application of this index for predicting the growth of industrial production for machine-building enterprises.

It was found that tax leverage helps to assess both financial risks and production risks and can reduce the risks of production activity in general. Thus, tax leverage related to the company's sustainability and determining the acceptable level of risk will further allow the authors to consider the company's performance in relation to taxes as one of the types of costs.

REFERENCES

1. Viktorova, N. *The Analysis of Tax Risks at the Macro- and Micro-Level: A*

Scientific Publication. SPbTEI, Saint Petersburg, 2010.
2. Vlasova, M., Laskina L. *Operational, Financial and Tax Leverage: Interpretation and interrelation.* Financial Analytics: Problems and Solutions, Moscow, 2014, 40 (226), 35-44.
3. Filobokova, L., Samkhanova, L. *Tax policy, Tax Risk and Overall Business Risk of Small Enterprises.* The Economic Analysis, (25) UEkS, 1/2011
4. Vlasova M. *The Development of a Mechanism to Manage the Economic Sustainability of a Mining Enterprise in Monitoring its Compulsory Payments:* Abstract of a PhD Thesis in Economics. Moscow, 2003, 22p.

APPENDIX 1

Author	Interpretation
N.A. Pavlenko	Monetary valuation of irrational actions of a particular officer responsible for tax liabilities of the company
D.N. Tikhonov L.G. Lipnik	Possible losses associated with the process of tax payment and optimization, as expressed in monetary equivalent
A.G. Ivanyan A.Yu. Che	Risk of unforeseen disposal of the taxpayer's funds due to the action (inaction) of state and (or) local authorities
E.V. Berezhnaya T.A. Porokhnya S.I. Kukota	Possible financial losses as a result of changes in tax policy and tax rates
V.G. Panskov	Possible financial losses incurred by all participants of tax relations
M.R. Pinskaya	Risks to the taxpayer, for example, in the form of re-classification of economic transactions by tax authorities
M.R. Pinskaya	Risk of increased tax burden due to the fact that a tax authority may invalidate the transaction concerned (declare it fraudulent or sham), declare illegal the tax charge regarded as legal by the taxpayer
E.V. Zamula I.A. Kuzmicheva	Risk to the tax subject of incurring taxation-associated financial losses; therefore, for the taxpayer, increased tax costs involve reduced property potential and less opportunities to solve future problems
M.S. Vlasova L.Yu. Laskina	Possible financial losses of the company as a result of a change in accounting policy for tax purposes and loss of tax sustainability

Table 1. Scientific approaches to the interpretation of the concept of "tax risk" [3]

ADDITIONAL DATA

1) Vlasova Marina, Laskina Lyubov, Musalimov Victor, Musalimova Liudmila
2) The Impact of Taxation on the Sustainability of Machine-Building Enterprises: A Case Study
3) 1. Vlasova Marina, cand. econ. sci.,
International Banking Institute ,
191023 Russia , Saint-Petersburg , Nevsky av. 60
vms68@yandex.ru

2. Laskina Lyubov, cand. econ. sci., associate professor,
ITMO University,
191002, Russia, St. Petersburg, Kronverkskiy,49 .
tel. +7-952-235-32-26., risk05@mail.ru
3. Musalimov Victor
ITMO University, Professor Department of Mechatronics,
191002, Russia, St. Petersburg, Kronverkskiy,49 .
tel. +79214217917, musvm@yandex.ru
4. Musalimova Liudmila
ITMO University, Professor Department of Mechatronics,
191002, Russia, St. Petersburg, Kronverkskiy, 49 .
tel +79214217917musvm@yandex.ru
4) Corresponding Author – Laskina Lyubov

10th International DAAAM Baltic Conference
"INDUSTRIAL ENGINEERING"
12-13th May 2015, Tallinn, Estonia)

PNEUMATIC FOOTBALL KICKING DEVICE BASED ON HUMAN ANATOMY IMITATION

Väisänen, J.; Hasu, O.; Hevonoja, T.; Mielonen, M.; Kiviluoma, P. & Kuosmanen, P.

Abstract: *Nature has developed several complicated and functional mechanical solutions during evolution of human body. The purpose of this project is to examine whether human anatomy can be imitated accurately and effectively by utilizing pneumatic muscles.*
The research platform is a football penalty kick training device, able to perform a penalty kick with adjustable force and direction. The device consists of steel pipe- and beam-based frame, industrial-grade pneumatic actuators and electrically controlled valves. Two fluidic muscles work as a quadriceps and a peroneus muscles, whereas opposing muscular force is created by springs. Raspberry Pi 2 computer is used for controlling and computational operations.
Our study proves that the highly efficient imitation of human anatomy can be achieved by pneumatics.

Keywords: fluidic muscle, sport training device

1. INTRODUCTION

Artificial imitation of human anatomy by robots is often considered to be difficult since human dynamics are very fine and complicated. It is an interesting to study and test, whether human anatomy can be imitated by utilizing modern mechatronic components, such as pneumatic muscles which closely resemble mechanical behavior of human muscle. Until now, human anatomy has been already studied, modelled and mechanically imitated in experimental projects. [1] Although devices have been structurally complicated and accurate, yet very high forces have not been applied in them. For instance, imitating human hand and fingers movement by pneumatic muscles is complicated task, but does not necessarily require high forces or speed. This raises a research problem: can human anatomy imitation be done effectively and accurately by utilizing pneumatic muscles? Football kick event was selected as a context for artificial imitation since it is a good example of high performance human anatomy execution. Yet it is not reasonable to build a human-like robot with all limbs and similar functional muscles, the construction can be simplified at feasible level. Robots with limbs are developed, but they are far too complex to be used on a field. [2] In this study an experimental robot which is able to perform precise performance footballs penalty kicks in acceptable level of human anatomy is built. Several football shooting machines have been built already, but most of them are based on two spinning wheels which are used to shoot the ball. [3] Adidas has developed the machine where the real shoe is used to shoot the ball as in our machine. [4] Possible outcomes of developing human anatomy artificially may encourage research community to apply human anatomy-alike construction in more demanding applications, such as in field of medical research (i.e. complicated prostheses or body replacements).

2 METHODS

2.1 Mechanical structure

Minimum technical requirements for the device are adjustable and firm mechanical structure and football launching mechanism which has enough force to deliver a kick similar to human players. [2] Mechanical structure will consist of pneumatically operated beam mechanism, where pneumatic muscles can be applied for generating necessary amount of force and precision. A thigh and ankle mechanisms are operated with pneumatic muscles. Return force in both mechanisms is simply implemented by pull springs, providing opposing force and returning movement. For executing strong and rapid impulse for football, pneumatic circuits and valves of high airflow capabilities are required. Robot must be able to be parametrically adjusted for a precise football shot, which requires adjustable mechanisms at least for a shooting direction and height parameters.

Mechanical design was carried out using PTC Creo 2.0. Several concepts were analyzed before the final design. The structure is based on steel beams and pipes which are attached together with bolted joints. All movements are implemented by pneumatic actuators. CAD model of the device is shown in Figure 1. For the bearings Y-bearing plummer block units and polyethylene plates were used. The most important parts of the device are shown in Figure 2 and parts are listed in Table 1.

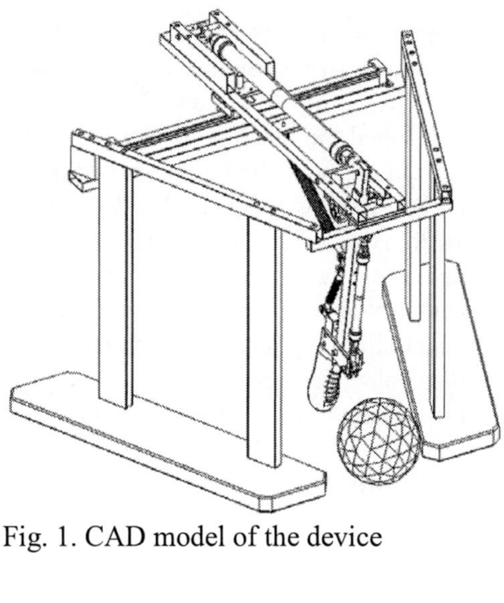

Fig. 1. CAD model of the device

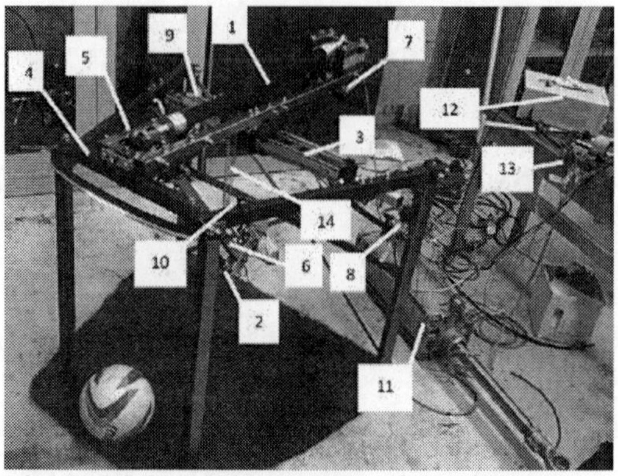

Fig. 2. Main components and structures

Pos	Description
1	40 mm pneumatic muscle for kicking action
2	20 mm pneumatic muscle for ankle angle
3	Rodless cylinder for alignment mechanism
4	Main frame assembly (solid steel beam structure)
5	Upper frame assembly (steel pipe structure)
6	Leg and ankle assembly
7	Proportional valve for 40 mm pneumatic muscle
8	5/3 valves (2 pcs.) for 20 mm pneumatic muscle and rodless cylinder
9	Linear sensor for rodless cylinder
10	Air pressure sensor for 20 mm muscle
11	Air tank
12	Power source (for 10 V and 24 V outputs)
13	Electronics
14	Returning springs

Table 1. List of main components

ADAMS multibody dynamics simulation software was used to analyze the most essential dynamics of the machine and define all critical values for the pneumatic components. A screen capture example of the analysis is shown in Figure 3. Required forces were also defined by dynamic analysis.

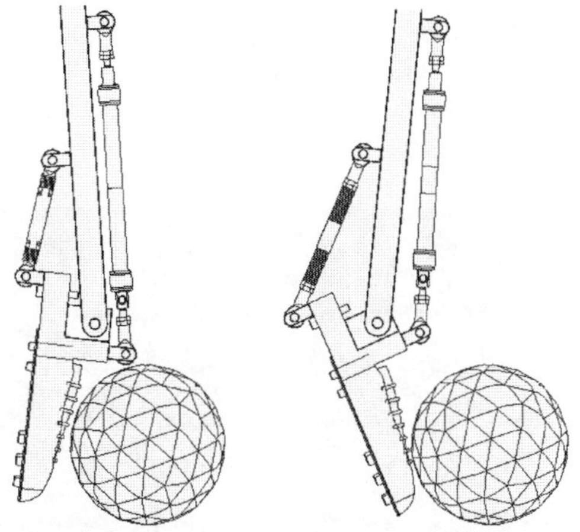

Fig. 3. Effect of ankle muscle pressure on ankle position

Final version of the device is shown in Figure 4. During the manufacturing process several changes were made. For example, some solid beams were replaced by pipes due to lighter weight. Also, placement of thigh and 40 mm muscle joint was moved closer to the axle to gain more speed for the kicking movement.

Fig. 4. Final device

2.2 Pneumatic system

All movements of the device are implemented by pneumatic actuators. Fluidic muscles are used for the kicking movements (kicking action and ankle adjustment). A rodless cylinder turns the upper frame where foot is mounted. This enables the foot to circle around the ball and thus allowing adjustable direction settings. Directional valves do control the kicking direction and proportional valve does control the speed of the kick. A pneumatic diagram is shown in Figure 5.

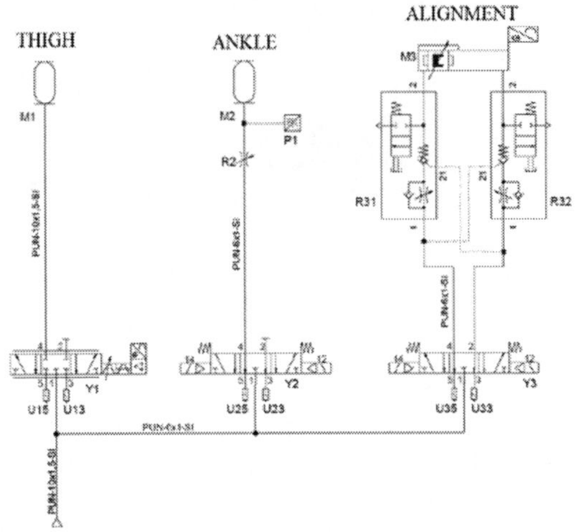

Fig. 5. Pneumatic system diagram

To maximize the airflow pulse to the main muscle, an air tank was used before the proportional valve. Feeding air directly from the air network is not enough to generate a rapid muscle movement essential for proper operation.

2.3. Control system

Raspberry Pi 2 computer is used for the overall control of the device. Arduino Nano microcontroller receives data from the pressure transmitter and linear drive potentiometer which measure the angle and direction of the foot. Received data is transmitted to the Raspberry Pi over serial connection. Raspberry Pi controls all the valves by its GPIO pins. For the proportional valve, PWM signal defines the control voltage. Control system diagram is shown in Figure 6. All programming was made by C and C++ languages and user can control the device by using command line or GUI. The device also has a front-mounted camera which can be used for defining the direction of the kick.

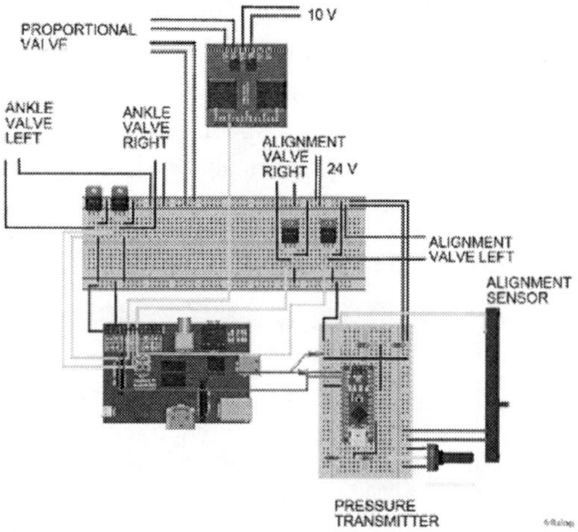

Figure 6. Control system diagram

3. RESULTS

In results section, two main performance values, accuracy and horizontal speed are evaluated. Tests were performed by maintaining air pressure at nominal 6 bar level. For the main pneumatic muscle (paired with a proportional valve) a separate air tank was used to store enough air to be delivered rapidly into the 40 mm muscle.

Speed performance was measured by utilizing Photron high-speed camera to record kick events with various ankle angle settings. Even though the proportional valve for main muscle was adjustable with different airflow values, it was always operated at maximum speed since lack of kick performance was apparent all the time. The speed test setup was configured by placing a one meter long ruler in front of the robot, and the high-speed camera on left side of the robot. Figure 7 shows the view captured by the camera. The kick event video clips were captured two times with six different ankle angle positions.

As seen in the result diagram, the overall speed varies between 41 - 45 km/h (shown in Figure 8), which is less than half of the goal value set in the beginning (100 km/h), and thus does not reach human performance. The results also indicate that the higher pressure applied in the ankle muscle, the lower is the horizontal speed. This is the result of higher vertical angle of the kick which reduces the horizontal component of speed vector. Horizontal speed can be perceived quite static in general.

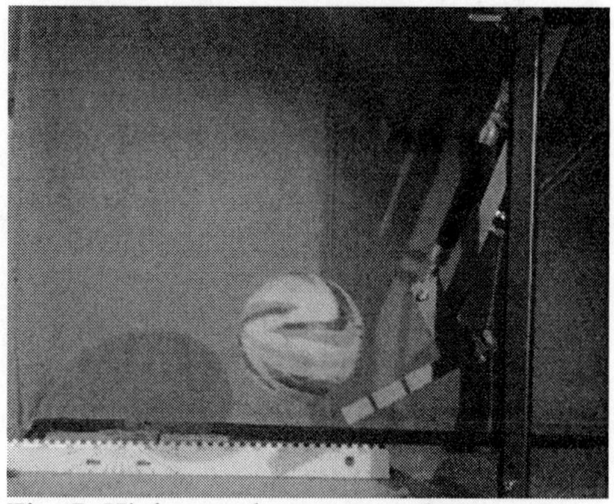

Fig. 7. High-speed camera analysis

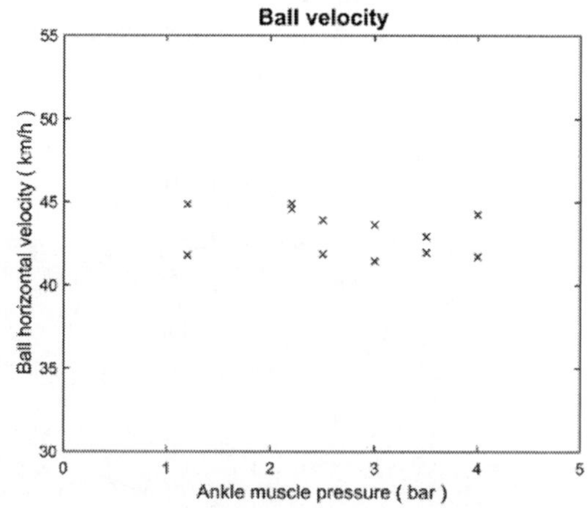

Fig 8. Speed data from the video analysis

Visual inspection also indicates that ankle muscle is flexible and thus works like a shock absorber absorbing kinetic energy. This phenomenon is clearly supported by high speed video recordings. Especially when the ankle muscle is operated with lower pressures (lower kick angles), the muscle is softer and more elastic.

Moreover, every time the foot hits the ball, the ankle is suspended a little bit backwards followed by a second major impact which delivers the most of the energy to the ball. This two-phased impact results in unclean kick and thus differs from human kick.

The accuracy test was carried out by performing subsequent football penalty kicks from 10 m distance into a football goal measuring 4 m of width and 2 m of height. Kicks were performed several times into same specific coordinates allowing evaluating mechanical accuracy of the device (maintaining direction and ankle angle adjustments unchanged). Scatter of impact points settled roughly inside a circle with radius of one meter, regardless of ankle angle and direction.

In addition to mechanical accuracy, also accuracy of control systems was evaluated. A laser pointer was mounted in front of the upper frame to help visual evaluation of directional accuracy. Although mechanical tolerances of the device were good (no significant gaps), parametric positioning of measurable pneumatic components (rodless cylinder and ankle muscle) was rather inaccurate and slow. When directional parameters were sent to the robot, the upper frame started oscillating around given location, sometimes for several seconds before setting in position. This was mainly caused by worn and stiff rodless cylinder with high friction. Inaccuracy of directional positioning was slightly compensated by creating smaller and more frequent air pulses to the cylinder when targeted location were close (inside 50 mm range). Rarely did the cylinder reach its dedicated coordinates accurately enough. Occasionally oscillating effect was also present when adjusting the ankle muscle, while ending setting was always precise enough to maintain static angular position.

4. DISCUSSION

Our research problem was to examine if human anatomy can be imitated precisely enough by pneumatic systems to perform a high performance football kick in terms of measurable speed and accuracy. First, it can be noted that human anatomy-based football kicking dynamics were able to be achieved at some level, yet the biggest issue is overall performance and lack of speed. The outcome reflects our hypothesis in a way that human anatomy was able to be reached at least at principle level, such artificially imitated and working dynamics, but did not meet performance goals.

The football kicking device does operate mechanically as expected. For example, dynamics did match our simulation and design of CAD-model, and the robot is able to perform adjustable kicks in different directions with controlled speed. Construction is also very firm with no significant mechanical gaps. Although mechanical accuracy can be considered to be at acceptable level, lack of kicking performance was very evident at all the time. Due to limited airflow into the muscle, or even probably limitations of the muscle performance itself, the device was not able to deliver the ball but only with half of the initial speed as planned. Another important issue in low performance is elastic behavior of pneumatic ankle muscle during the event of impact, where significant amount of kinetic energy is absorbed. High-speed video captures also shows that the impact occurs in two-phased event, dissociating it from human-performed football kick. The placement of the football on artificial grass was also inaccurate at some level, since there was no precise joint point or any kind of tee to support the ball exactly in one permanent location. It was also analyzed whether the profile of shoe could improve the kicking accuracy. In other words, our current foot solution, a 30 mm wide steel pipe does not exactly equal the interaction compared to human foot and shoe, being more narrow, blocky and thus less accurate.

Even though our research goals were not entirely reached, the study does still show that the efficient imitation of human anatomy will be possible to achieve by utilizing pneumatics muscles. By developing and redesigning mechanics, pneumatic and control systems, more optimal result is expected to be achieved with same basic concept.

The football kicking device does operate acceptably enough at principle level. However, achieving human performance needs considerably further refinement and testing. In mechanical point of view the construction of the robot can be considered to be firm and reliable enough to be equipped with more advanced designs and components.

During the project, several new questions were raised and left open to be solved in future; could robot perform considerably better if leg profile or the whole kicking mechanism is built from different design perspective to deliver more energy to the ball? Since our goal of reaching 100 km/h initial speed was not reached, what modifications or equipment is needed to increase the force? One of the key answers might be finding solution to increase pressure and airflow, as well as to study if pneumatic muscles were the bottleneck in performance. Since pneumatic muscles are very strong, could there be any mechanical solutions to increase transmission ratio? How to prevent pneumatic muscle in the ankle mechanism from suspending and absorbing significant amount of kinetic energy, which should be delivered to the ball? One possible option might be revising the ankle mechanism so that the muscle reaches its positions with higher pressure levels. Finally, could the robot or the whole concept of human anatomy imitation be productized for real sport training purposes by equipping it with machine vision and decision making capabilities?

6. ACKNOWLEDGEMENTS

The valuable support of researching Professor Jyrki Kajaste from Department of Engineering Design and Production, Hydraulics Group, is gratefully acknowledged.

7. REFERENCES:

[1] Festo, Project 2007, Airic's Arm [Online]: Available: http://www.festo.com/cms/de_corp/9785_10400.htm#id_10400 [2015, May 4]

[2] Schempf, H. 1995, ROBOLEG: A robotic soccer-ball kicking leg, Robotics and Automation IEEE vol 2 p. 1314-1318

[3] Griffith L. 1980 Soccer ball practice machine Patent US 4352348 A

[4] Haddadin S et al. 2009, Kick it with elasticity: Safety and performance in human–robot soccer, Robotics and Autonomous Systems vol 57:8 p. 761-775

8. CORRESPONDING ADDRESS

Panu Kiviluoma, D.Sc. (Tech.), Senior University Lecturer
Aalto University School of Engineering
Department of Engineering Design and Production
P.O.Box 14100, 00076 Aalto, Finland
Phone: +358504338661
E-mail: panu.kiviluoma@aalto.fi
http://edp.aalto.fi/en/

9. ADDITIONAL DATA ABOUT AUTHORS

Väisänen, Joni
E-mail: joni.vaisanen@aalto.fi

Hasu, Olli
E-mail: olli.hasu@aalto.fi

Hevonoja, Toni
E-mail: toni.hevonoja@aalto.fi

Mielonen, Matti
E-mail: matti.mielonen@aalto.fi

Kuosmanen, Petri, D.Sc. (Tech.), Professor
Phone: +358 500 448 481
E-mail: petri.kuosmanen@aalto.fi

III SUPPLY CHAIN MANAGEMENT

10th International DAAAM Baltic Conference
"INDUSTRIAL ENGINEERING" -
12-13 May 2015, Tallinn, Estonia

VALUE STREAM MAPPING AS A TOOL IN OPTIMISING PRODUCTION LOGISTICS. CASE: HE TELETECHNICS

Hurt, U.; Tomba, A. & Koppel, O.

Abstract: *Contemporary production logistics might involve newest IT-solutions, wireless communication and technological supplies for running the automation, but it still needs surveillance of process speed and productivity. Value stream mapping (VSM) among other LEAN methods becomes even more necessary, but also more productive as a tool in Industry 4.0. conditions. The current paper explains the usage of VSM in production logistics optimisation. Based on an example case, the paper also analyses the strengths and weaknesses of the method. Finally, it suggests methods of better time cycle and performance analysis for production logistics.*
Key words: automation, Industry 4.0., optimisation, production logistics, value chain, value stream mapping.

1. INTRODUCTION

The paper focuses on the specifics and perks of using value stream mapping (VSM) in production logistics optimisation. Since Toyota first introduced methods of LEAN in production management, the aim of the sustainable production is to diminish the waste of resources through better planning and timing of production as well as reducing the amount of stock (both material, semi-finished and finished products) [1].
All tools of applying sustainable production such as 5S, just-in-time, VSM derive from the principles of sustainability and the company's interest of minimising the waste of resources.
It has been a recent trend and a practice at the department of Logistics and Transport at Tallinn University of Technology to provide the companies of the sector a service of optimisation calculations often using value stream mapping as a method.
HE Teletechnics Ltd [2] requested such an analysis of their production logistics and material flow. The problem of the current Estonian production sector is that LEAN and other optimisation methods (incl. VSM) are relatively new tools in development processes. Miina [3] states that even though LEAN ideas have been known and studied extensively, there is still yet much to be studied for development of better imple-mentation methods.
The systematic methods of optimisation such as VSM are applicable and even more effective in Industry 4.0. conditions as the data is available online in real time.

2. VALUE STREAM MAPPING (VSM)

2.1. VSM and LEAN

Optimisation is a concept in need of tools and methods in order to be prepared and implemented. LEAN thinking is a wider structure that has supported new emerging methods and tools for analysing and suggesting improvement actions/changes in production process in order to gain savings or higher rate of meeting deadlines.
LEAN thinking has been the driver of systematic elimination of waste in the supply chain. It has first been introduced in

Toyota's production system and discussed and studied by many researchers.

LEAN as well as its methods (incl. VSM) aim at elimination of waste and non-value-adding activities [4]. But Bicheno [5] stresses on the process of avoiding rather than eliminating non-value-adding activities and waste.

The target of the implementation of better lean tools is the reduction problems of lead time in every stage of production. According to Womack *et al* [6] it leads to lower costs and higher quality as well as improvement of safety and morale.

According to Ohno [7] the wastes can be structured into eight groups and explained as following.

- Overproduction: excess production to storage leads to excess inventory as well as waste of resources in production process.
- Waiting: all time the production process meets waiting is considered waste.
- Excess transport: transport of raw materials, components or final products in between the process cycles or territory.
- Ineffective production methods or technologies: tools or product design specialties leading to excess movements or actions during the production process.
- Unnecessary or excess inventory: raw materials, components, products in production, completed products in storage.
- Unnecessary movements: all excess movements that the employees have to make to reach material, tools or their materials.
- Fault management: system of defining and managing faulty products.
- Low use of knowledge and experience of employees: system allowing improvements suggested by members of the process.

2.2. VSM in production logistics

Originating from Toyota's production management optimisation, VSM is a tool that allows visualisation of transport, material and information flow throughout all processes in the supply chain or a part of it. Mapping the current situation, the total time of production and total value adding time can be assessed [1] as well as attention to hidden problems or waste of time/resources can be discovered.

Following the mapping of current value stream and defining of the bottlenecks, the suggestions for avoiding waste can be suggested. Eliminating the sources of waste is a process following the defining of the structure and details on selected types of wastes [8].

3. METHODS OF IMPLEMENTATION

The methods of implementing VSM as an optimisation tool involve modeling the movement/flow of activities, information and material [8].

Among the mapping, assessment of processes involves measuring time use of the activities as well as stating whether the process is a value adding or non-value-adding process [9]; the first involving material unpacking or delivery of materials and the latter involving mainly assembling, folding, painting [10]. The results are in addition to calculations, presented it visually.

The method involves mapping the current value stream and suggesting a future value stream. The map of the future value stream can be compiled through consultations with all stakeholders, such as production participants, operations managers and company's management [1].

Goals to be achieved with the future map are [1, 9-10]:

- information on the cycle time and rhythm of the production;
- information on planning and management of production;

- one process decided to be the pace-maker (speed setter);
- if possible, constant flow of material must be achieved, if not possible, the pulling principle must be implemented;
- the productivity of the production flow must increase and all methods of diminishing time-waste must be counted in;
- data on information flow for the production management.

Indicating the wastes and suggesting ways of eliminating or minimizing the causes of wastes is the main goal to be achieved within the method.

The mapping has visual tools and key icons for the visualised map are shown in Fig.1.

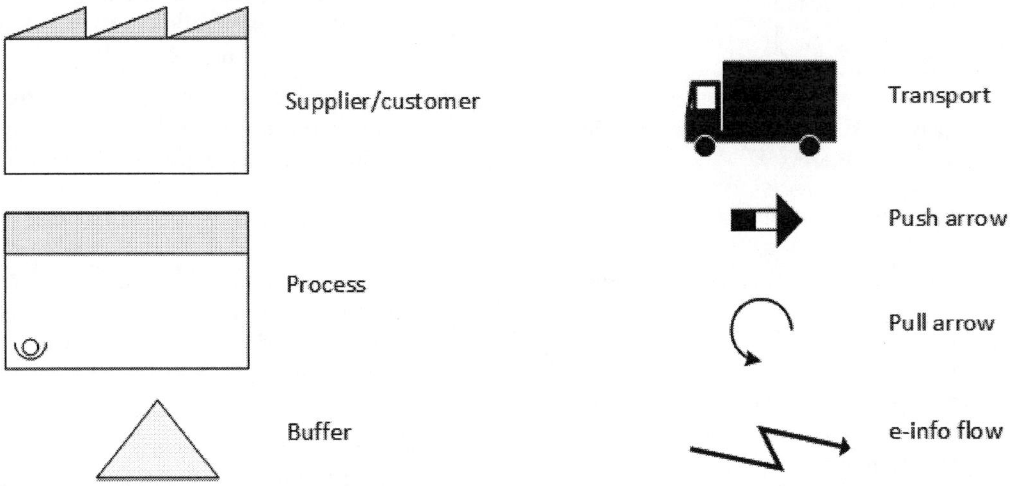

Fig. 1. Visual tools and key icons for VSM

4. RESEARCH RESULTS

4.1. The current value stream.

On-site measurements and production information system data analysis combined led to mapping the current value stream and observations, calculations, interviews and consultations led to mapping the future value stream. During the research, four different high-runner products were thoroughly analysed.

The flow was analysed separately on all the products aiming at the effectiveness one-piece flow analysis [12].

The future value stream was composed based on the research performed. A simulation as also suggested by McDonald *et al* [13] of the lead time was calculated and alternatives suggested.

The map stated bottlenecks and wastes in the processes for four different products. The current value stream as an example of one product is visualised on Fig. 2.

The material for the selected products are delivered in two patches - one to the storage of raw materials in the beginning of the process flow, the other to the buffer storage before final assembling. Unpredictability extends the amount of time materials "wait" in the buffer. Even though "safety stock" is a concept often used to explain the over-production of components [14], in current case, the level of safety stock was not unified for similar products as well as safety stock level was not set.

In addition, the production in some cases tends to over-produce some details "just in case" as the setting times of machines for detail-cutting are relatively long.

There are no shortages in details, rather than raw material as it has been produced into details waiting in storage for products that are not yet ordered by customers.

The main task was to define the pace-maker of the material flow and production logistics for the selected products and its components throughout the production process. Painting was defined as the most inflexible process as well as concluded to be the pace-maker.

4.2. Defined wastes

The following wastes were indicated and prioritized for actions of minimization (see also in Table 1).

Fig. 2. Current value stream

Factor	Importance/ impact
overproduction	***
waiting time	***
transport	*
work arrangement	*
inventory	***
unnecessary movements	*
defects	*
employee knowledge	*

Table 1. Wastes and their importance

- Overproduction: during the analysis period, it could be stated that the focus of the stamping work-station is to use one sheet of metal fully, not according to the needs.
- Excess waiting time derives from overproduction of semi-finished products in the first workstations bringing the waiting time of products even up to 60-100 minutes in the process.
- Excess transport does exist between the workstations, but as they are overly close and structured in a following layout, the factor is not maximum-critical.
- As for unnecessary/unsuitable arrangement of work the factors' impact was not significant on the material flow analysis.
- Excess inventory is in of the key indicators in the case researched as due to overproduction, excess inventory is gathered before every process/work-station.
- Unnecessary movements, could be considered low in amount separate from the excess movements made due to overproduction.
- Defects are well handled and stopped before reaching the next workstation.
- Use of human capital is at a good level in the case researched and well communicated.

4.3. The future value stream

The future value stream was composed and accompanied with calculation methods. The future value stream is visualised on Fig. 3.

The future value stream shows eliminating buffers and overproduction of semi-finished products.

Mainly, the focus is on lowering excess

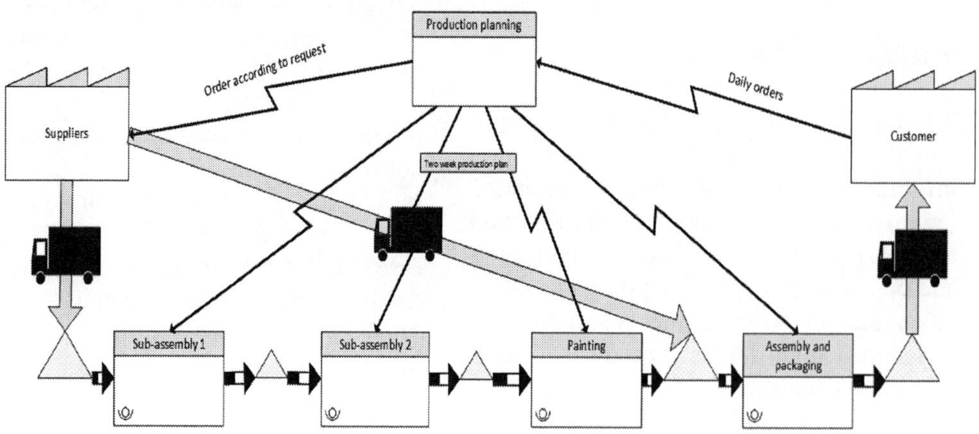

inventory in every stage of production and eliminating the current overproduction leading to components waiting between processes.

Three main suggestions and changes were underlined for HE Teletechnics in optimising their production process using tools of VSM. First, together with the setting time, the "pulling" principle needs to be implemented to avoid excess waiting time. For that, the "pulling" principle best discussed by Jonsson *et al* [[15]] was suggested.

Second, the material usage structure was suggested to be rearranged to avoid overproducing details not needed in production yet.

Thirdly, deriving from lower amount of details overproduced the number of buffers and amount of semi-finished products in buffers and buffer storage is lowered. The minimising of the inventory was suggested to be fully observed and re-calculated with the help of the production information system.

5. CONCLUSION

The methods of VSM find extensive use as a tool of optimisation in production process. It still needs discussion as the university has also contributed in promoting the method.

In Estonia and in the case of HE Teletechnics, both LEAN and VSM have become useful tools that have proven its applicability.

Based on the research, the company acknowledged the real pace-maker in the production process and re-arranged (at least got suggestions of doing so) the production flow based on the tact time of the painting unit.

Today, as not all the companies are up to date, the visualisation as well as data gathering and comparison, unfortunately is still time-consuming.

In progress of more and more automation as well as computerized and network-managed Industry 4.0. production systems, the optimisation methods such as VSM can be used in a perfected way as precise data availability is higher.

Finally, the methods are components of teaching methodologies in the Tallinn University of Technology Department of Logistics and Transport as well as a hands-on learning case at the Laboratory of

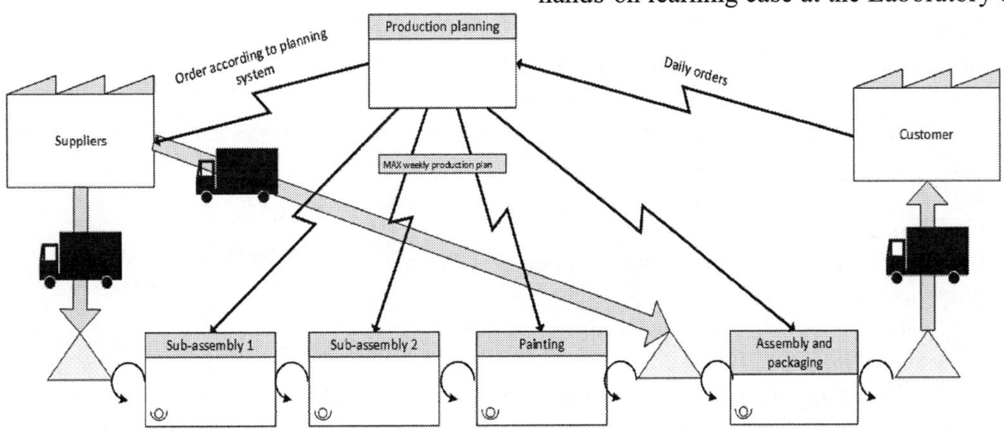

Fig. 3. Future value stream

Logistics and Supply Chain Engineering in the TUT Mektory Centre.

6. REFERENCES

[1] Liker, J. K. *The Toyota way: 14 management principles from the world's greatest manufacturer.* McGraw-Hill, New York, 2004.

[2] HE Teletechnics. [WWW] http://www.harjuelekter.ee/teletehnika (2012-2015).

[3] Miina, A. *Critical Success Factors of Lean Thinking Implementation in Estonian Manufacturing Companies.* TUT Press, Tallinn, 2012.

[4] Santos, J., Wysk, R., Torres, J. M. *Improving production with Lean Thinking.* John Wiley and Sons, Hoboken, 2006.

[5] Bicheno, J. *New Lean Toolbox: Towards Fast, Flexible Flow.* PICSIE Books, Buckingham, 2004.

[6] Womack, J. P., Jones, D. T., Roos, D. *Lean Thinking: banish waste and create wealth in your corporation.* Free Press, New York, 1990.

[7] Ohno, T. *Toyota production system: beyond large-scale production.* Business Diamond, Tokyo, 1978.

[8] Liker, J. K. *The Toyota Way Fieldbook: A Practical Guide for Implementing Toyota's 4Ps.* McGraw-Hill, New York, 2006.

[9] Rother, M, Shook, J. *Learning to see: value stream mapping to add value and eliminating muda: A Lean tool kit method and workbook.* Lean Enterprise Institute, 2003.

[10] Lasa, I., de Castro, R., Laburu C. O. Extent of the use of Lean concepts proposed for a value stream mapping application. *Production planning & Control,* 2009, **20**, 82-98.

[11] Monden, Y. *Toyota production System: An Integrated Approach to Just-in-Time Engineering and Management Press.* IEEE, Norcross, 1998.

[12] Nyhuis, P, Vogel, M. Adaptation of logistic operation curves to one-piece flow processes. *Int. J. Productivity and Performance Manage-ment,* 2006, **55,** 284-299.

[13] McDonald, T, Van Aken, E. M, Rentes, A. F. Utilising simulation to enhance value stream mapping: a manufacturing case application. *Int. J. Logistics,* 2002, **5,** 213-232.

[14] Piasecki, D. *Optimizing safety stock.* [WWW] http://www.inventoryops.com (18.04.2015).

[15] Jonsson, P., Mattson, S. A. *Manufacturing planning and control.* McGraw-Hill, New York, 2009.

7. ADDITIONAL DATA ABOUT AUTHORS

1) Ulrika Hurt (MA), corresponding author, PhD researcher at Tallinn University of Technology, ulrika.hurt@ttu.ee, phone +372 52 14251, Department of Logistics and Transport, Tallinn University of Technology, Akadeemia tee 15A, 12616 Tallinn, Estonia;

2) Andrei Tomba (MSc), supply chain engineering professional in private sector;

3) Ott Koppel (PhD), visiting professor at Tallinn University of Technology.

10th International DAAAM Baltic Conference
"INDUSTRIAL ENGINEERING" -
12-13 May 2015, Tallinn, Estonia

DECISION-MAKING FRAMEWORK FOR INDUSTRIAL-SIZE DATACENTERS

Leppiman, A.; Kotka, T.; Kõrbe Kaare, K. & Koppel, O.

Abstract: *Daily life in modern society creates data, with predicted exponential growth rate in coming years. Three aspects of the business environment have captured considerable attention in recent years: information technology, globalization, and the service economy. Hosting industrial scale datacenters in its territory could be appealing to countries due to economic benefits. Many countries are researching economic and technological synergies to attract industrial-size datacenter projects.*

In this paper, the authors compare classical foreign direct investment decision determinants with key factors in choosing locations for datacenters. The purpose of this paper is to relate theories from the field of international economics and logistics, to give a roadmap of related measures and to pose several research questions as challenges to compose models to support the strategic planning process of datacenters.

This decision framework and research helps managers allocate investments and assess alternative locations for industrial-size datacenters.

Key words: decision framework, industrial-size datacenters, location decisions, power consumption.

1. INTRODUCTION

The IT industry in general and datacenters in particular are subject to a very dynamic development. Datacenters are the information warehouses and distribution centers of most private and public institutions. Several converging trends, such as increased number of users, more devices and a lot more data (Big Data, Data Lakes) have pushed the storage environment to a new level.

Now, these new technologies are not only driving the cloud – they are pushing forward all of the technologies that support cloud computing. Datacenter is the central point where all information is gathered, and then distributed to other datacenters or to the end-user.

Because cloud computing will only continue to advance, there will be new demands placed around storage. Storing data has created the need for additional datacenters and warehouses and the demand is predicted to be exponential [1].

Global enterprises developing new datacenter strategies throughout the next years will require a holistic approach and a new type of analyzes. The demand for new, high-quality facilities, geopolitical and socio-economic factors, rapid changes in technologies and new financial models will come together forcing users to think differently about the future of their enterprise datacenters.

In September 2012, the European Commission adopted a strategy for "Unleashing the Potential of Cloud Computing in Europe". The strategy outlines actions to deliver "a net gain of 2,5 million new European jobs, and an annual boost of 160 billion euros to the European Union GDP (around 1%), by 2020" [2].

The strategy is designed to speed up and increase the use of cloud computing across all economic sectors. Although it is too early to evaluate the actual size of those numbers, the cost savings, scalability and high availability achievable with virtuali-

zation, it is clear that there is a huge potential to benefit from using cloud computing [3]. According to the IDC, the amount of data that will need to be stored in 2020 will be 50 times larger than in 2010 [4]. This underpins the need to build new datacenters.
The purpose of current paper is to relate theories from the field of international logistics to pose several research questions as challenges to compose models to support the strategic (especially location) planning process of datacenters.

2. INDUSTRIAL-SIZE DATACENTERS

Datacenters do not create a lot of jobs. The servicing of a 20 000...40 000 square meter datacenter does not require more than 50 to 100 employees. Jobs will, however, be primarily created in the software industry which will have to update and amend existing solutions to make them cloud compatible. Therefore, the interested government show in attracting new datacenters to their territories is based on other arguments than just job creation.
High demand for new datacenters creates additional burden and challenges to other sectors. In addition to the economic benefits and sectorial synergies, investors for datacenters have requirements on site, namely sufficient size in tens of hectares, no high restrictions, zoning, requirements on ICT infrastructure as dark fiber links as well as requirements on business environment.
Still, as the datacenters profitability relies on the highest input in the business model – energy price – industrial datacenters competitiveness will be influenced on the proper location in terms of stabile and low price power supply.
However, the explosion of digital content, big data, e-commerce, and Internet traffic is also making datacenters one of the fastest-growing consumers of electricity in other developed countries, and one of the key drivers in the construction of new power plants [5].
Data centers are approximately 50 times more energy-intensive than conventional office buildings, where ICT equipment consumes about 50% of total electricity and cooling energy is roughly 35% of total energy use [6].
Normally developers of industrial scale datacenters require dual feed, high voltage power connection up to 100 MW. Electricity cost for large consumer should be in competitive level, but more often higher renewables breakdown of countries energy portfolio has influence attracting investments.
On site selection process depends on also re-use of energy, to use excess heat from servers in a datacenter in nearby industries or feeding it into the district heating networks. Especially energy use and re-use optimization has high impact on business model.
To ensure constant power supply, a variety of back-up power systems to help protect the client may require extra investments, so that lack of power does not cause service slowdown or failure. Cooling is accomplished by letting cold air or liquid flow between the racks of servers [6].
Therefore, the key term which is important to understand the energy balance in datacenters are Power Usage Effectiveness (PUE) value, free air cooling and forced air cooling. PUE value is a measure of datacenters efficiency of energy use and the calculating power achieved.
In ideal case PUE = 1,0, so all incoming energy feeds the processor only to produce calculating power, which would ecologically also be the most sustainable outcome. Traditionally, PUE values in datacenters depending on the age of facility vary from 1,2 to 3,0, whereas the newest data centers can achieve values as low as 1,1 [7].
As datacenters have constant energy needs such clients are expected customers for energy traders. Still with escalating demand and rising energy prices, it is essential for the owners and operators of

these mission-critical facilities to assess and improve datacenter performance using energy efficiency and greenhouse gas emission metrics [8].

3. DECISION FRAMEWORK FOR DATACENTERS

The international data storage market is largely dominated by corporations based in the United States (Amazon, Google, Microsoft, etc). These companies do not only establish datacenters in the United States, but choose locations globally, with consideration given to climate, the speed and distance of data transfers, as well as the price and cleanliness of electricity [9].

For industrial-size datacenters it is rare to be storing one single clients data, most business models are created for the datacenter to function as a warehouse where all interested parties can store their data. When in-house datacenters reach their capacity or become financially not competitive with large-scale datacenters decide whether to outsource date warehousing [10].

As outsourcing becomes more widespread, customers, the general public and politicians have increasingly questioned and criticized firms' outsourcing decisions to foreign locations. After the decision for outsource the next question is where to outsource.

The strategic supply chain design problem describes the numbers, locations, and capacities of manufacturing, assembly, and distribution facilities and then the flow of the materials from suppliers to customers. Global supply chain design involves international trade rules and financial issues and allows suppliers and facilities to be located in multiple countries [11].

„Location theory addresses several important questions: who produces what goods or services in which locations and why." [11]. Therefore, one of the most important decision making process of administrators is the location decision of the distribution center [12].

Richardson and Marshall [13] identified eight major location considerations in modern industries: availability of advanced telecommunications; telecommunication costs; labor pool of sufficient quality; labor costs; government financial incentives; attractive living environment; low occupancy costs, and access to good transporttation. Pennington [14] describes several barriers to international location decisions (especially in IT-industry), including competitive pressure, hacking and other criminal risks, and cultural differences.

Location-specific advantages are based on resources, networks or institutional structures that are specific to a geographic entity. Internalization advantages accrue to a firm when it eliminates the transaction costs associated with market interaction [10].

Dunning [15] identified three determinants of location. These include infrastructure, country risk and government policy. Several authors [10-12, 16-18] have found empirical support for this theoretical framework for foreign direct investments decisions, especially regarding manufacturing investment.

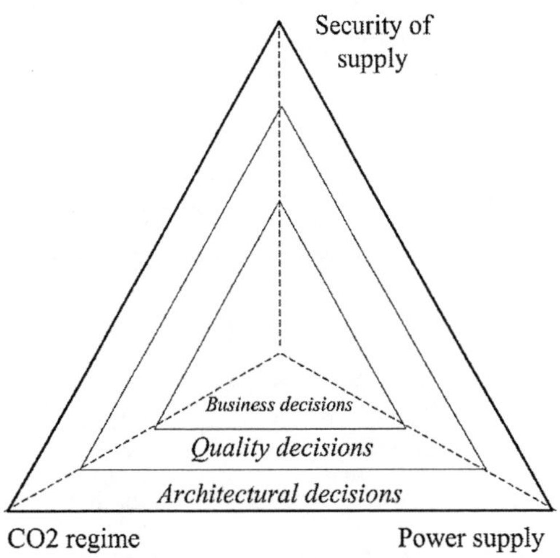

Fig. 1. Datacenters decisions hierarchy and key determinants

Datacenters can also explored as a type of distribution centers. Datacenters investments differ from traditional manufacturing investments in the importance of security risk in assessing country risk. When in most cases price (overall cost, operating cost) something money related is the key determinant – in case of datacenters it is security [10].

Taking into account all the above mentioned, the location criteria of datacenters can be interpreted as depicted on Fig. 1 on previous page.

Schniederjans *et al.* [18], Kumar [19], Kaisler *et al.* [20] and others have brought out and categorized the critical factors that influence business, quality and architecttural decisions in establishing industrial-size datacenters.

- Processor/system design/power (e.g. the use of utility and grid architecttures);
- Architectural topology and real-time infrastructure (e.g. the need to manage a heterogeneous hardware environment);
- Operational processes and tools (e.g. the use of monitoring and measuring tools);
- Disaster recovery and business continuity (e.g. dealing with social threats that disrupt technology use);
- Capacity growth (e.g. growth of new datacenters);
- Operating system and application changes (e.g. effects of business process optimization on application development);
- Consolidation and rationalization (e.g. managing physical datacenter consolidation projects);
- Facilities modification (e.g. environmental and energy designs for new hardware and datacenters).

Combining decision hierarchy of Fig. 1 with the critical factors that have been brought out the authors propose a modified decision framework for choosing datacenters location (see Fig. 2).

4. CONCLUSION

In this paper, author's present a modified decision chain for choosing locations for industrial-size datacenters. Results of the research are presented below.

- Based on literature the datacenter location decision making process can be divided in four stages: business decisions (especially return on investment), quality decisions, architectural decisions and location decisions).

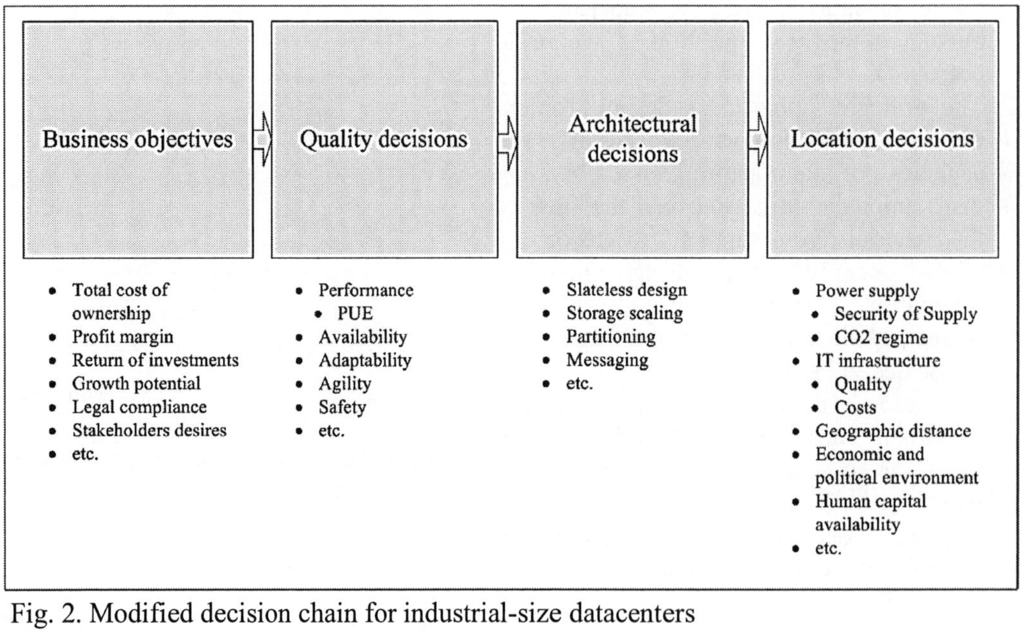

Fig. 2. Modified decision chain for industrial-size datacenters

- The theoretical bases of datacenter location decision is a combination of classical investment decision principles and distribution center decision principles known from logistics theory.
- The major differences in the above-mentioned principles are caused by extreme security and geopolitical stability requirements and also the large energy consumption of data centers. In international dimensions it can be said that the main criteria in the location decision process are: geopolitical stability, security of supply of energy and energy efficiency.
- However, the challenges to meet the expectations of all interested parties and society as a whole will continue to be a major concern for decision-makers of datacenters.
- The effects of global socioeconomic changes on disaster recovery should not be underestimated. Enterprise datacenter strategies have come down to a balancing act between managing the costs of a defined level of IT service delivery and managing risks of the failure of that delivery.

In conclusion, further research involves testing the presented model on countries starting with a case study on Estonia.

5. REFERENCES

1. *Big data: The next frontier for innovation, competition, and productivity.* McKinsey & Company, Washington, 2011.
2. *Unleashing the Potential of Cloud Computing in Europe.* COM(2012) 529 final. European Commission, Brussels, 2012.
3. Biscotti, F., Skorupa, J., Contu, R. et al. *The Impact of the Internet of Things on Data Centers.* Gartner, Stamford, 2014.
4. *IDC Digital Universe Study: Big Data Is Here, Now What?* EMC^2, Silicon Valley, 2011.
5. *Scaling Up Energy Efficiency Across the Data Center Industry: Evaluating Key Drivers and Barriers.* Natural Resources Defense Council (NRDC), New York, 2014.
6. Cho, J., Yang, J., Lee, C., Lee, J. Development of an energy evaluation and design tool for dedicated cooling systems of data centers: Sensing data center cooling energy efficiency. *Energy and Buildings*, 2015, **96**, 357-372.
7. Aaltonen, P. H. M., Aaltonen, P. Re-using old infrastructure to host datacenters: eco-innovation as an examptation process. In *DRUID Society Conference 2014.* CBS, Copenhagen, 2014, 1-33.
8. *Harmonizing Global Metrics for Data Center Energy Efficiency.* 2014 [WWW] http://www.thegreengrid.org/~/media/Regulatory/HarmonizingGlobalMetricsforDataCenterEnergyEfficiency.pdf?lang=en.
9. Marston, S., Li, S., Bandyopadhyaya, S., Zhang, J., Ghalsasi, A. Cloud computing – The business perspective. *Decision Support Systems*, 2011, **51(1)**, 176-189.
10. Graf, S., Mudambi, S. M. The outsourcing of IT-enabled business processes: A conceptual model of the location decision. *J. Int. Management*, 2005, **11**, 253-268.
11. Lee, C., Wilhelm, W. On integrating theories of international economics in the strategic planning of global supply chains and facility location. *Int. J. Production Economics*, 2010, **124**, 225-240.
12. Özcan, T., Çelebi, S., Esnaf, S. Comparative analysis of multi-criteria decision making methodologies and implementation of a warehouse location selection problem. *Expert Systems with Applications*, 2011, **38**, 9773-9779.
13. Richardson, R., Marshall, J. N. Teleservices, call centres and urban and regional development. *The Services Industry Journal*, 1999, **19(1)**, 96-116.
14. Pennington, E. Approaching the location of international call centers from a secure perspective. *Call Center CRM Solutions*, 2000, **19(4)**, 86-90.

15. Dunning, J. H. *Explaining International Production*. Unwin Hyman, London, 1988.
16. Hätönen, J. Making the locational choice: A case approach to the development of a theory of offshore outsourcing and internationalization. *J. Int. Management*, 2009, **15(1)**, 61-76.
17. Koong, K. S., Liu, L. S., Wang, Y. J. Taxonomy development and assessment of global information technology outsourcing decisions. *Industrial Management & Data Systems*, 2007, **107(3)**, 397-414.
18. Schniederjans, M. J., Hamaker, J. L., Schniederjans, A. M. *Information Technology Investment: Decision-making Methodology*. World Scientific, Singapore, 2010.
19. Kumar, R. *Eight Critical Forces Shape Enterprise Data Center Strategies*. Gartner, Stamford, 2007.
20. Kaisler, S., Money, W. H., Cohen, S. J. A Decision Framework for Cloud Computing. In *45th Hawaii Int. Conf. System Sciences*. IEEE, 2012, 1553-1562.

6. CORRESPONDING ADDRESS

Associate Prof. Kati Kõrbe Kaare
TUT, Department of Logistics and Transport
Akadeemia tee 15A, 12616 Tallinn, Estonia
Phone: 372+620 2605,
E-mail: kati.korbe@ttu.ee

10th International DAAAM Baltic Conference
"INDUSTRIAL ENGINEERING -
12-13 May 2015, Tallinn, Estonia

THE IMPACT OF TECHNOLOGY TRENDS ON SKILLS OF LOGISTICS ENGINEERS – A NOVEL COMPETENCE APPROACH

Niine, T. & Koppel, O.

Abstract: *Logistics is heavily influenced by technologies and engineered solutions. Identification, tracking, process control, automation and sustainability-oriented technologies are advancing rapidly. These areas should be studied in classroom by all future logisticians and in-depth by logisticians with engineering focus.*
The paper summarizes relevant technologies in modern-day and near-future logistics and analyses the presence of these elements in current competence models in logistics (by APICS, AST&L, ELA and SOLE). A novel competence model titled "logistics systems engineer" is presented as research outcome. The paper explains model structure and promotes it as significantly modernized way to integrate crucial technological viewpoints into logistics engineer's competence profile to overcome gaps present in current models. The model can be used as curriculum development guideline.
Key words: logistics technologies, logistics engineering, logistics skill areas, logistics competence models.

1. INTRODUCTION

The environment of logistics education is swiftly evolving due to technological progress and economic advancement. In many areas, capability of rapid innovation and technological modernization is a key success factor. Often this is coupled by emphasised role of logistics in the mix of competitive advantages to ensure agile and reliable global deliveries.
Universities need to thoroughly understand essential competences of future logistics professionals today [1, 2, 3]. Logistics education has to be interdisciplinary, as society needs broad knowledge and know-how to manage interrelated functions of logistics systems [4, 5].
This paper argues that the technological element is not represented to sufficient extent and depth in some renowned competence models of logistics professionals. It has been suggested that logistics has over time turned more business management focused [6] and that there is a lack of engineering students in logistics because of it [7].
Firstly, the paper reviews literature on major technology trends in logistics and identifies ten key technology areas. In the next section, five international models of logistics competence, which are often used as standards for curriculum development, are reviewed in terms of references to these technologies. As gaps are notable, the authors present a novel competence model titled *„logistics systems engineer"* designed in Tallinn University of Technology as a modern interdisciplinary and systemic view to logistics engineering education.

2. TECHNOLOGY TRENDS IN LOGISTICS

Three major studies of technology advancements with biggest impact for future logistics systems, all from 2014, are summarized in table 1 [8-10].
Notable driving force of logistics technologies are "green" sustainability oriented solutions, pushing towards lower environmental impact, improved services, lower costs and greater efficiency and also

resulting in increased reliability, service innovation and increases in revenue and reputation [11]. A study on the evolution of supply chains by SMI [12] has concluded that cutting carbon emissions will be the greatest challenge, followed by fuel supply in the future of oil scarcity.

Logistics Trends 2020 [8]	Logistics Trend radar [9]	Material handling and logistics: US roadmap [10]
application of telematics	big open data	e-commerce
deeper penetration of logistics systems with ICT	cloud logistics	mass personalisation
implementation of GPS systems	autonomous logistics	mobile, wearable computing
traffic information, real time routing	3D printing	robotics, automation and driverless vehicles
e-marketplaces for logistics services	robotics and automation	sensors and the internet of things
networking and integration in IT	internet of things	big data and predictive analysis
real time transport information systems	localization and local intelligence	new methods of distribution
application of mobile computing	wearable technology	tracking integration
traffic information systems	augmented reality	cloud-based visibility
logistics simulation models	low-cost sensor technology	sensor data standards
	crypto-currencies and -payment	process optimisation tools

Table 1. Notable technology trends in logistics

Innovative tools make logistics operations cleaner and resource-efficient through advances in vehicle emissions, energy efficiency and technologies for smart cities [13]. The smart city concept includes vehicle sensors and intelligent transport systems (ITS) for controlled traffic, but indirectly also integrated information solutions for businesses and online marketplaces. Promising trends in ITS are collision avoiding systems, lane keeping systems, RFID tracking, driving monitoring systems and real time travel data analysis [14].
As an underlying trend, the SMI study [12] foresees that the customers of the future "*continue to demand greater control over the logistics process, and will more actively intervene in the delivery process of the goods they do order. This will increase the complexity of logistics processes, making necessary a highly sophisticated technical infrastructure.*" To provide control, advances in both information as well as physical delivery capabilities are required, which are the main pillars of progress in logistics.
McKinnon *et al* [15] have treated the impact of modern vehicle technologies through three pillars: *carrying capacity, energy efficiency* and *externalities.* Improving truck aerodynamics can notably improve fuel efficiency [16]. Another efficiency gain comes from lightweight materials, which reduce fuel consumption, increase capacity and as a result requires less road space.
In another study, the fields of smart cities, e-mobility and zero emission technologies belong to the top of global economic mega-trends [17]. Wide-scale implementation of automatic technologies in cargo handling, packaging and robotic transport in industry applications is also forecasted. Similarly, the SMI study [12] proposes considerable growth in autonomous systems and increases in capacities across modes, ultra-large container vessels, aircraft and LHVs.
Driverless transportation systems can provide cognitivity, safe navigation and notably altered cost patterns, which may become a reality by 2030: "*Autonomous vehicles with radar, navigation and ultra-sonic sensors can steer themselves and enable dynamic real-time traffic-dependent routing*" [9]. Another area of autonomous logistics is UAVs (unmanned aerial vehicles, drones), which is in first phases of testing commercial applications across courier express parcel sector.
Advanced cognitivity is also impacting warehouse systems where processes can be facilitated by magnetic or optical guidance. The solutions are supported by advances in software, such as swarm intelligence platforms. Intermec [18] has listed voice recognition, RFID, digital imaging and resulting remote management among their top technology trends.
In a Delphi study of logistics realities in 2050 [19], one possible scenario, "mega-efficiency in megacities", is described as

"transition to the automation age" and embraces green paradigm shift, smart urban logistics to deal with externalities, high efficiency traffic concepts, robotics-based logistics, underground networks, global grid of large-scale transport, information logistics, open trade, and global governance.

Another scenario, "customized lifestyles", predicts the emergence of 3D printing and localized production – only raw materials and data would flow globally and managing "last mile" transport would become critical, while global flows decline. The study identified 14 key trends, of which three are technological: ICT and robotics, materials and urban development solutions.

The potential of RFID-tracking is also strong in retail, improving efficiency and saving costs. Four key impacts of RFID are meeting demand, sharing real-time data, creating delivery value and error reduction [20]. Additional outcomes are transparency, improved availability and labour savings.

Logistics is innovative application industry for IT. Cloud computing has been promoted as a means for fundamental redesign of logistics systems [21]. As the real-time aspect of logistics information grows, the future constraint is not obtaining the data but rather distributing data along supply chain, which requires inter-company integration and efforts in analysing the data.

In another vision for next decade supply chain advances [22], improved supply chain infrastructure and affordable technologies and big data are prominent. The key elements of infrastructure relate to faster deliveries across modes, unitization technologies, continuing modal shift to intermodal solutions and semi-automatic handling. The report suggests "information explosion is a certainty" with widespread internet and mobile coverage and huge increases in data generation and storage. The report foresees internal data from ERP systems complemented with external such as geospatial data and point-of-sale terminals. As data storage costs dropping, partly due to cloud computing, more data fuels growth in flexible, reliable and affordable data analytics architectures [22].

The authors of this paper have consolidated a list of ten major technology areas, which should be essential for future logistics engineering specialists:

1. Electric vehicles, alternative fuels and clean technologies;
2. Telematics, real-time tracking and intelligent transport systems;
3. Auto-pilot and autonomous vehicles: UAV, self-driving cars, ships etc.;
4. Vehicle design, materials and systems of safety, costs and performance;
5. Robotics, sensors and ID-solutions in cargo handling and security;
6. Mobile and cloud computing applications, wireless communication;
7. Logistics process and network simulation and optimisation software;
8. Electronic marketplaces, e-commerce and smart networking;
9. Big data, augmented reality, automatic data analysis and integration;
10. Additive manufacturing (3D printing) applications.

In conclusion, the authors agree with Zelewski *et al* [8] who remarked that no particular trend stand out, which suggests the fields synergize to realize the concept of continuous shipment tracking including corresponding added value services.

3. TECHNOLOGIES IN LOGISTICS COMPETENCE MODELS

The authors have carried out a comparison of four internationally recognised models of logistics professional competences in an ongoing curriculum development process:

- Distribution and logistics managers' competency model by APICS [23];
- Certified in transport and logistics (CTL) by AST&L [24];
- Certified master logistician (CML) by SOLE[25];
- Logistics professional by European Logistics Association ELA [26].

All the models are aimed at describing logistician competences on the level of higher education and suggest a relatively broad scope in their title. The study goal was to identify to what extent are the models in line with the envisioned ten technology areas and would these models then prove suitable for logistics engineering curriculum benchmarking. The results, however, demonstrated notable gaps and mismatches across board.

APICS model assumes students to "*demonstrate an understanding of the factors that are considered important to the branch of knowledge or technology*" and "*to implement new technology*" [23] with no *ad hoc* list of technologies. However in various sections, the following technology elements are mentioned: materials and distribution requirements planning, ERP-systems, advanced planning systems, renewable materials, energy reduction, warehouse management systems and electronic data interchange (EDI).

AST&L model [24] is approaching logistics, and more specifically transport, more in terms of exploitation and management rather than design and development, by requiring the competence of "*how the operating and service characteristics of each mode affect cost, performance, and the products moved*". The only time innovation is mentioned in the model is under "creative component".

SOLE [25] lists some educational areas connected to technologies and to related design and implementation life-cycle aspects: conceptual system design, civil engineering, safety and reliability engineering and user tests. However, no technology area is treated in detail.

ELA model [26] states: "*Due to the constant progress in ICT, specific technologies are not defined in the modules. It is a prerequisite, though, that current technologies must be applied in all relevant fields. ICT competences are implicit in every module.*" The model mentions vendor managed inventory, e-procurement, APS, WMS, transport management system, customer relationship management (CRM) systems and software testing. No reference is made to other technologies, except: "*Understands the impact of technological innovation on supply chain design*".

In summary it appears that while the models are mostly capable of describing soft skills and business workplace process-oriented competences, the models are poor guidelines in terms of developing a logistics curriculum that would be founded on natural sciences and technologies.

AST&L perspective to logistics is mostly that of transport management. ELA and APICS are more interdisciplinary but not in terms of technologies. Even SOLE, originally named Society of Logistics Engineers, does not draw dedicated focus to the building blocks of modern logistics engineering.

4. LOGISTICS SYSTEMS ENGINEER COMPETENCE MODEL

To reinforce technology element in logistics competences, the authors have created a novel competency approach titled "logistics systems engineer". The guiding idea has been to infuse the modern understanding of logistics with dedicated focus to technologies and to the design and implementation process (the focus of systems engineering).

The model structure is shown on Figure 1. It consists of six layers, starting from foundational engineering competences. In this view, systems engineering treatments form the conceptual basis. This is followed by a layer for specific technologies that need to be engineered and maintained in logistics, and a core layer of more conventional logistics topic areas. The technology layer is essentially an abbreviated version of the ten technology areas. However, all business information system related topics are intentionally not included on this layer – rather they form a separate segment on the next layer, as they integrate all logistics data aspects.

Fig. 1. Competency areas and structure in "logistics systems engineer" training profile

The core layer views logistics as a system with inventory and information as key variables. All technology solutions in logistics essentially imbue information and physical inventory handling and related operations. The top two layers add logistics management elements to the technology foundation and point out that all applied solutions should be viewed as parts in a value creation system of an enterprise.

In brief, the key technological competences in the model are the following.

1. Understands the characteristic, design, applications and limitations of technology solutions in transport, warehousing, tracking and handling.
2. Analyses modern technologies and application environments in terms of capabilities, costs, implementation requirements, constraints and risks.
3. Analyses current organisation processes and workflow and identifies suitable technology solutions.
4. Analyses the impact of various material flow technologies to logistics system and supply chain performance.
5. Understands synergetic relations of material flow technologies, IS and information flow configurations and utilises it in systems development.
6. Plans, manages and controls new technology implementation projects throughout the life cycle.
7. Defines human, IS and technology interfaces and integrates physical technologies with information systems and workflow in an optimal way.
8. Initiates and carries out feasibility, risk and impact studies and cost-benefit analysis, including aspects of safety, security and environmental impact.
9. Cooperates with specialists in partner companies to create systems reaching across organizational boundaries.
10. Analysis technological competencies in company and assists in defining training, serves as a technological expert and supports innovation.
11. Is aware of the boundary of human and machine-based operations and the conditions when human labour can be replaced with machines.
12. Understands delicate relation between lengthy implementation projects and changing environments which can impose risks regardless of initial plans.

The entire competence model includes over 250 defined competency elements and is far too extensive to present here. The details are available from corresponding author on contact.

5. CONCLUSIONS

In conclusion, some competence models in logistics do not emphasise technologies from modern logistics frontier. The authors have designed a new competency approach more fitting for future logistics engineers. The new model serves as a curriculum development guideline for all universities wanting to develop logistics programs with technology orientation. The entirety of the model is suitable for 5-year integrated masters' studies, but it can be partially applied also on bachelor level.

6. REFERENCES

1. Kisperska-Moron D. Evolution of competencies of logistics and supply chain managers. *LogForum: El. Sci. J. of Log.*, 2010, **6**, 21-31.

2. Klumpp, M., Clausen, U., Hompel, M. Logistics research and the logistics world of 2050. In *Efficiency and Logistics* (Klumpp, M., ed.) Springer, Heidelberg, 2013, 1-7.
3. Lancioni, R., Forman, H., Smith, M. F. Logistics and supply chain education: roadblocks and challenges. *Int. J. Phys. Distrib.*, 2001, **31(10)**, 733-745.
4. Rahman, S., Yang, L. Skill requirements for logistics managers in China: an empirical assessment. *IIMB Man. Rev.*, 2009, **21(2),** 140-148.
5. Wu Y. J. Contemporary logistics education: an international perspective. *Int. J. Phys. Distrib.*, 2007, **37(7),** 504-528.
6. Murphy, P., Poist, R. F. Skill requirements of senior-level logisticians: a longitudinal assessment. *Supply Chain Manag.*, 2007, **12(6),** 423-431.
7. Jian, M., Ma, J., Chen, Q. Statistics and analysis on education of logistics undergraduate. In *Proc. of ICLEM 2010 Int. Conf. Log. Eng.*. ICLEM, Chengdu, 2010, 1035-1041.
8. Zelewski S., Münchow-Küster, A., Föhring, R. Logistics Trends 2020: a national Delphi study concerning the German logistics sector. In *Next Generation Supply Chains* (Kersten, W., Blecker, T., Ringle, C. M., eds.) HICL, Hamburg, 2014, 69-85.
9. DHL Trend Research. *Logistics Trend Radar: delivering insight today, creating value tomorrow.* DHL Customer Solutions & Innovation, Troisdorf, 2014.
10. Gue, K., Akcali, E., Erera, A., Ferrell, W., Forger, G. *Material Handling and Logistics: US Roadmap.* Auburn University, Auburn, 2014.
11. APICS & PricewaterhouseCoopers. *Sustainable Supply Chains: making value a priority.* PricewaterhouseCoopers, Delaware, 2014.
12. SMI & PricewaterhouseCoopers. *Transportation and Logistics 2030: How will supply chains evolve in an energy-constrained, low-carbon world*? PricewaterhouseCoopers, London, 2009.
13. Golinska, P., Kawa, A. *Technology Management for Sustainable Production and Logistics: Environmental Issues in Logistics and Manufacturing.* Springer, Cham, 2015.
14. Taniguchi, E., Thompson, R. G. *City Logistics: mapping the future.* CRC Press, Boca Raton, 2015.
15. McKinnon, A., Browne, M., Piecyk, M., Whiteing, A. *Green Logistics: improving the environmental sustainability of logistics.* Kogan Page, London, 2015.
16. Deutsche Post AG. *Delivering Tomorrow: Towards Sustainable Logistics.* Deutsche Post AG, Bonn, 2010.
17. Frost & Sullivan. *World's Top Global Mega Trends to 2020 and Implications to Business, Society and Cultures.* Frost & Sullivan, San Antonio, 2010.
18. Intermec. *Top Ten Supply Chain Technology Trends.* Intermec Technologies, Washington, 2007.
19. Deutsche Post AG. *Delivering tomorrow: Logistics 2050. A Scenario Study.* Deutsche Post AG, Bonn, 2012.
20. Fu, H., Chang, T., Du, Z., Lin, A., Hsu, K. Key factors for the adoption of RFID in the logistics industry in Taiwan. *Int. J. Log. Man.*, 2015, **26(1),** 1-33.
21. Hompel, M., Rehof, J., Wolf, O. *Cloud Computing in Logistics.* Springer, Cham, 2015.
22. A.T. Kearney & CSCMP. *Supply Chain 2025 – Trends and implications for India.* A.T. Kearney, Chicago, 2014.
23. The Association for Operations Management APICS. *Distribution and Logistics Managers Competency Model.* APICS, Chicago, 2014.
24. American Society of Transportation and Logistics AST&L. *Certification in Transportation and Logistics.* AST&L, Chicago, 2014.
25. International Society of Logistics SOLE. *A Study Guide for the CML Examination Program.* SOLE, Laurel, 2005.
26. European Logistics Association ELA. *Qualification Standards for Logistics Professionals.* ELA, Brussels, 2014.

7. CORRESPONDING AUTHOR

Ph. D. student Tarvo Niine,
TUT, Dpt. of Logistics and Transport
Ehitajate tee 5, 19086 Tallinn, Estonia
E-mail: tarvo.niine@ttu.ee